Advances in African Economic, Social and Political Development

Series Editors

Diery Seck, CREPOL—Center for Research on Political Economy, Dakar, Senegal

Juliet U. Elu, Morehouse College, Atlanta, GA, USA

Yaw Nyarko, New York University, New York, NY, USA

Africa is emerging as a rapidly growing region, still facing major challenges, but with a potential for significant progress—a transformation that necessitates vigorous efforts in research and policy thinking. This book series focuses on three intricately related key aspects of modern-day Africa: economic, social and political development. Making use of recent theoretical and empirical advances, the series aims to provide fresh answers to Africa's development challenges. All the socio-political dimensions of today's Africa are incorporated as they unfold and new policy options are presented. The series aims to provide a broad and interactive forum of science at work for policymaking and to bring together African and international researchers and experts. The series welcomes monographs and contributed volumes for an academic and professional audience, as well as tightly edited conference proceedings. Relevant topics include, but are not limited to, economic policy and trade, regional integration, labor market policies, demographic development, social issues, political economy and political systems, and environmental and energy issues.

More information about this series at http://www.springer.com/series/11885

Calvin Atewamba · Dorothé Yong Ngondjeb
Editors

Inclusive Green Growth

Challenges and Opportunities for Green Business in Rural Africa

UNITED NATIONS
UNIVERSITY
UNU-INRA
Institute for Natural Resources in Africa

Springer

Editors
Calvin Atewamba
Institute for Natural Resources in Africa
United Nations University
Brampton, ON, Canada

Dorothé Yong Ngondjeb
Institute for Natural Resources in Africa
United Nations University
Ottawa, ON, Canada

ISSN 2198-7262 ISSN 2198-7270 (electronic)
Advances in African Economic, Social and Political Development
ISBN 978-3-030-44179-1 ISBN 978-3-030-44180-7 (eBook)
https://doi.org/10.1007/978-3-030-44180-7

This Springer imprint is published by the registered company Springer Nature Switzerland AG
The registered company address is: Gewerbestrasse 11, 6330 Cham, Switzerland

Foreword

Green growth is a growth strategy that is low carbon, resource efficient, resilient to shocks and socially inclusive. According to the African Development Bank, promoting green growth in Africa means addressing existing and emerging development challenges without locking into pathways that deplete Africa's natural capital and leave economies and livelihoods more vulnerable to climate change and other environmental, social and economic risks. Whereas this new paradigm for development has the potential to create tremendous business opportunities, there are also challenges. To understand these challenges and opportunities in rural Africa, the United Nations University Institute for Natural Resources in Africa implemented a three-year collaborative research project entitled "Unleashing the potential of rural economies through green growth" from 2013 to 2016. The project brought together more than 60 researchers from African universities and other research institutions to investigate the potential impacts of green growth strategies through countries and sectorial case studies.

The book on *Inclusive Green Growth: Challenges and Opportunities for Green Business in Rural Africa* is one of a series of five books compiling the findings of that research project. A book by African researchers is an important contribution to the green economy literature in Africa which, in its current state, is quite limited. The overall objective of this book is to provide empirical evidence on the conditions for the emergence of the green business in Africa. These empirical evidences are generated mainly from case studies in various parts of Africa covering a wide range of countries at different levels of development. Despite their differences, most African countries face similar development challenges, such as access to credit, low level of education and experience, low innovation and entrepreneurship, weak competitiveness of businesses, to name a few.

Inclusive Green Growth: Challenges and Opportunities for Green Business in Rural Africa provides evidence on how small businesses and informal businesses can play a role in promoting sustainable development. The chapters demonstrate how green growth strategies can, inter alia, influence employment, income generation, innovation, competitiveness, food security and social inclusion in Africa. A number of chapters in this book identify both the determinants of small- and medium-size enterprises (SMEs) engagement in sustainable development and the factors that hinder innovation in business and entrepreneurial activities. Some chapters focus on studying awareness, acceptability and constraints related to the use of environmentally friendly techniques. The studies in this book make use of several scientific methodologies to study how SMEs contribute to the green economy in Africa. The book enriches the literature on green economy for Africa. It will play a significant role in developing appropriate regulations and policies to stimulate the development of green business in the continent. Investments in education, training and technology are some of the measures that can help raise awareness of green production technologies, increase labor productivity and help to provide options in resolving the problem of unemployment and poverty in Africa. The findings of the various chapters will benefit the community of policymakers, researchers and practitioners by providing evidence on the rural implications of the transition to green growth; on the diversification of green growth strategies pursued by local populations involved in microenterprises and the importance of these enterprises to generate jobs and incomes in rural Africa; on the transition to a circular economy; and on the relationship between environmental efficiency, labor productivity and business competitiveness.

This contributed volume is a knowledge product from a collaborative research project that benefited from the financial support of the International Development Research Centre (IDRC). UNU-INRA is grateful to IDRC and to all researchers that participated in this research project.

Kigali, Rwanda

Elias T. Ayuk
Director

Acknowledgements

This book is a knowledge product from a collaborative research project of the United Nations University Institute for Natural Resources in Africa (UNU-INRA). We wish to thank UNU-INRA whose international research fellowships made this work possible.

Contents

Contributors

Clifford Afoakwah Centre for Applied Health Economics (CAHE), Griffith University, Brisbane, QLD, Australia

Julius Ajah Department of Agricultural Economics and Extension, Faculty of Agriculture, University of Abuja, Abuja, Nigeria

Koffi Paul Assandé Université Félix Houphouet Boigny D'Abidjan, Abidjan, Côte d'Ivoire

Calvin Atewamba United Nations University Institute for Natural Resources in Africa, Accra, Ghana

Yusuf Kiwala College of Business and Management Science, Makerere University, Kampala, Uganda

Isaac Koomson Faculty of Science, Agriculture, Business and Law, UNE Business School, University of New England, Armidale, NSW, Australia;
Network for Socioeconomic Research and Advancement (NESRA), Accra, Ghana

Moussa Macalou University of Ghana, Acrra, Ghana

Chioma Ifeyinwa Madueke Department of Metallurgical and Materials Engineering, Federal University Oye-Ekiti, Ekiti, Nigeria

Yirviel Janvier Métouolé Méda Professional University Institute, University of Dédougou, Dédougou, Burkina Faso

Mmaduabuchukwu Mkpado Department of Agricultural Economic and Extension, Federal University Oye-Ekiti, Ekiti, Nigeria

Chipo Mukonza Institute for Corporate Citizenship: Exxaro Business and Climate Change, UNISA, Pretoria, South Africa

Assémian Faustin Naouré Université Félix Houphouet Boigny D'Abidjan, Abidjan, Côte d'Ivoire

Dorothe Yong Ngondjeb United Nations University Institute for Natural Resources in Africa, Accra, Ghana;
Faculty of Economics and Management, University of Yaoundé II, Yaoundé, Cameroon

Chidiebere Ofoegbu University of Pretoria, Pretoria, South Africa

Oluwafunmiso Adeola Olajide Department of Agricultural Economics, University of Ibadan, Ibadan, Nigeria

James Atta Peprah Department of Applied Economics, School of Economics, University of Cape Coast, Cape Coast, Ghana

Maruf Sanni National Centre for Technology Management, Federal Ministry of Science and Technology, Obafemi Awolowo University, Ile Ife, Osun State, Nigeria

Isaac M. B. Shinyekwa Economic Policy Research Center (EPRC), Makerere University, Kampala, Uganda

Fatai Abiola Sowunmi Department of Agricultural Economics, University of Ibadan, Ibadan, Nigeria

Yann Cédric Armel Vangah Université Lorignon Guédé de Daloa, Daloa, Côte d'Ivoire

Chapter 1
Green Businesses for an Inclusive and Sustainable Future in Africa

Dorothe Yong Ngondjeb and Calvin Atewamba

1.1 Introduction

The emergence of the green economy is attracting a lot of attention around the world and more so in Africa. This economic model, promoted by several major international organizations, derives from the urgency of solving environmental problems such as climate change and biodiversity loss. It is also seen as a real alternative to the current models of economic growth and is positioned as the one that would allow the responsible use of resources while at the same time ensuring economic development. This emergence of the green economy can also be explained by the market and job creation opportunities to be seized, which stem from environmental issues. There is a strong demand for clean technologies and associated expertise. This demand, which comes mainly from emerging economies (including China and India), contains significant business opportunities and quality job creation (Écotech 2012). Finally, the green economy is gradually being borne out of the failures of the current economic system such as its inability to solve the problems of global poverty, the degradation of natural capital, the financial crisis of 2008 and soaring food prices (UNEP 2011).

The theme "green economy" emerged in 1989 in a report entitled Blueprint for a Green Economy (Pearce et al. 1989). This widely circulated report was written by a group of economists for the UK Government. The term was later taken up in various international documents, including the popular and recent 2009 Global Green New

D. Y. Ngondjeb (✉) · C. Atewamba
United Nations University Institute for Natural Resources in Africa, Accra, Ghana
e-mail: njedora@gmail.com

C. Atewamba
e-mail: atewamba@gmail.com

D. Y. Ngondjeb
Faculty of Economics and Management, University of Yaoundé II, 1365, Yaoundé, Cameroon

C. Atewamba and D. Yong Ngondjeb (eds.), *Inclusive Green Growth*, Advances in African Economic, Social and Political Development,
https://doi.org/10.1007/978-3-030-44180-7_1

Deal, which proposed that states invest heavily in green stimulus to restart the economy and exit the financial crisis of 2008 (PNUE 2009). Even though the "green economy" theme is relatively recent, several components of this concept (internalization of environmental costs, use of economic instruments, elimination of unsustainable production and consumption patterns, etc.) existed earlier and were enshrined in the 1992 Rio Declaration, Agenda 21 and the Johannesburg Plan of Implementation (UNDESA 2012). For UNEP (2011), the green economy is "an economy that results in improved human well-being and social equity while significantly reducing environmental risks and resource scarcity. In its simplest form, it is characterized by low carbon emissions, rational use of resources and social inclusion". Lacarrière (2011) defines it as "a development model less dependent on carbon energies, but without renouncing the lifestyles and habits of consumption that characterize them". This type of economy represents a perfect reconciliation of environmental sustainability and economic growth to form what Jurgensen (2009) calls "econology".

According to Perret (2010), the green economy is based on six main principles: Favor renewable resources; use scarce resources efficiently; reuse, repair, recycle; prioritize use of local resources to avoid energy costs associated with transportation; maintain diversity by avoiding, "one-way best, specialization and standardization"; take account of interdependencies and foster cooperation. According to the United Nations (UN) and other international organizations, environmental, social and financial crises are due to a bad investment of capital, that is to say that the public and private actors invest mainly, and for a long time, in sectors that deplete natural capital. They are also the result of development policies that take little or no account of environmental and social externalities. In the same vein, the price of consumer products is also pointed out because it does not reflect the real cost it brings to society in terms of environmental impact. This failure to take into account externalities in policies and in the price of products leads to the issuance of poor signals to producers and consumers. In this context, the major international organizations urge countries to join the global transition movement toward the green economy in order to mitigate the growing risks of crises and shocks inherent in the existing economic model and to allow the protection and restoration of natural capital.

In Africa, the continent where small and medium enterprises (SMEs) and the informal sector are the engines of economies, the promotion of eco-initiatives poses as many challenges as opportunities. It then seems appropriate to look at examples that show how the green economy (despite its potential) creates many challenges for the continent. This book highlights, through studies in some African countries, the factors that condition the promotion of the green economy in Africa. It must be said that poverty in most of Africa can hamper any eco-initiative since the very people who are supposed to be the vectors of this new form of the economy are having difficulties to change the way in which their activities operate to meet the conditions for sustainable societal development. As Verreault (2011) points out, the transition to the green economy must be financed both by public and private actors.

African SMEs, which are one of the driving forces of the economy in many African countries, suffer from limited skills, lack of investment and sometimes insufficient technologies to guarantee wealth creation within the traditional growth model. Moreover, much of the continent's wealth is created by informal structures that occupy a significant place in the current African economies. Therefore, relying on these SMEs and to some extent, on the African informal sector to adopt and support eco-initiatives while expecting growth rates higher than those currently experienced by African countries, is a real challenge. The green economy as understood requires that the constraints and barriers to Africa's emergence be lifted and overcome.

This book highlights the factors explaining the success or not of the green economy in Africa through case studies carried out in select African countries with unique socio-economic and even political characteristics. Most of this work is trying to identify the factors that hinder the emergence of the green economy or its drivers. This research shows that raising awareness of the green economy is a major factor in the emergence of eco-initiatives. At the SME level, barriers include transaction costs, market factors, access to finance, technological capacity, lack of knowledge and access to information. In a more geographical approach, agricultural enterprises in forest or coastal zones have more potential to engage in green initiatives than those in the Sahelian zones. Rural agriculture also suffers from access to credit. It can also be noted that the regulatory environment in the countries studied is a relevant factor in the adoption of eco-initiatives.

This introductory chapter comprises six sections. The section following the introduction will present some perspectives of the green economy in the literature with a focus on innovation, green entrepreneurship, business diversification, corporate social responsibility and environmental finance. The third section highlights the challenges and opportunities of the green economy for Africa. The fourth section reviews the purpose of this book, its importance and its justification. The contribution of the other chapters is discussed in the fifth section and the conclusion in the last section.

1.2 Prospects for Green Business

Promoting the economy is like looking at opportunities and constraints for development by people and businesses. These constraints and opportunities may be related to the introduction of a new business model that respects the environment or to a new form of entrepreneurship that can create markets for environmental products and services. They may also be related to business diversification in order to internalize environmental externalities or to improve corporate social responsibility. It is also about understanding how corporate financing constraints and opportunities encourage not only investment in productivity improvement, but also investment in environmental protection.

1.2.1 Green Business Model and Innovation

The term "business" and the various concepts associated with it, such as "entrepreneurs, entrepreneurship," are used to describe a social activity with an economic vocation. The modern meaning of these notions takes shape from the eighteenth century (Peredo and McLean 2006). Precisely, they refer to associating and action of control with an economic activity allowing either the creation of economic wealth pertaining to the industry, the creation of social wealth for the purpose of equity, or finally the creation of political wealth according to Say (Fontan 2011). The achievements of a company are sources of positive or negative externalities. They generate events that reflect situations of imperfection or incompleteness in terms of the regulation of society. This usually leads the company to readjust its production plans, hence the concept of innovation. Defined as a process of creative destruction according to Schumpeter (1935), innovation in the enterprise is the realization of an economic activity capable of creating new institutional arrangements, which take the form of new products, new processes, new ways of organizing work or new uses for existing products (Fontan 2011).

When green innovation or eco-innovation occurs in a business, the new business model result is then called the green business model and innovation. According to FORA (2011), the green business and innovation model is a new way of doing business that refers to non-technological "green" innovation between businesses, which pushes core businesses to produce services rather than integrate sustainability into the operating and management process. The green business and innovation model is also seen as a new way of approaching products and service-oriented R&D, value chain, organizational models, operations management, incentive contracts, marketing, etc. (Rajala et al 2016; Morioka et al. 2016).

Innovation is therefore increasingly driving the progress of the economy and society at the national (macro) level, as well as driving business success and competitive advantage at the enterprise (micro) level. Porter and Kramer (2006) also state that "innovation is the central problem of economic prosperity". Many governments are now considering eco-innovation as a part of their growth strategy. In light of the great global challenges of economic slowdown, environmental degradation and resource scarcity, eco-innovation is seen as a way of reconciling economic and environmental priorities—and opening new sustainable avenues for the industry.

1.2.2 Green Entrepreneurship

Entrepreneurship from the Schumpeterian perspective is defined as an attitude to help innovative ideas become reality by establishing new business models and replacing conventional business systems and making them obsolete (Korres et al. 2011). The concept of "green entrepreneurship" stems from the combination of the main characteristics of entrepreneurship itself: innovation, risk, a new business idea and the ecological and social commitment of those who do business.

In order to reconcile the notion of entrepreneurship and sustainable development within the company, a hybrid form of business should emerge to redefine the boundaries of for-profit and non-profit organizations (Porter and Kramer 2011). Several characteristics of entrepreneurship are therefore stated by theoreticians. First, sustainable entrepreneurship, which is defined as a process of discovering, evaluating and exploiting economic opportunities that are presented in market failures and that go against sustainability (Dean and Mccullen 2007). Or even more recently authors like Shepherd and Patzeld (2011) and Chen and Sintov (2016) define it as a mode of business management focused on the preservation of the nature and supports to the life of the community by seeking opportunities to create goods, services and processes that produce economic and non-economic gains for individuals, the economy and society. Next, transition-oriented entrepreneurship, which also suggests that the transitioning firm puts into practice business strategies focused on improving social and environmental conditions rather than maximizing profit or just job creation (Hopkins 2016). Finally, green transition entrepreneurship, which represents the type of business that reduces and minimizes waste, cancels or improves the negative externalities of existing products and services (Farinelli et al. 2011). Located outside the circles of administrative pressure, the ecological transition firm is able to perceive the extent of the pressures exerted by the ecological crisis and climate change at the landscape level.

1.2.3 Business Diversification and Sustainable Business Growth

History shows us that the model of extensive industrial growth that allowed rich countries to develop during the twentieth century has adverse effects. These effects are considered externalities, insofar as they are situations in which the economic activity produces harmful effects for certain agents, without being involved in the action or compensated by the author of the effects. These externalities include cases of pollution and overexploitation of certain resources, leading to their rapid depletion and health risks. The urgency of these environmental challenges has prompted politicians to think about a new development model in which the aspects of social equity and environmental sustainability, as well as economic growth, will be of equal importance. This refers to the concept of sustainability that "contributes to sustainable development by simultaneously bringing economic, social and environmental benefits—the so-called triple bottom line".

The sustainability of diversified enterprises in economies can be defined as the ability of a group of individuals or firms to cope with the internal and external pressures of cultural, economic, institutional and resource aspects through change and innovation of specific products or services. Diversification strategies are widely known to have positive and negative implications for business efficiency (Kaulich 2012; Jara-Bertin et al. 2015). Tax benefits, redistribution of funds, reduction in cash flow volatility, reduction in business risk, synergy and economies of scope are the benefits that are extracted from diversification. In addition, the presence of diversified firms in an economy most often refers to an increase in profitability, risk reduction, increased debt capacity and growth and the extension of the business life cycle and effective use of human and financial resources because of its long-standing relationship with other institutions. These advantages are cited in an article by Jara-Bertin et al. (2015).

Some studies indicate that diversification destroys the value of the business in developing economies (Lins and Servaes 2002; Martin and Lorenzen 2016). In developing countries, diversified firms face many challenges due to the highly dynamic nature of institutional change, external shocks and crises, rival political links (Zhu and Chung 2014). These issues influence the market value and performance of these companies. In other words, we can say that these are the "sustainability issues," which include environmental, social and economic aspects.

1.2.4 Social Responsibility and Sustainable Business Growth

The productivity growth model that the world has experienced in recent years has not been without impact. Economic growth with the pollution it generates through economic activity leads to a deterioration of the natural capital stock and the related social well-being. The world has gradually realized that economic growth does not necessarily lead to social progress and even threatens to jeopardize the natural balance of the planet. A new model has emerged—sustainable development. Concerns about protecting the environment are now more intense than ever, just as social and economic concerns are. At a time when sustainable development is an integral part of the global economy, every company must consider its impact on the environment. Whether it is through its transport logistics that generate greenhouse gases or by the management of its industrial waste that causes pollution, a company causes an impact that can be harmful. In order for the sustainable development policy to be effective, social responsibility is increasingly part of the business model of companies (Rahdari et al. 2016). Corporate social responsibility is defined as *"the way in which companies integrate, on a voluntary basis, social, environmental and ethical preoccupations in their economic activities as well as in their interactions with all stakeholders, whether they are internal (managers, employees, shareholders,* etc.*) or external (suppliers, customers,* etc.*)"* (Bocken et al 2015).

More clearly and simply stated, it is the responsibility of an organization to manage the effects it has on society (European Commission 2011). The most committed companies in this area are also those that are the most innovative, those with higher profitability, those with the highest productivities with a real ability to attract and retain the best employees and the best reputation. In France, corporate social responsibility has been regulated. Article 116 of the law of May 15, 2001, on new regulations stipulates that publicly traded companies shall include in their annual report a series of information relating to the social and environmental consequences of their activities. In Africa, the ISO 26000 standard on corporate social responsibility came into force in November 2010. Indeed, in a globalized and highly competitive economy, the notion of societal responsibility is becoming a strategic positioning tool in the conquest of market shares.

1.2.5 Environmental Finance and Sustainable Business Growth

The environment as an inseparable part of the development was recognized at the United Nations Conference on the Environment in Stockholm and confirmed in 1992 by the United Nations Conference on Environment and Development. At the end of the conference, states reaffirmed that "the right to development must be achieved in a way that equitably meets the development and environmental needs of present and future generations" (Principle 3, Rio Declaration) and that "in order to achieve sustainable development, the protection of the environment must be an integral part of the development process and cannot be considered in isolation" (Principle 4, Rio Declaration). A healthy and sustainably managed environment is thus recognized as an essential condition for the development of a society.

Economic growth can pose a threat to the environment. It can increase the burden on the environment to intolerable limits. However, it is able to generate more resources to finance environmental policy, investments in clean technologies, the realization of necessary infrastructure and so on. On the other hand, it would be unrealistic to think that we can deal with the main current environmental problems in a situation of economic stagnation, unless we accept radical changes in the allocation of resources. The problem of financing the protection of the environment cannot be mentioned without recalling the question of the efficiency with which the available resources are used. In general, the first measures tend to be the most effective: Investing 1 million francs to reduce the pollution of an old-generation thermal power plant will have a much greater impact than the same amount invested in a modern power station with combined cycle. Therefore, there is no doubt that in some areas, it would be more efficient to use the resources devoted to combat environmental degradation in the former communist countries and in the developing countries of the world rather than in industrialized countries (Polzin et al. 2016).

In this overall situation, financing to eliminate poverty and ensure sustainable development remains very important. These needs include: (a) the financial resources to eradicate poverty and hunger, improve public health and education, ensure access to affordable energy resources and promote gender equality; (b) financing needs for national sustainable development investment, including infrastructure, rural development, climate change adaptation and resilience and energy; and (c) the financing of global public goods, including the protection of the global environment and the fight against climate change and its effects. The cost of sustainable development also depends on the efficient use of resources. The estimates of the funds required therefore vary considerably. Environmental protection research can lead to a substantial reduction in costs thanks to energy savings, efficiency gains and less waste. Thus, in a "win-win" perspective, often called Porter's hypothesis, the reduction in pollution tends to reduce the quantities of materials and energy used and thus to increase productivity (Boiral 2005). In addition, banks are increasingly considering environmental and social risk in their lending procedures (Barabel et al. 2007; Anderson 2016). A company that is truly committed to sustainable development will have easier access to sources of financing at a lower cost. The company is therefore faced with the need to first identify and then manage in the medium and long term the risks and opportunities related to its activities, with regard to their impact on different groups of its stakeholders. The sustainable approach leads to a redesign of the risk map and an integration of risk prevention (Notat 2003), whether it concerns environmental risk (impact of ecological accidents, challenges by new standards), social risk (for example, degradation of the image of the company following a denunciation of working conditions at a subcontract) or in terms of product or process safety.

1.3 Green Business, Challenges and Opportunities for Africa

Solid arguments encourage economic actors to engage in green business, the new economic transformation. For governments, this means setting more favorable rules for green products, i.e., proactively giving up subsidies from another time, reforming policies, adopting new incentives, strengthening market infrastructure and economic mechanisms, reorienting public investment and greening government procurement. In other words, it is a question of proposing a regulatory framework able to direct investments toward environmentally friendly and socially inclusive activities. It is also a question of reorienting public spending, limiting those that contribute to stimulating activities that are harmful to the planet and promoting those that fuel the transition to a green economy. For the private sector, this helps to understand and measure the real chance of transitioning to a green economy in many key sectors and to respond to policy reforms and price signals by increasing funding and investment levels. The 2013 UNEP report "GEO-5 for Business: The Impact of a Changing Environment on the Business Sector," for example, states that the private sector could contribute more than 80% of the capital required to cope with the consequences of climate change. Beyond an obvious need, the report also highlights the expected benefits for companies that will anticipate and be part of a sustainable responsible approach. In fact, the ability of companies to innovate and develop goods and services that are low in natural resources and have low environmental and climatic impacts will increasingly be one of the essential criteria for evaluation and reputation. Finally, citizens, through their choices and their consumption practices, can be real catalysts for the transition to a green economy. This is one of the major axes developed by the Ten-Year Programming Framework on Sustainable Consumption and Production (10YFP), launched in 2012 on the occasion of the Rio + 20 Conference. Indeed, "consumer information" is one of the five initial programs of the 10YFP, aimed at enabling citizens to make more sustainable consumption choices. The green economy thus appears not only as a social, economic and ecological imperative but also as a unique opportunity to generate growth, create jobs and help to eradicate poverty through investments in natural capital, all by preserving it to ensure the survival of the planet in the long run. In this perspective, the green economy appears as one of the main treasures for the development of African countries and would undoubtedly participate in the stability of the continent constructing peace and prosperity.

The opportunities are many and concern all sectors of activity: agriculture, energy, industry, tourism, transport, building, urban planning and the everyday life of the population. With regard to agriculture, the stakes are high given what this sector represents for Africa as well as the many challenges to overcome in the context of global warming. In this context, the transition to green agriculture seems inseparable to ensure the sustainable future of the region. The energy issue is also one of the main challenges while emphasizing that it brings with it multiple opportunities. Energy efficiency in buildings, the optimization of public transport, the deployment of renewable energies with considerable potential for development on the African continent, improvement of industrial processes and so on are all levers of action that make the prospects for a green and socially inclusive economy realizable.

The transition to the green economy in Africa faces many challenges; Africa faces a literature deficit in understanding the movement of the green economy within it. Since this understanding requires studies of the variables that constitute vectors or obstacles to the smooth running of eco-initiatives, it is essential to have a scientific platform that deals with this subject. More specifically, there is no consistent literature to understand how African SMEs, engines of economic activity are able to implement appropriate green initiatives. It is, however, essential to look at green economy incentives for SMEs, as they face enormous challenges. Large companies or large groups also have a role to play in the emergence of the green economy in Africa. Whether in the oil industry, manufacturing, and agribusiness, to mention only a few sectors, there is a screaming need for works or studies able to shed light on how multinationals, transnational corporations or simply large industrial groups and national authorities are able to contribute to the judicious exploitation of natural resources and respect for the environment. As for SMEs, the informal sector is no longer the subject of abundant literature in terms of how it could contribute to the emergence and development of green initiatives. It is nevertheless recognized as one of the major players in the economic life of African countries, by the number of people it employs, the income it generates and the many areas in which it is deployed. Africa is a continent still predominantly rural, and rural people, farmers, pastoralists or simply rural dwellers have a huge impact on the environment through their economic activities or simply subsistence. However, at this level too, there is a lack of work elucidating the role of rural people in adopting green initiatives.

1.4 Objectives, Importance and Justification of this Book

This book seeks to highlight the constraints and opportunities for promoting green business in rural Africa. More specifically, it seeks to highlight, through studies done in select African countries, the factors that condition the promotion of the green economy in Africa.

It looks at the conditions, variables and factors that determine the contribution of each of the actors/sectors mentioned above in the emergence of the green economy in Africa. The studies presented in this book attempt to identify the factors that hinder not only innovation and innovation opportunities in green enterprises, but especially green entrepreneurship in general. They also try to highlight the difference in intensity in the adoption of green initiatives between rural and urban populations. They examine the sensitization, acceptability, uses and constraints associated with environmentally friendly infrastructure. They seek to identify the determinants of SME engagement in corporate social responsibility (CSR). They highlight and examine the barriers that limit the ability of SMEs to introduce eco-innovation into their activities. Finally, they provide a conceptual framework for transforming traditional farm business models in Africa into inclusive and green business models through the inclusive business innovation model and green innovation business model.

The enrichment of the economic literature on the subject of the conditions of the emergence of the green economy is the purpose of this book. This contribution seems essential in many respects. Researchers and other scientists interested in the evolution of the green economy in Africa will find in this book an early answer to some questions they are asking about this concept in African countries. This book also helps to understand the barriers that prevent green economy pioneers from developing their businesses. The timidity of small, medium and large businesses to adopt eco-initiatives is partly explained in this book. The behavior of populations and state actors is also one of the issues that case studies in this book address.

This book presents studies in several African countries with different socio-economic and political characteristics to help explain the factors determining the emergence or otherwise of the green economy in Africa. This approach seems to be necessary as countries are socially, economically and politically different, and the question of the green economy is different depending on the country being studied. This makes it possible to avoid falling into a sort of generalization and clichés that often constitute the error that occurs when analyzing African economies. Africa is a vast, multi-faceted continent with highly differentiated regions, and its economic characteristics are different from one country to another and even from one region to another within the same country. In other words, what is true in Ghana is not necessarily true in Cameroon. And sometimes even what is true in southern Cameroon is not true in the North. However, such different studies may offer the possibility of extending the results to countries or regions with characteristics similar to the region or country under study.

The conditions for the success of the green economy could then be understood not by country or region, but rather by economic domains with the same characteristics. In fact, the characteristics encountered in an agricultural sector in southern Sudan may correspond to those of an agricultural sector in Chad or Mali. One could then explain the difficulty or the ease of adopting eco-initiatives according to the characteristics of these two sectors rather than comparing them by country or region. Africa is a vast laboratory where studies can lead to interesting and useful results in other parts of the world, provided that these studies are developed without clichés or preliminary considerations that do not always reflect reality.

1.5 Contribution of the Other Chapters

This book presents 11 studies conducted in Nigeria, South Africa, Ghana, Côte d'Ivoire and Uganda.

The second chapter "*Constraints and Opportunities for Innovation in Green Enterprises: Implications for Land and Water Management in Rural Uganda*" examines the factors that hinder innovation and innovation opportunities in green enterprises in rural Uganda. A descriptive analysis of the secondary data of 423 companies was made. In addition, a comparative analysis of the case studies of three green firms was also conducted to allow for more in-depth and contextualized investigation. The results show that cost and market factors are the main obstacles to innovation in green enterprises.

The third chapter, "*An Analysis of Factors Influencing Green Entrepreneurship Activities in South Africa,*" examines the factors affecting green entrepreneurship activities in South Africa. A mixed method approach is used. A total of 103 green entrepreneurs are asked about the factors that influence green entrepreneurship activities. In-depth interviews, key informants, observations and a comprehensive review of the literature were conducted for triangulation purposes. Access to finance, knowledge, skills, access to information and support from the government and the private sector are all factors that affect green entrepreneurship activities in South Africa. The study notes, however, that for green entrepreneurship to be a driving force in the global transition to a green economy, there is a need for legislation, government regulation and agencies that support the industry.

The fourth chapter, "*Analysis of crop yield volatility among smallholder farmers in Ghana,*" examines the instability of crop yields among small rural and urban farmers in Ghana. Data from the study come from the sixth round of the Ghana Living Standards Survey. The study used three separate models of ordinary least squares. It shows that education on agricultural technology and access to credit would help reduce the volatility of agricultural output of small farmers in coastal and forest areas.

The fifth chapter "*Prospects and Constraints to Investments in Environmentally Friendly Infrastructure in Western Nigeria*" examines the awareness, acceptability, uses and constraints of using environmentally friendly technologies in rural areas of the country in western Nigeria. The study analyzes field survey data from 160 households in the states of Ekiti and Ondo. Using descriptive statistics, the Likert scale, the Chi-square and the regression, the study establishes that the level of awareness of climate change, age and income level are the determinants of the adoption of green technologies or environmentally friendly infrastructure in these states.

The sixth chapter "*Sustainable development and SMEs: How to meet the challenges of mobilizing Ivorian SMEs?*" identifies the determinants of the commitment of Ivorian SMEs in sustainable development and corporate social responsibility. The study combines a qualitative and quantitative approach based on data from fifty-seven Ivorian SMEs from various sectors of activity. It follows that staff management, the environment and the organizational climate are areas that reflect the commitment of these Ivorian SMEs for sustainable development and corporate social responsibility.

The seventh chapter *"Environmental and Economic Burden of Sand Dredging on Artisanal Fishing in Lagos State, Nigeria"* examines the environmental effectiveness of fishermen in sand dredging and non-dredging areas. Using cost/benefit analysis, the study shows that education, fishing experience and sand dredging affect the environmental effectiveness of the fishermen in the sand dredging area, in contrast to the fishermen in the non-dredging area.

The eighth chapter "*Reducing the Impact of Grazing Livestock on Crop Production and its Consequences in Nigeria: The Way Out*" highlights the impact of livestock grazing on crop production in Abuja, Nigeria. Questionnaires were used to interview farmers on small-scale crops in agrarian communities. The analysis of the survey reveals several problems that require a viable policy to reduce the negative impact.

The ninth chapter, "*Climate Change-Induced Migration and Its Implication for Green SMEs Development and Rural Livelihoods in Nigeria*," examines climate change caused by migration (CC-IM) and the potential it provides for the development of green SMEs and rural livelihoods in Nigeria. An agricultural and rural system approach was used, as well as a multi-stage random sampling technique on 120 households in rural Bende local government area of Abia State. Three systems were analyzed, the Resource Poor Labor Intensive System (RPLI), the Resource Rich Labor Intensive System (RRLI) and the Resource Poor Non-Labor Intensive System (RPNLI). The study shows that the "Resource Rich Labor Intensive System (RRLI)" is the most economically viable of the three systems with a relatively better standard of living.

The tenth chapter "*An Assessment of Factors Shaping Green Growth Uptake in the Forest Sector at Rural Community Level in South Africa*" is a study that evaluates the current pattern of forest use and management at the rural community level in South Africa in locating space for the growth of green initiatives and factors that could facilitate or hinder the adoption of such initiatives. The study was conducted in twenty rural communities with a total of 366 households selected and interviewed. The study shows that the informal forest sector is more likely than the formal forest sector to contribute to the development of green initiatives. However, competence and technical capacity remain an obstacle to the adoption of green growth initiatives.

The eleventh chapter "*Effects of Barriers to Eco-Innovation within Small and Medium-Sized Manufacturing Firms in the Peripheral Regions of Lagos*" is a study that examines the obstacles to the viability of small and medium-sized manufacturing companies to introduce eco-innovation in the peripheral regions of the Lagos metropolitan area. Using the perspectives of the evolution of the economy and the theory of innovation, the study establishes that the technological capacity, the regulatory environment, the search for external knowledge and the sector of the enterprise are the key factors that improve eco-innovation in SMEs.

The twelfth chapter "*Inclusive Green Agricultural Business Model Innovation for Rural Africa: A Conceptual Framework*" is a study that proposes a conceptual framework for transforming traditional agribusiness in Africa into green and inclusive agribusiness through a model of inclusive innovation business and a green innovation business model. The main contribution is the development of an integrated framework of the two models in response to both types of innovations to address the socio-economic and environmental challenges facing small farmers in Africa.

1.6 Conclusion

This introductory chapter aimed to present the contribution of the book to the challenges and opportunities of operationalizing green economy in rural Africa. The chapter first looked at some green economy perspectives in the literature with a focus on innovation, green entrepreneurship, business diversification, corporate social responsibility and environmental finance. The purpose of this review was to find out if this new analytical framework offers potential for applications in the typical rural environment of Africa. Since the green economy model was designed from a general perspective, it could be applied to any economic sector, including agriculture. However, the emerging literature on developed green economy concepts pays more attention to the service and manufacturing sectors than to agriculture. Thus, they are not comprehensive enough to address the challenges facing rural Africa. This chapter then looked at the challenges and opportunities of the green economy for Africa. In particular, it highlighted the temptation that is often great for the authors to provide general conclusions when it comes to studying a concept in Africa, forgetting the specificities of each region. It then presented the importance and approach of this book as a collection of studies in different countries and regions to try to identify the factors that promote or hinder the growth of green initiatives in Africa. Through the studies presented in this book, the reader can understand why the adoption of green technologies in agriculture is more pronounced in one region than in another. Therefore, do not dwell on a factor observed in one country to generalize it across the continent.

References

Anderson J (2016) Environmental finance. Handbook of environmental and sustainable finance. Academic Press, San Diego, pp 307–333

Barabel M, Huault I, Meier O (2007) Changing nature and sustainability of the industrial district model: the case of technic Valley in France. Growth Change 38(4):595–620

Bocken N, Rana P, Short SW (2015) Value mapping for sustainable business thinking. J Ind Prod Eng 32(1):67–81

Boiral O (2005) The impact of operator involvement in pollution reduction: case studies in canadian chemical companies. Bus Strat Environ 14(6):339–360. https://doi.org/10.1002/bse.431

Chen B, Sintov N (2016) Bridging the gap between sustainable technology adoption and protecting natural resources: predicting intentions to adopt energy management technologies in California. Energy Res Soc Sci 22:210–223

Dean TJ, McMullen JS (2007) Toward a theory of sustainable entrepreneurship: reducing environmental degradation through entrepreneurial action. J Bus Ventur 27:50–76

Écotech Q (2012) Les technologies propres au Québec, Diagnostic de la chaîne de financement. In Écotech Québec la grappe des technologies propres. Publications. Études et mémoires. (Page consultée le 15 mars 2013). http://www.ecotechquebec.com/publications/etudes-et-memoires/

European Commission (2011) Youth opportunities initiative (COM [2011] 933). Author, Brussels, Belgium. Retrieved from http://ec.europa.eu/social/BlobServlet?docId=7276&lan gId=en

Farinelli F, Bottini M, Akkoyunlu S, Aerni P (2011) Green entrepreneurship: the missing link towards a greener economy. ATDP J 8:42–48

Fontan J-M (2011) Entreprenariat social et entreprenariat collectif: synthèse et constats. Can J Nonprofit Soc Econ Res/Revue Canadienne de recherche sur les OSBL et l'économie sociale 2(2):37–56

FORA (2011) Project proposal green business model innovation [EB/OL]. www.ebst.dk/file/170960/projectproposal.pdf

Hopkins V (2016) Institutions, incentives, and policy entrepreneurship. Policy Stud J 44(3):332–348

Jara-Bertin M, López-Iturriaga FJ, Espinosa C (2015) Diversification and control in emerging markets: the case of Chilean firms. BRQ Bus Res Q 1–16 (online). http://doi.org/10.1016/j.brq.2015.01.002 (in press)

Jurgensen Philippe (2009) L'économie verte: Comment sauver notre planète. Odile Jacob, Paris, p 313

Kaulich F (2012) Diversification vs. specialization as alternative strategies for economic development: can we settle a debate by looking at the empirical evidence? Vienna. Retrieved from: http://www.unido.org/fileadmin/user_media/Publications/Research_and_statistics/Branch_publications/Research_and_Policy/Files/Working_Papers/2012/WP032012_Ebook.pdf

Korres GM, Papanis E, Kokkinou A, Giavrimis P (2011) Measuring entrepreneurship and innovation activities in E.U. Interdisc J Contemp Res Bus 3(3):1155–1167

Lacarrière S (2011) La croissance verte: un mythe salutaire pour un monde solidaire ? , Revue internationale et stratégique 1(81):183–188

Lins KV, Servaes H (2002) Is corporate diversification beneficial in emerging markets? Financ Manage 31(2):5–31

Martin SM, Lorenzen K (2016) Livelihood diversification in rural laos. World Dev 83:231–243

Morioka SN, Evans S, De Carvalho MM (2016) Sustainable business model innovation: exploring evidences in sustainability reporting. Procedia CIRP 40:659–667

Notat N (2003) La responsabilité sociale des entreprises, Futuribles, n°288, pp. 11-28. Orse, 2003. « Comment élaborer un rapport de développement durable? » Synthèse des réunions du groupe de travail ORSE-EPE 2002, janvier

Pearce D, Markandya A, Barbier EB (1989) Blueprint for a green economy. Earthscan Publications, London, p 192

Peredo AM, McLean M (2006) Social entrepreneurship: a critical review of the concept (July 2, 2006). J World Bus 41(1):56–65

Perret B (2010) Croissance verte ou développement humain? Revue Projet 4:49–55. (n° 317)
Polzin F, von Flotow P, Klerkx L (2016) Addressing barriers to eco-innovation: exploring the finance mobilization functions of institutional innovation intermediaries. Technol Forecast Soc Chang 103:34–46
Porter ME, Kramer MR (2006) Strategy and society: the link between competitive advantage and corporate social responsibility. Harvard Bus Rev 84(12):78–92
Porter M, Kramer M (2011) Creating shared value. Har Bus Rev 89(1/2):62–77
Programme des Nations Unies pour l'Environnement (PNUE) (2009) Global green new deal. Policy brief. In: PNUE. Publications. Search UNEP Books. http://www.unep.org/pdf/GGND_Final_Report.pdf
Rahdari A, Sepasi S, Moradi M (2016) Achieving sustainability through Schumpeterian social entrepreneurship: the role of social enterprises. J Clean Prod 137:347–360
Rajala R, Westerlund M, Lampikoski T (2016) Environmental sustainability in industrial manufacturing: re-examining the greening of interface's business model. J Clean Prod 115:52–61
Shepherd DA, Patzelt H (2011) The new field of sustainable entrepreneurship: studying entrepreneurial action linking "what is to be sustained" with "what is to be developed". Entrepr Theory Pract 35(1):137–163
Schumpeter JA (1935) Theorie de l'Evolution Economique, Dalloz
United Nations Department of Economic and Social Affairs (UNDESA) (2012) A guidebook to the green economy. Issue 2: exploring green economy principles. In UNDESA—sustainable development knowledge platform. Publications series, Green Economy Guidebooks
United Nations Environmental Programme-UNEP (2011) Towards a green economy. Pathways to sustainable development and poverty reduction. Available online at https://www.unep.org/green economy. Accessed 11/03/2015. ISBN 978-92-801-3143-9
Verreault L (2011) L'émergence de l'économie verte: quel rôle pour les acteurs publics ? In ÉNAP. Publications. http://leppm.enap.ca/leppm/1091/Environnement.enap. (Page le 29 septembre 2016)
Zhu H, Chung C-N (2014) Portfolios of political ties and business group strategy in emerging economies: evidence from Taiwan. Adm Sci Q 59(4):599–638

Chapter 2
Constraints and Opportunities for Innovation in Green Enterprises: Implications for Land and Water Management in Rural Uganda

Isaac M. B. Shinyekwa and Yusuf Kiwala

2.1 Introduction

The future of green enterprises is local and entrepreneurial. This study gives insights into opportunities for and constraints to innovation for sustainable entrepreneurs in rural Uganda. The study combines literature from innovation theory, sustainable development practices, and small and medium-sized enterprises (SMEs) studies to establish a basis on which we examine the constraints and opportunities for innovation in green enterprises. This is important in understanding how SMEs can change toward sustainable behavior. The relationship between a firm's ability to innovate and value created for customers is well documented (Jacobides et al. 2006; Walker and Preuss 2008; Huarng and Yu 2011). This debate is now beyond the economic aspects of value creation; it embraces wider stakeholders' perspectives which consider environmental and social benefits. This line of business thinking envisions a future, where the renewal of products and improvement of practices, technology, and services lead to economic, social, and environmental performance. This is termed as green innovation. The term green innovation encompasses constructs coined to mean innovations which improve environmental performance, such as sustainable-driven innovation, eco-innovation, and environmental innovation. These terms are used interchangeably; but for the purpose of this study, we adopt eco-innovation because it is well defined in the literature. The study aims to explore SMEs' innovation trends in order to identify the constraints and potential opportunities for eco-innovation in the agricultural sector, as well as implications on land and water management.

I. M. B. Shinyekwa (✉)
Economic Policy Research Center (EPRC), Makerere University, Kampala, Uganda
e-mail: ishinyekwa@eprcug.org

Y. Kiwala
College of Business and Management Science, Makerere University, Kampala, Uganda
e-mail: kyosseuf2@gmail.com

C. Atewamba and D. Yong Ngondjeb (eds.), *Inclusive Green Growth*, Advances in African Economic, Social and Political Development,
https://doi.org/10.1007/978-3-030-44180-7_2

2.1.1 Context of the Study

This research is set within the context of the agricultural sector in Uganda and focuses on SMEs that are involved in different enterprises within the sector. It is important to study innovation generated by green enterprises within the agriculture sector because agriculture is a part of the environmental problem and a contributor to nonpoint pollution and environmental degradation (Dumanski 1997). A number of small entrepreneurs in Uganda are engaged in agriculture and operate informally, but research on green innovation within the agricultural sector is still nascent and limited. The agricultural sector, unlike other resource-based sectors, involves millions of small-scale entrepreneurs, who make individual decisions on the investment of capital and management of natural resources. Although the land- and water-use decisions of an individual farmer seem insignificant, these decisions, when repeated over time collectively, may have a global impact on the environment. In order to minimize the pressure on resources, small-scale entrepreneurs have to embrace eco-innovation in their strategic agenda. The constraints to innovation among SMEs threaten environmental performance. The factors that limit SMEs in Africa include access to resources and market opportunities; cost of doing business; policy gaps; poor infrastructure; inadequate workforce skills; adverse business start-up climate; low technological environment and innovation. However, many of these challenges happen to be more prevalent in Uganda than any other country in the region (Muhanguzi and Kyobe 2014). The study explores potential opportunities for innovative SMEs that can be used to integrate sustainable activities into business practices and the implications this has on water and land use since innovativeness is highly context-dependent (Morand 2008); yet, the constraints that affect SMEs are sector-specific, it is important to focus on SMEs in a single sector. This is in line with Loader's (2010) recommendation that the future research on barriers that SMEs face should be investigated by sector.

2.2 Literature Review

2.2.1 Eco-Innovation: Conceptual Definitions

There is hardly any universal definition of eco-innovation, although several attempts have been made in the literature. According to Kemp and Pearson (2008), eco-innovation is the production, assimilation or exploitation of a novelty in products, production processes, services or in management and business methods, which aims, throughout its life cycle, to prevent or substantially reduce environmental risk, pollution, and other negative impacts of resource (including energy) use. Oltra and Jean (2009) broadly define eco-innovation as innovations that consist of new or modified processes, practices, systems, and products, which benefit the environment and thus contribute to environmental sustainability. Kemp and Pearson (2008) define innovation as the production, application, or exploitation of a good, service, production process, organizational structure, management or business methods that are novel to the firm or user and which results, throughout its life cycle, in a reduction of environmental risk, pollution, and the negative impacts of resource use (including energy), compared to relevant alternatives. Kemp and Pearson's (2008) definition pays attention to the adoption of innovation previously introduced by others and focuses on environmental effects contrary to the conventional economic perspective of innovation (Schumpeter 1934), which regards innovation as the first introduction of a new product, process, or organizational structure. Therefore, it is important in the African context, because SMEs on the continent mainly depend on innovations already developed by other firms. This research explores both the constraints that are related to the creation of new products, processes, and organizational methods, as well as those related to the adoption of eco-innovation practices already developed by other firms or individuals. Another distinctive feature of eco-innovation is that it is not limited to innovation in products, processes, and organizational methods, but includes innovation in social and institutional structures (OECD 2009). It is in this context that this paper adopts Kemp and Pearson's (2008) definition of eco-innovation because it is more relevant to the study context.

In general, there is consensus among authors that eco-innovation reduces negative environmental impacts caused by consumption and production activities, whether intended or not. However, according to Carrillo-Hermosilla et al. (2010), the definitions that emphasize the intention of the innovator are problematic, as it becomes difficult to establish the relationship between one's intention and environmental performance. Further, authors allude to problems with definitions that focus on environmental impact, because it is quite challenging to deduce which innovation actually reduces the environmental impact of products and production. In sum, this paper considers the main distinguishing feature of eco-innovation as its likelihood to improve environmental performance. Moreover, it does not matter whether the initial motivation of an entrepreneur to innovate had economic or social reasons, provided the outcome of the activity improves environmental performance.

From a theoretical perspective, eco-innovation is acknowledged as bringing incremental or radical changes to products or systems. However, it is also known that most innovations by SMEs within Africa are a result of incremental changes (Carrillo-Hermosilla et al. 2010). The authors define incremental changes as gradual and continuous competence-enhancing modifications that preserve existing production systems and sustain the existing networks, creating added value in the existing system in which innovations are rooted. This is in contrast to radical changes, which are competence-destroying, discontinuous changes that seek the replacement of existing components or entire systems and the creation of new networks, creating value added. In relation to the foregoing distinction, Christensen (1997) draws a line between sustaining innovation and disruptive innovation, by describing the latter as innovation that renders obsolete existing systems and structures akin to radical innovation. Unlike incremental innovation, system innovation or radical technology change has often been related to environmental performance (Kemp and Pearson 2007; Nill and Kemp 2009). It is observed that more systematic changes can potentially yield higher environmental improvement in the long run, compared to simple modifications in products and process (OECD 2009). This study examines both mechanisms of innovation to get a clear understanding of the nature of constraints embedded along each path.

The growing importance of eco-innovation demonstrates different lines of research. Whereas the first line focuses on motivations for eco-innovation and related environmental outcomes, the second examines the measurements of eco-innovation and the third explores the dimensions of eco-innovation (Cheng and Shiu 2012). In order to understand constraints, it is important to focus on the activities that happen in the different dimensions of eco-innovation. In line with this, eco-innovation activities are studied along three main dimensions: target, mechanism, and impact of innovation on the environment (Oncioiu 2015). Target is described as the focus area of eco-innovation, which includes processes, product, marketing methods, organizations, and institutions. Mechanism refers to the way in which changes are made in the target, which may be modification, redesign, alternative, and creation, while impact is defined as the effect of eco-innovation on the environment. The discourse on eco-innovation activities suggests that constraints are at every stage of the innovation life cycle. It is believed that, enabling SMEs to change their environmental behaviors requires weakening the resistant forces to eco-innovation, such as poor eco-literacy, strengthening of the driving forces, such as effective research or preferably a combination of both to encourage a strategic response (Tilley 1999). Factors driving eco-innovation in SMEs are examined in the next section.

2.2.2 Drivers of Eco-Innovation in SMEs

It is well known that SMEs, irrespective of their location, are better placed to adopt eco-innovation because they are more agile and flexible; therefore, they are a focus for eco-innovation. For instance, decision-making in SMEs is often less bureaucratic and often depends on a single person. According to Oxborrow and Brindley (2013), eco-innovation requires a firm's internal capacity to acquire information, absorb knowledge, commitment from teams and organizational learning. Conversely, Cuerva et al. (2014) study shows that innovation propensity depends on the availability, or the lack of resources and capacity to innovate by firms. Regarding this, knowledge resources, human skills and provision, as well as access to finance are essential drivers of eco-innovation for firms (Cuerva et al. 2014). The role played by a firm's internal capacity to acquire information was also emphasized in Pinget et al. (2015), which reveals the need for effective research, education, and training as major drivers to eco-innovation for small firms. SMEs in Uganda, especially those in the agricultural sector, lack financial resources and consequently rarely employ skilled labor, yet they find difficulty in accessing public extension workers. The current agricultural knowledge system gap in Uganda provides opportunities for entrepreneurs to close this gap. Furthermore, a firm's capacity to innovate also depends on technology capability. Since research and development (R&D) enhances technology capabilities in green firms, SMEs without it are likely to face a cost disadvantage in developing innovations. Many SMEs in Uganda are not undertaking R&D; similarly, their human skills and technological capability are lacking. SMEs often depend on technologies already developed by universities, public research centers, and foreign companies, but where such technologies are not easily accessed; the rate of adoption is often negatively affected. The lateness associated with the release of biotechnology policy in Uganda may account for low levels of eco-innovation activities in the country.

According to Lewis and Cassells (2010), the factors that drive SMEs to adopt green practices include government legislation; the need to expand market share; employee motivation and eco-efficiency. Eco-efficiency is described in relation to cost reduction due to the efficient use of such resources as water, materials, improvements in the quality of products and more effective management of risks. Since SMEs have a more reactive stance to innovation, they are more inclined toward eco-efficiency, as a first step in eco-innovation (Klewitz, et al, 2012), in a development that affirms the importance of economic benefits as critical drivers of eco-innovation in SMEs. In light of this, since SMEs adapt easily to opportunities in their external environment, they are more likely to easily adjust to positive environmental behavior, if they are saving a cost or making a gain, as opposed to improving images or becoming legitimate in communities.

The analysis of the SMEs sector in Uganda suggests that some of the drivers highlighted above may not apply in the context where SMEs operate informally. Therefore, any effort to influence SMEs' positive behavior toward environmental performance should focus on the drivers relevant to sub-Saharan Africa (SSA). For instance, Uganda, like many SSA countries, lacks a dedicated eco-innovation policy and measures to promote eco-innovation which are often framed under the national environmental policy. SMEs that operate informally may not easily respond to government regulations, especially where there is no dedicated policy to guide enforcement. Therefore, using environmental laws to initiate eco-innovation is bound to fail. For this reason, Uganda's legal environment may not effectively influence SMEs' behavior toward green practices. It is on this basis that this paper explores additional information to understand the nature of opportunities to exploit in accelerating adoption of eco-innovation for SMEs.

Another body of knowledge demonstrates the relationship between the drivers of eco-innovation and environmental performance, but most of the work has been done in formal economies. For example, there is evidence that government regulation is particularly important with regard to rebating air, water, and noise pollution, increasing recyclability and avoiding hazardous substances (Horbach et al. 2012). Furthermore, cost-saving is important in reducing energy and material use and customer requirements is important with regard to products with improved environmental performance and process innovation, increases material efficiency, reduces energy consumption, waste, and the use of dangerous substance. This discourse suggests that the drivers for eco-innovation are linked to the type of innovation (product, process, organization, and institution). For example, market-based forces are related to products and process innovation. Likewise, regulatory drivers are linked to resource use hence product innovation. By demonstrating a clear linkage between the drivers of eco-innovation and the type of innovation, there is no doubt that different drivers of eco-innovation ultimately have different areas of environmental impact.

2.3 Methodology

Since research on the constraints to and opportunities for innovation in green SMEs within the agricultural sector in Uganda is limited, the study employs mixed methods. First, using existing survey data and extensive literature review, various constraints to innovations of Ugandan SMEs are examined. Subsequently, a comparative case study analysis of three green enterprises is done to allow for a more detailed and contextualized investigation of the opportunities for innovation and implications for water and land management. The use of case studies is in line with Yin (2003), who advocates for case studies in the early stages of developing a theory. Comparative case studies are particularly useful for understanding and explaining how context influences the success of a phenomenon and how better to tailor the phenomenon to the specific context to achieve intended outcomes. Therefore, the case study approach provides an intimate feel of owner/managers' experience, a deeper understanding of opportunities for each of the cases, and the implications for resources management.

2.3.1 Research Sample

The case study participants were selected based on the three motivations of corporate sustainability: eco-centric orientation, value creation, and compliance (Keijzers 2002, 2005). According to Hamann et al. (2015), a manager who is motivated by the compliance driver minds about cleanup or health and safety efforts of a company. Managers who hold the eco-centric view mind about adequate environmental and social management, while those who are driven by value creation care about integrating all ecological and social issues into all business decision-making (Hamann et al. 2015). Using purposive sampling, three companies with characteristics that fit the above criteria were selected and they produce organic products. These companies were recommended at the Food Technology and Business Incubation Center in Makerere University.

2.3.2 Data Collection

2.3.2.1 Secondary Data

Secondary data retrieved from Uganda National Council for Science and Technology (UNCST) 2014 for the Uganda National Innovation Survey were employed for the study. The survey was conducted by the Uganda National Bureau of Statistics (UBOS), together with UNCST. The questions were adopted from the Community Innovation Survey (European Commission 2000). The Uganda National Innovation Survey was designed to cover all sectors, including agriculture. The questionnaire raised questions from companies about their innovation activities and expenditures; types of innovation; sources of information; collaborative partners for innovation; effects of innovation; factors hampering innovation and intellectual property rights; and the impact of innovation on environmental performance. The latter was important in assessing the extent to which the innovation in the companies selected can be described as eco-innovation.

2.3.2.2 Primary Data

The data from the three case studies was collected in two stages, using interviews, observation, document analysis (industry statistics, reports, media, and government environmental regulation reports), targeting the most knowledgeable informants. A combination of groups and in-depth interviews for the business owners was employed for the study. Six in-depth interviews were carried out among managers and workers to understand the nature and constraints to innovation in the selected firms. The interview sessions were recorded to get an accurate summary of the interview and to help in transcribing the answers. In the first stage of the interview, emphasis was mainly on examining the opportunities and barriers, while in the second phase monitoring the impacts that innovative activities of green enterprises have on water and land management was conducted. The study used semi-structured interviews within pre-defined topics. The interview was guided by the stages of eco-innovation along its life cycle, including idea generation, innovation, investment decision, invention, as well as adaptation process and commercialization. Therefore, constraints were examined at each of these stages.

2.3.3 *Data Analysis*

The first set of empirical analyses for which results are presented is based on data collected during the 2014 National Innovation Survey. In the 2014 survey, 423 firms were surveyed based on random sampling. Tables, charts, and figures are generated based on the analysis of the data.

A thematic approach to data analysis was used in order to identify commonalities and differences between responses of people in the same position (Stebbins 2001). An iterative approach was followed in order to identify any emerging patterns and themes in the data. The data were assigned codes and counted to determine how often each code was repeated by the respondents. This process was carried out by the second author and verified by the first author to achieve greater objectivity. In order to assess the implications that the activities of green SMEs have on water and land management, an evaluation approach was adopted, using a set of short-term indicators but with a sound theoretical path since the research project was only six months. The evaluation model followed a clear scientific protocol to ascertain the implications that green innovations have on land and water management.

2.4 Research Findings

The research findings are structured in two parts: firstly, a quantitative analysis of secondary data to get the nature of SMEs' practices in innovation and the constraints they face. It is important to note that the constraints that SMEs face in innovation are contextual. This implies that green enterprises, in the same study area as other SMEs, share the same constraints that other SMEs meet along the innovation path. In fact, eco-innovation only differs from traditional innovation (Schumpeter 1934) on the pretext that eco-innovation has a positive impact on the environment. Therefore, the quantitative analysis provided a good basis to interrogate the constraints to innovation, using a large sample size. Secondly, a qualitative contextualized analysis was made to provide an accurate position of opportunities for eco-innovation and the implications for water and land management. The presentation of results from secondary data starts with the characteristics of firms covered.

2.4.1 Characteristics of Study Companies

2.4.1.1 Period of Establishment

The key inclusion and definition of SME were that the company has not more than 250 employees. In total, 423 companies were retrieved from secondary data. Majority of these companies were established between 2001 and 2011 (51%), with those established before 2001 constituting about 35%. On average, the companies had been operating for 14.4 years (95% ci = 13.7–15.1 years). This implies that the majority of the companies have a fairly long life to have ventured into innovation. Therefore, their inability to innovate is more likely to be caused by constraints and not necessarily time. Table 2.1 shows the details of the proportions of companies by origin.

About three out of any ten companies were part of a larger group, with operations outside Uganda (29%). This is typical of most SMEs in Uganda given that majority operate on an informal basis, which in itself limits innovation. Considering the companies which are part of a larger group, majority of them (74%) had their head offices in Uganda, with the remaining 26% in other countries (other country in the

Table 2.1 Proportion of companies by origin

	Thru to 1990	1991–2000	2001–2011	Not stated	Total
	$N = 38$	$N = 148$	$N = 214$	$N = 23$	$N = 423$
Whether the enterprise is part of a larger group					
Yes	28.9	31.8	26.2	34.8	28.8
No	71.1	68.2	73.8	65.2	71.2
Total	100	100	100	100	100
If part of larger groups, country where the head office is located					
	$N = 11$	$N = 47$	$N = 56$	$N = 8$	$N = 122$
Uganda	90.9	72.3	69.6	97.5	73.8
Other EAC country	0.0	10.6	7.1	0.0	7.4
Other African countries	0.0	0.0	7.1	0.0	3.3
USA	9.1	6.4	7.1	0.0	6.6
Others	0.0	10.6	8.9	0.5	9.0
Total	100	100	100	100	100
Country where the head office is located–all companies considered					
Uganda	97.4	91.2	92.1	95.7	92.4
Other EAC country	0.0	3.4	1.9	0.0	2.1
Other African countries	0.0	0.0	1.9	0.0	0.9
USA	2.6	2.0	1.9	0.0	1.9
Others	0.0	3.4	2.3	4.3	2.6
Total	100	100	100	100	100

Data Source UNCST and UBOS, 2014

EAC[1] = 7.4%, other African country[2] = 3.3%, USA = 6.6%, and other countries in the world outside Africa and USA = 9%). This suggests that the majority of all the sampled companies had their head offices in Uganda.

2.4.1.2 Geographical Markets Supplied by Ugandan SMEs

The main geographical markets, where the companies sold products in the period of 2011–2014, included Uganda (14%), East African markets (13%), Common Market for Eastern and Southern Africa (COMESA) (12%), other African countries other than those in East Africa and COMESA (13%), European countries (12%), USA (12%), Asia (12%), and other countries (11%). The structure of the markets for the products is spread among the different segments of the market with the largest concentration in regional markets.

[1]This was mainly Kenya.

[2]Countries mainly included South Africa, Ghana, and Nigeria.

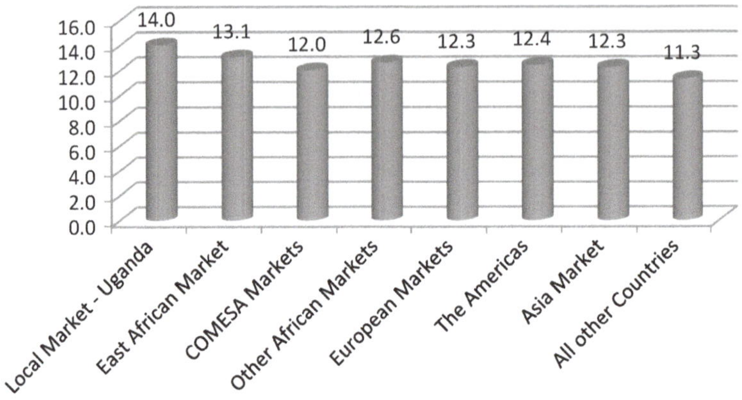

Fig. 2.1 Proportion of companies supplying different markets. *Data Source* UNCST and UBOS, 2014

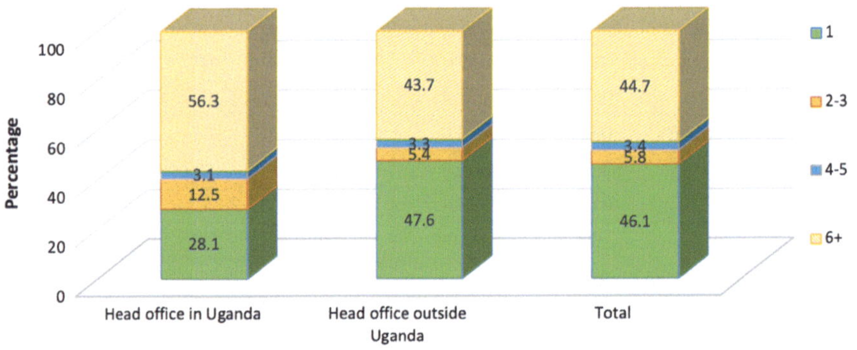

Fig. 2.2 Number of geographical markets served. *Data Source* UNCST and UBOS, 2014

Furthermore, Fig. 2.1 indicates that majority of the companies (46%) supplied only one of the geographical markets while about 45% supplied at least six of the geographical markets. The number of geographical markets supplied had neither significant relationship with the number of years the companies had been in existence nor whether the company had their head offices in Uganda or outside (Fig. 2.2).

2.4.1.3 Employee Composition and Characteristics

In total, the companies had about 33, 300 employees (males = 70.4%, females = 29.6%) but with the numbers varying across companies. The results presented in Fig. 2.3 show that 31% of the companies have no more than 10 employees and 27% had 11–20 employees, while 14% had 21–30 employees. Only 15% of the companies had more than 50 employees. This is typical of SMEs in developing countries.

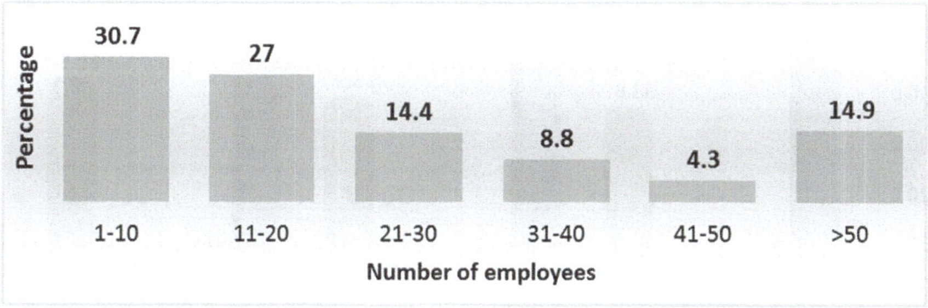

Fig. 2.3 Number of employees in the companies. *Data Source* UNCST and UBOS, 2014

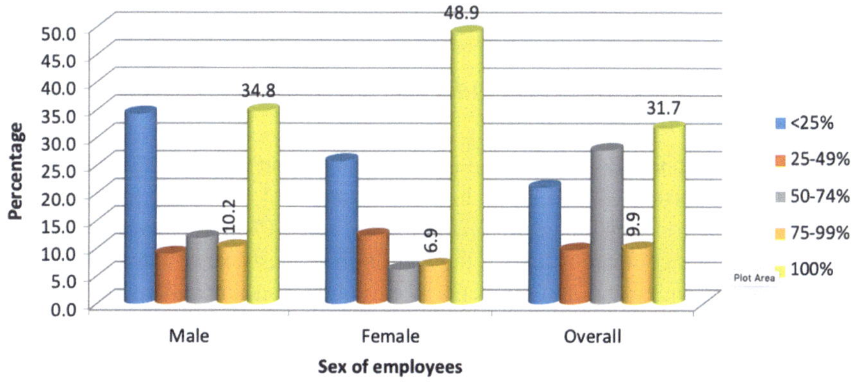

Fig. 2.4 Number of employees with university education. *Data Source* UNCST and UBOS, 2014

The employee's level of education influences the company's capacity to absorb information, hence innovation. On exploring the proportion of company employees with a university degree, results presented in Fig. 2.4 illustrate that about 37% of the companies had all their employees holding a university degree. Overall, about 69% of the companies had at least half of their employees holding university degrees.

2.4.1.4 Turnover

Figure 2.5 depicts a growth in turnover[3] of companies. The proportion of companies with a turnover exceeding UGX100m (the exchange rate in 2014 was UGX2800 per US$1) increased from 34% in 2012 to 36% in 2013 and 40% in 2014. The companies had an average growth rate in their turnover of about 15.4% between 2012 and 2013, which reduced to 12.4% between 2013 and 2014. This suggests a moderate, if unimpressive, growth in turnover.

[3]Turnover is defined as the market sales of goods and services (Include all taxes except VAT).

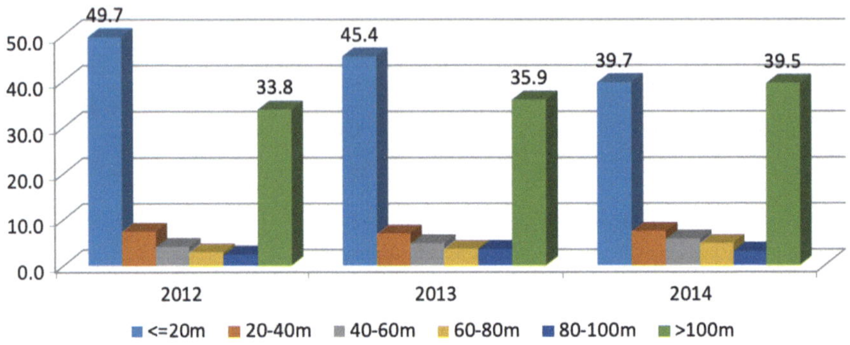

Fig. 2.5 Distribution of sales revenue for the three-year period. *Data Source* UNCST and UBOS, 2014

2.4.2 *Types of Innovation Undertaken by Uganda SMEs*

Companies had innovation during the period of 2011 and 2014. Of the total companies, the proportion of SMEs that had organizational,[4] marketing, process, and product[5] innovation constituted 81%, 75%, 65%, and 61%, respectively. Whereas about 88% of the companies had an innovation in at least one of the areas (product, process, organization, and marketing), only 48% had innovations in all these areas. Given the fact that innovation is multidimensional, the statistics suggest that inadequate innovation is experienced by the sampled firms.

2.4.2.1 Product Innovation

Product innovation is associated with the introduction to market of a new or significantly improved good or service with respect to its capabilities, such as improved user-friendliness, components, software, or subsystems. Whereas about 41% of the companies introduced either a service (41%) or product innovation (42%), only 22% had both product and service innovation during the three years 2011–2014 (Table 2.2). The innovations were mainly developed/generated within the enterprises (60%), with the enterprise group constituting 20%, other enterprises 13%, and institutions/SME in partnership with other institutions 7%. The statistics suggest that firms conduct own innovation with limited interaction from firms outside their systems. Given the

[4]Included new business practices for organizing procedures, new methods of organizing work responsibilities, and decision-making or new methods of organizing external relations with other firms or public institutions.

[5]A product innovation is the introduction to market of a new or significantly improved good or service with respect to its capabilities, such as improved user-friendliness, components, software, or subsystems. The innovation (new or improved) must be new to your enterprise, but it does not need to be new to your industry sector or market. It does not matter if the innovation was originally developed by your enterprise or by other enterprises.

Table 2.2 Proportion of firms that introduced product and services innovations

Theme	Particulars	%
Proportion of SMEs that had new or significantly improved	Goods	41.8
	Services	41.1
	Both product and service	22.2
By who were these products (goods and services) innovations developed?	Mainly your enterprise	59.7
	Mainly your enterprise group	19.8
	Mainly your enterprise by adapting or modifying goods or services originally developed by other enterprises or institutions	13.1
	Your enterprise together with other enterprises or institutions	7.4
	Total	100
Did these innovations originate mainly in Uganda or abroad?	In Uganda	81.5
	From outside Uganda	18.5
	Total	100
Whether innovations were new to your market or new to your firm	New to the market	43.5
	Only new to your firm	56.5
	Total	100

Data Source UNCST and UBOS, 2014

constraints to innovation, this trend implies that limited innovation takes place among these firms. This is further underpinned by the fact that about 81% of the enterprises had innovation ideas originated from within Uganda while 19% originated from outside the country. Given that firms in Uganda hardly conduct R&D, they are not likely to innovate with landmark outcomes and effects, hence the limited innovation. However, most of the firms noted that the ideas were not new to the markets but rather new to their firms.

2.4.2.2 Process Innovation

Process innovation is the use of new or significantly improved methods for the production or supply of goods or services. The innovation (new or improved) must be new to the enterprise, but it does not need to be new to the industry sector or market. It does not matter if the innovation was originally developed by the owner's enterprise or by other enterprises. The results presented in Table 2.3 show that within the reference period of 2011–2014, more than half of the SMEs had new or significantly improved on their methods of manufacturing or producing goods or services (52%). About 46% improved on logistics, delivery or distribution methods for inputs, goods or services, while 43% introduced or significantly improved operating systems for

Table 2.3 Proportion of firms that introduced process innovation

Item	Innovation attributes	%
Proportion of SMEs that have new or significantly improved	Methods of manufacturing or producing goods or services	51.5
	Logistics, delivery, or distribution methods for your inputs, goods or service	45.6
	Supporting activities for processes, such as maintenance and operating systems for purchasing, accounting, or computing	42.8
	All the three process innovations	27
By who were these products (goods and services) innovations developed?	Mainly your enterprise	62.9
	Mainly your enterprise group	18.4
	Mainly enterprise by adapting or modifying goods or services originally developed by other enterprises or institutions	11.4
	Your enterprise together with other enterprises or institutions	7.4
	Total	100
Did these innovations originate mainly in Uganda or abroad?	In Uganda	81.5
	From outside Uganda	18.4
	Total	100

Data Source UNCST and UBOS, 2014

purchasing, accounting, and computing. Only 27% had new or significant improvements in all the three process innovations. The process innovations were mainly originated by the enterprises themselves (63%), enterprise groups (11%), and other enterprises (11%), and in partnership with other institutions (7%).

Majority of the process innovation ideas originated from within Uganda (82%), with those originating from outside countries constituting about 18%. Like the case of product innovation, process innovation was dominated by ideas originating from within Uganda, with limited interaction from outside the country.

2.4.2.3 Organizational Innovation

An organizational innovation is a new organizational method in the enterprise's business practices (including knowledge management), workplace organization or external relations that have not been previously used by an enterprise. It must be the result of strategic decisions taken by management. Results in Table 2.4 reveal that about 64, 74, and 48% of the SMEs introduced new or significantly improved on their business practices, methods for organizing work responsibilities, decision-making, and methods for organizing external relationships with other firms, respectively. It is

Table 2.4 Proportion of firms that introduced organizational innovation

Proportion of SMEs that introduced new or significantly improved …	% of SMEs
Business practices for organizing procedures (i.e., supply chain management, business re-engineering, knowledge management, lean production, quality management, etc.)	64.3
Methods of organizing work responsibilities and decision-making (i.e., first-time use of a new system of employee responsibilities, teamwork, decentralization, integration or de-integration of departments, education/training systems, etc.)	73.5
Methods of organizing external relations with other firms or public institutions (i.e., first-time use of alliances, partnerships, outsourcing, or subcontracting)	48.2
Innovation in all the three organizational areas	38.5

Data Source UNCST and UBOS, 2014

noted that only 39% of the firms carried out all the three organizational innovations, a rather small proportion. Organizational innovation is imperative in the current competitive world of business, where telecommunications technology is at the core of management discourse and therefore critical to the survival of firms.

The performance index of organization innovations on various outcomes, including increase/maintenance of market share; improvement in the quality of goods and services; reduction in the costs of production per unit output; and improvement in employee satisfaction sheds more lights. The main areas where the organizational innovations were rated to have contributed significantly included improving quality of goods and services (29%); increasing or maintaining market share (26); improvement of employee satisfaction (24%); and reduction of production costs per unit output (21%). Note that firms that reported organizational innovations did not reduce time taken to serve customers or client needs. This implies that firms were only able to serve a few customers, which affects sales margins.

2.4.2.4 Marketing Innovations

A market innovation is the implementation of a new marketing concept or strategy that differs significantly from the enterprise's existing marketing methods, and which has not been used before. It requires significant changes in product design or packaging, product placement, product promotion or pricing. As illustrated in Table 2.5, within the reference period of 2011–2014, 57% of the SMEs changed or improved on their packaging for commodities; 40% considered better methods for product promotion; 39% improved on their distribution channel; and 57% adjusted their pricing for goods and services in order to place themselves in a more competitive position. It is noted that because of the costs associated with placement and promotion innovations, only 20% were able to implement all the four areas of marketing innovations, an extremely small proportion of the firms in the sample.

Table 2.5 Proportion of firms that introduced marketing innovation

Proportion of SMEs that new or significantly improved …	% of SMEs
The aesthetic design or packaging of a good or service (exclude changes that alter the product's functional or user characteristics—these are product innovations)	56.5
Media or techniques for product promotion (i.e., the first-time use of a new advertising media, a new brand image, introduction of loyalty cards, etc.)	39.7
Methods for product placement or sales channels (i.e., first-time use of franchising or distribution licenses, direct selling, exclusive retailing, new concepts for product presentation, etc.)	39
Methods of pricing goods or services (i.e., first-time use of variable pricing by demand, discount systems, etc.)	56.5
In all the above market innovations	20.3

Data Source UNCST and UBOS, 2014

2.4.2.5 Sources of Information Used for Product and Service Innovations

The results presented in Table 2.6 indicate the ratings for the various information sources on innovation. Clearly, the results reveal limited use of education and research institutions, conferences and trade fairs, scientific publications and professional and industry associations in acquiring information on process and product innovations.

Information sources mainly used include internal sources within the enterprise and its networks; clients/customers; suppliers and competitors. There is limited consideration for interactive approaches to innovate; instead, firms concentrated on internal approaches which may not be the best strategy. This development may arise from lack of opportunities or funds.

For the sources of information outside a firm, only one in three SMEs collaborated in any of its innovation activities with other enterprises or institutions during the reference period of 2011–2014, as illustrated in Table 2.7.

The main partners associated with included clients/customers (20%); suppliers (17%); other enterprises in the group (16%); competitors (15%); consultants (12%); government and research institutes (10%); and universities or institutes of higher learning (10%). It should be noted that the most valued partners associated with collaboration were clients or customers, as illustrated in Table 2.7. Majority (55%) of the enterprises considered clients/customers as their most valuable partners, while 13% considered the other enterprises in the group to be the most valuable for their enterprises and 11% considered suppliers.

2.4.2.6 Effects of Product and Process Innovations

The objective of innovation introduced by a firm is related to the outcome (Horbach et al. 2012). Table 2.8 presents the rating scores of effects arising from the product and service innovations that were taken during the reference period of 2011 and 2014. The

Table 2.6 Proportion of firms by sources of information used for innovation

	Source	High	Medium	Low	Not used	Total
Internal sources	Sources within your enterprise or enterprise group	43.3	24.8	5.2	26.7	100
Market sources	Clients or customers	42.3	22.9	7.8	27.0	100
	Suppliers of equipment, materials, components, or software	21.5	27.4	13.5	37.6	100
	Competitors or other enterprises in industry	19.9	24.3	20.6	35.2	100
	Consultants and commercial laboratories	12.5	12.8	13.9	60.8	100
Education and research institutes	Government, public, or private research institutes	5.0	10.2	15.1	69.7	100
	Universities or other higher education institutions	4.5	9.9	17.7	67.8	100
Other sources	Conferences, trade fairs, exhibitions	13.7	21.0	14.4	50.8	100
	Professional and industry associations	11.1	14.4	16.8	57.7	100
	Scientific journals and trade/technical publications	7.3	18.0	16.3	58.4	100

Data Source UNCST and UBOS, 2014

leading effects from the product and process innovations include improvement on the quality of goods and services (13%); increase in the range of goods and services (11.3%); increased capacity for producing goods and services (10.3%); improved flexibility in production (10.3%); increased market share (9.8%); and penetration into new markets (10.3%). Other benefits include the reduction of production costs (8.8%) and replacement of outdated products or processes (9.2%) as well as reduction of environmental impact (7%). The fact that only 7% of the effects reduced negative environmental impacts demonstrates that innovations are not primarily geared toward mitigating negative environmental impacts.

2.4.2.7 Factors Hampering Innovation Activities

Constraints are considered as evolving and changing at the various stages of the eco-innovation process. The constraints experienced by a firm at any of the stages may

Table 2.7 Most valued partner for the enterprise

Institution/partner	Partner cooperated with (%)	Most valuable partner for the enterprise (%)
Clients or customers	19.5	55.2
Suppliers of equipment, materials, components, or software	16.7	11.0
Other enterprises within your enterprise group	16.2	13.1
Competitors	15.2	7.6
Consultants, commercial laboratories	11.7	5.5
Government, public, or private research institutes	10.4	4.1
Universities or other higher education institutions	10.2	3.5
Total	100.0	100

Data Source UNCST and UBOS, 2014

Table 2.8 Importance of the effects of the products and process innovations during the period of 2011–2014

	Effect	Percentage
1	Improve the quality of goods or services	12.8
2	Increased range of goods or services	11.4
3	Entered new markets	10.3
4	Improve flexibility for producing goods or services	10.3
5	Improve capacity for producing goods and services	10.3
6	Increased market share	9.8
7	Improved working conditions on health and safety	9.2
8	Reduce production costs per unit output	8.8
9	Reduced environmental impacts	7.0
10	Meet governmental regulatory requirements	10.0
	Total	100.0

Data Source UNCST and UBOS, 2014

mean a company has to abandon the innovation or may face serious delay as a result of failure in addressing a particular constraint. This may result into dropping the idea at the concept stage or dropping an idea that is ongoing or sometimes delayed. The results reveal that within the reference period of 2011–2014, 32% of the SMEs delayed some of their innovation activities, while 15% abandoned some activities after initiating their implementation, with a similar proportion abandoning activities while still in their conceptualization stages. Although it is evident that more firms are able to overcome the constraints compared to those that fail, it is also true that abandonment and delay in execution of innovations are of reasonable proportion.

The factors that limit enterprises from introducing an innovation or influence decisions to abandon innovation activities are divided into four: cost, knowledge, market, and other factors. These factors were further divided into sixteen subcategories (Table 2.9). The most severely deterring factor to innovation is the "cost factor," more specifically lack of finance from sources outside a firm. The financial system in Uganda imposes serious barriers for SMEs that need funding for innovation. In addition to cost factors, market factors also pose serious challenges to Ugandan SMEs. The knowledge related factors, such as difficulty in finding collaborative partners for innovation, lack of qualified personnel are also often mentioned as important factors deterring SME innovation in Uganda.

The data also reveal reasons for a firm not to innovate. This demonstrates the proportion of firms that did not venture into innovation. A small number of firms

Table 2.9 Importance of obstacles hampering innovation for Ugandan SMEs

Hampering factors	%	Cumulative %	
Cost factors	Lack of funds within your enterprise or group	10.1	35.4
	Innovation costs too high	9.6	
	Lack of finance from sources outside your enterprise	8.1	
	Excessive perceived economic risks	7.5	
Knowledge factors	Difficulty in finding cooperation partners for innovation	6.7	25.4
	Lack of information on technology	6.5	
	Lack of qualified personnel	6.2	
	Lack of information on markets	6.0	
Market factors	Market dominated by established enterprises	7.9	20.5
	Uncertain demand for innovative goods or services	7.1	
	Innovation is easy to imitate	5.5	
Reasons not to innovate	No need because of no demand for innovations	4.0	7.7
	No need due to prior innovations	3.7	
Other factors	Limitations of science and technology public policies	5.9	10.9
	Organizational rigidities within the enterprise	5.0	
	Total	100.0	100

Data Source UNCST and UBOS, 2014

indicated lack of involvement in innovation, which is an indication that majority of Ugandan SMEs try to be innovative.

The factors presented were categorized as internal and external factors to the organization. Internal factors relate to firm characteristics, management structure, and human resources, while external factors include institutional factors, access to finance, access to information on technologies, among others. The top three factors constraining innovation among SMEs are "cost factors". Cost factors include lack of funds within the enterprise; high innovation costs; lack of finances from sources outside the enterprise; and excessive perceived economic risks constituted the highest percentage of factors hindering innovation (35%). Cost factors were followed by knowledge factors, such as difficulty in finding cooperation partners for innovation; lack of information on technology; lack of skilled personnel; and lack of information on markets, which cumulatively accounted for about 25%. These constraints are attributable to the weak education system that is not producing graduates, with the requisite skills to promote innovation, cost and time of collecting information on technologies.

The external factors that were perceived as important in deterring innovation relate to markets and other institutional factors. Under the category of markets, dominance of established firms, uncertain demand for innovative goods and services, and the tendency by firms to imitate innovation developed by other firms were perceived as major constraints for innovation. The factors related to institutional barriers, include limitation of science and technology, public policies, and insufficient flexibility of regulation or standards. A question was asked regarding whether organization rigidities within enterprise constrain innovation among firms. The fact that SMEs are more flexible, and organization rigidities had minor effect on firm innovation behavior.

2.4.3 Case Study Analysis

2.4.3.1 Characteristics of Firms

The firms that participated in the interview are described using codes. Company one (C1) is owned by female entrepreneurs, who are involved in producing organic mushrooms. The firm produces different products, including mushroom powder, confectionery, jelly, and nutritious porridge for babies. The market for their product is mainly local. Besides, C1 offers two main services, that is, training and massaging using organic products. Company two (C2) is owned by a group of female entrepreneurs in their youthful age. The company produces two products: organic fertilizer made out of urine and cooking bananas, which they sell on the local market. Also, company two (C2) offers training to its clients. Company three (C3) is owned by a male entrepreneur and produces mainly organic vegetables for the local and international market. Two people in each of the companies were interviewed: the owner/manager and one senior staff. The number of employees in each of these companies was 9. 12 and 25, respectively.

2.4.3.2 Opportunities for Innovation

There are factors that present opportunities to entrepreneurs in Uganda to participate in eco-friendly business practices. Interviews for the study reveal a number of factors, including the new customers who are demanding eco-friendly products; suppliers mainly from the Brazil, Russia, India, China, and South Africa (BRICS); cost-saving technologies; and indigenous knowledge. Table 2.10 identifies opportunities for innovation and the decision taken by the entrepreneurs to exploit the opportunity. Two respondents were interviewed from each firm.

2.4.3.3 New Clients Demanding Eco-Efficient Products

From the analysis of secondary data, it is evident that SMEs mainly depend on customers for innovation. The same trend is associated with the case studies. However, for green enterprises, not all customers matter. The main market for green products in Uganda is largely customers from Europe and Asia. This implies that the customers from foreign countries have great potential to influence the behaviors of firms toward eco-innovation. With the increasing number of foreign companies, especially from Brazil, Russia, India, China, and South Africa (BRICS), local demand for organic products greatly increased, yet supply did not grow at the same rate. There are imports of eco-friendly products, mainly from Kenya and South Africa, but local products still have space on the shelves. According to the owners of company C3, one of their customers requested them to open up a weekly market for fresh vegetables at his restaurant. The idea improved company C3's sales and was very much appreciated. In addition, the owner/manager (C3) also realized that their clients preferred Asian vegetables; therefore, they adjusted the production mix by introducing the new varieties to match customer preferences. On the other hand, company C1 diversified into medicinal products, having interacted with medical doctors at one of the hospitals, where they sell their products. According to an employee in C1', mushrooms are very nutritious and are good as "*antibiotics*". This is the main reason why their customers opted for the medicinal products. For company C2, the owner was proud about the high demand for urine—organic fertilizers—which continued to grow with the new packaging. The new clients, especially from Asia created a market niche, which the companies are taking over by introducing new products, packaging, and management strategies. Overall, most of the innovations were initiated by customer markets, which, again, confirm earlier findings from the secondary data.

2.4.3.4 New Suppliers Offering Eco-Efficient Inputs

The results of this study suggest that most of the innovations in Ugandan SMEs are incremental and mainly based on adaptation and learning. There is a great opportunity for local entrepreneurs to learn from both foreign and local suppliers that produce low-cost inputs. The new innovative suppliers, who have introduced more efficient

Table 2.10 Opportunities for innovation in green enterprises

Identified opportunities	Examples
New clients demanding eco-efficient products	Guided by customers to start weekly farmer market at the restaurant for the fresh organic vegetables (C3) Adopted Asian vegetables because customers from China prefer that type (C3) Interest by new customers in the medicinal benefits of using mushrooms (C1) The growth in demand for organic fertilizer on a daily basis (C2)
New suppliers offering eco-efficient materials	Use organic pesticides prepared by our supplier (C3) Use cockpit husks from Asian suppliers to improve the germination rate (C3) Dependency on a company from China for already made mushroom kits (C1) The emergence of urine separating toilets to maintain big clients (C2)
Local Indigenous knowledge	Use of cheap wood ash and red chili to control banana weevils (C3) Use of tobacco to control snakes that carry a virus which destroys mushroom (C1) Use of urine than industrial fertilize which is cheaper and available (C3) The locally fabricated solar dryers are cheaper and easy to get (C1) Use of dogs to scare away monkeys from our gardens (C2, C3)
Farmer led demonstration units and local production networks	Money from training on mushrooms significantly improved revenues (C1) Working with horticulture associations to develop new products (C3) A member of the association offered to demonstrate for the group to learn (C2) Groups in our association meet once a month to share knowledge (C3) We scaled down production, improved the demonstration units to focus more on training and advisory services (C1)
Potential for local fertilizer market and cottage industry	The substrate used in growing mushrooms is decomposed for selling (C1) Urine is the single option for nitrogen, potassium, and phosphorus (C2) Use of animal urine and compost manure to improve soil fertility (C3) Making compost from cow dung and wood ash (C3)

technologies, that cut short the long production cycles, are reaping a lot from local SMEs. For example, a new technology—use of *hydrated lime*—for neutralizing the substrate for growing mushroom adapted by C1 was introduced by a supplier from China. This technology has been widely adopted by mushroom growers in Uganda. There are more examples of local innovative suppliers, such as the producers of solar dryers used by C1 and suppliers of organic pesticides used by C3 in the management of strawberries and vegetables. In view of the above, the study suggests that SMEs have a great potential to exploit partnerships with suppliers, both local and foreign, who have invested in affordable eco-friendly solutions.

2.4.3.5 Indigenous Knowledge and Technologies

Indigenous knowledge is an important national resource that offers new models for businesses' performance. All the entrepreneurs visited introduced locally available technologies in their production process. The three companies used indigenous knowledge to stimulate process innovation and experiments because it was a cheaper option than other alternatives. For example, companies C2 and C3 found the cost of organic fertilizer made out of urine cheaper than industrial fertilizer. Tobacco and cats were introduced to control snakes and rodents, respectively. According to the owners of company C1, rodents and snakes carry a virus that destroys mushroom gardens. According to the owner/manager of C3, "snakes feed on strawberries, and other fruits; so, it was necessary to get rid of them by planting tobacco, which is more permanent solution to maintain productivity". In addition, two of the companies owned dogs they used to control monkeys from destroying their gardens. These innovations reduced the cost of production, while, at the same time, conserving nature. The role of indigenous knowledge is not new; several authors have acknowledged its role in conserving biodiversity (Warren 1996; Ford 2000; Berkes 2004). The companies covered in this study show a clear demonstration of how this knowledge is being used to create economic value and improve food security and environmental performance. According to the feedback from interviews, the use of indigenous knowledge tackles both the cost and knowledge constraints to innovation.

2.4.3.6 High Demand for Knowledge on Organic Farming

The study revealed that there is potential to exploit the knowledge gap on green farming practices in the farming communities of Uganda. This follows an earlier decision by the Government of Uganda to finance the private sector led agricultural extension system which created model farmers (farmers that formed a nucleus for other farmers to learn from) across the country. The model farmers reinforced confidence in farmer-to-farmer learning; therefore, the recent decision by the government to revert to public extension system created a vacuum in the agricultural knowledge system that is being exploited by entrepreneurs. All the entrepreneurs interviewed had a knowledge transfer component as part of their business. Company C1 had provided more services to its client compared to the rest. Company C1's services include massaging, using mushroom therapy, training on mushroom growing, and agribusiness. Company C3 advised their clients on agronomy for vegetables and fruits, general management, and irrigation. Company C2's services were mainly on general management for bananas and application of organic fertilizer made out of urine. So, the high demand for knowledge on organic farming practices implies that there is potential to harness service innovation in farming practices of local entrepreneurs.

2.4.3.7 Potential for Local Fertilizer Market and Cottage Industry

Uganda is ranked as the lowest consumer of industrial fertilizer in the East African region. The low consumption of artificial fertilizer is attributed to the dominance of subsistence farming, where majority of farmers cannot afford to buy fertilizer for their gardens. However, with the land in most parts of the country over plowed, productivity has declined, which is a threat to food security. To reclaim soil fertility, subsistence farmers have to try out the affordable options that are locally available. As a result, the demand for organic fertilizer is steadily growing. The entrepreneurs introduced low-cost alternatives, including earthworms, compost manure, and human excreta in trying to improve soil fertility. According to C3, the use of human excreta in the food industry, especially vegetables, had a lot of resistance among local communities. However, we found that the fertilizer from human urine had potential demand from banana farmers and in the flower industry because of its compatibility with the greenhouse fertigation system. The emergence of urine separating toilets is a great opportunity for entrepreneurs, who want to use urine on large commercial farms. Due to the increased demand for manure, company C1 produces compost manure from the substrates that remain after harvesting mushrooms. The heaps of residues were left to decompose for a while, before farmers bought it as compost manure.

2.4.4 Implications for Water and Land Management

2.4.4.1 Application of Urine as a Fertilizer

The organic fertilizer that is made out of urine has a large component of water that is required by plants. For example, in the greenhouses, farmers who use urine in fertigation systems require fewer liters of water than what is required when solid fertilizer is used. The study found that less water is needed to dilute urine during the process of applying fertilizer for plants in the greenhouse.

2.4.4.2 Application of Organic Pesticides

The use of indigenous technologies, such as tobacco, red chili, and wood ash, does not have negative effects on land. The use of local technologies did not disrupt microorganisms in the ecosystem, thereby maintaining soil structure and fertility. In fact, farmers reported that spraying urine maintained soil fertility and controlled banana weevils and grasshopper infestation. However, there were fears among farmers about the long-term impact that may result from continuous application of urine on their farms. Anderson (2015) noted that continuous application of urine creates high deposits of phosphorous and nitrogen content, hence suggesting a residual buildup of these nutrients following urine application.

2.4.4.3 The Use of Hydrated Lime for Neutralization

The use of *hydrated lime* for neutralizing the mushroom substrate improved soil productivity. In addition, the used-up substrate was left for some time to decompose. This was later sold to farmers as composed manure. Alternatively, some farmers used the substrate to make charcoal briquette for cooking. These innovations not only served the purposes, but significantly reduced the costs that firms incurred.

2.4.4.4 The Use of Plastic Bags for Mushrooms Growing

The study established that the use of plastic bags in making the growing bags for mushrooms is very risky for the environment. We found that when plastic dries up, it breaks down into small pieces which remain in the residue that is decomposed and sold as compost manure. Since plastic cannot be decomposed by microorganisms, plastic pollution undermines lands and waterways. Plastic bags are typically used for a short period of time in mushrooms growing but take hundreds of years to break down in the garden. The large pieces of plastic bags were picked out for burning, with implications for increased greenhouse gases. The burning of plastic releases *carbon dioxide* into the atmosphere, which degrades the environment; therefore, it is important to look for replacement materials, such as the new forms of *polymer*s made from renewable materials that are digestible by microorganisms.

2.5 Discussion of Findings

Ugandan SMEs encounter serious obstacles to innovation but the majority do not give up. The main hindrances to innovation in SMEs include costs, knowledge, and market factors. The cost factors relate to limited resources to finance innovation; yet access to credit is still a problem for small companies in sub-Saharan Africa. There are uncertainties regarding the outcomes of innovation, which obstruct decisions to invest in new knowledge and technology. The factors that deter innovation mirror the most common type of innovation undertaken by Ugandan SMEs. The most common type of innovation was organizational innovation, followed by marketing because both types of innovation are less costly, compared to process and product innovations. According to the findings, product innovation was very uncommon, followed by process innovation. It is acknowledged that product innovation stems from new technologies and skills, which are lacking in Ugandan SMEs. Elsewhere, SMEs that lack resources often depend on collaborating partners, with potential to invest in innovation, but Ugandan SMEs find a lot of difficulty in finding cooperating partners for innovation. This implies that green SMEs have to depend on locally available knowledge and affordable sources of information, such as customers and suppliers as a strategy to promote eco-innovation. This significantly limits meaningful innovation that ensures competitiveness beyond national borders.

Knowledge constraints, such as the lack of skills and information about technologies, are significant because very few Ugandan SMEs are partnering with universities, consultants, and research institutions, where they could access new knowledge. The tendency to use trade fairs and journals as sources of information for innovation is also low. The lack of resources to initiate innovations within firms means that Ugandan SMEs have to rely on external parties. However, the most valued external partner for innovation among Ugandan SMEs is the customer. Customer requirements have been found to influence eco-innovation (Horbach et al. 2012), but where firms lack resources to invest in innovation, they tend to respond to customer needs by imitating other firms. Imitative innovation is largely incremental innovation. Ugandan SMEs need partners with new technologies to improve product and process innovation; however, the new technologies should be affordable. The lack of information on new eco-friendly technologies is therefore a great opportunity for partnerships to support knowledge creation critical to the national innovations system. It is therefore important to support the creation of knowledge on organic farming, as part of the national innovation system.

The research reveals that green SMEs in the agricultural sector depend on indigenous knowledge and technologies to undertake innovation. This is mainly because local knowledge and technologies are cheaper to adapt. Indigenous knowledge has become a major component of process and product innovation, because it reduces the costs of production and improves environmental performance; therefore, innovation among green SMEs in low-income countries can be supported by tapping into existing knowledge and know-how.

It is evident from the study that the biggest barrier for market innovation is dominance of established companies. Firms that lack skills and new knowledge struggle to undertake market innovation; consequently, they are more likely to imitate innovations introduced by other companies. This also explains why most innovation activities undertaken by Ugandan SMEs are incremental and based on experimentation. However, incremental innovation is not often linked to environmental performance (Smith et al. 2005; Nill and Kemp 2009). However, the multiplier effect of these innovations ultimately has cumulative impact on the environmental outcome. The study shows that incremental innovations are easier and used by poor people because they are socially and economically appropriate.

The study also found that most of the technologies employed by green SMEs have a positive impact on the environment. However, the use of plastic bags for mushrooms growing was accompanied by traces of *polymer* materials, which were introduced as part of process innovation. They inevitably have a negative impact on soil and waterways. This requires more innovative technologies, like the digestible *polymer* to replace plastic, which takes years to decompose. On the other hand, the use of urine and other materials as organic fertilizer not only demonstrate a high demand for organic option but contribute water that is required by plants. The adoption of tobacco, red chili, and wood ash as pesticides was found to be a cheaper option for controlling pest infestation. In addition, mushroom farmers used *hydrated lime* and maize bran to neutralize the substrate which improved productivity.

2.6 Conclusion and Policy Implications

In conclusion, to enable SMEs to change their environmental behaviors require strengthening the driving forces or weakening the resistant forces to eco-innovation, or a combination of both. The study concludes that Ugandan SMEs are involved in incremental innovation because of the high multiplier effect of individual small farmer's actions. To this end, incremental innovation is likely to have large and gradual effects that significantly impact on environmental performance, contrary to earlier findings (Smith et al. 2005; Nill and Kemp 2009).

The study illustrates that customer requirements form a driving force for innovation in Ugandan SMEs. There is a need to strengthen this driver through training customers and promotion of eco-literacy, as a potential option to stimulate eco-innovation among SMEs in Uganda.

The main hindrances to innovation in SMEs are the high costs, inadequate knowledge, and market factors. This has led to the perpetuation of the least cost type of innovation undertaken by SMEs. The government should consider using its existing research institutions to assist SME to innovate.

It is noted that product innovation is very uncommon, followed by process innovation, owing to limited new technologies and skills. This implies that investments into new technologies and development of skills are critical to firms undertaking product innovation.

Since SMEs lack resources and may have to depend on collaboration with partners to invest in innovation, private sector apex and umbrella institutions, like Uganda National Farmers Federation, Uganda Private Sector Foundation, among others, should come in to foster these linkages and cooperation.

Given that local and indigenous knowledge plays a critical role and is accessible to farmers undertaking innovation, strategies should be developed to promote its exploitation and investment into, through policies, frameworks, and regulations.

In addition, there is a great opportunity for potential suppliers, especially foreign firms to tap into existing knowledge systems, including indigenous knowledge to develop more affordable and efficient technologies, such as degradable polymers to replace plastic bags.

In order to mitigate the knowledge constraints, such as the lack of skills and information about technologies, SME should partner with universities, consultants, and research institutions, where they could access new knowledge.

The lack of information on new eco-friendly technologies creates an opportunity for partnerships to support knowledge creation on organic farming, as part of the national innovation system.

Although most of the technologies employed by green SMEs have a positive impact on the environment, the traces of *polymer* materials that were introduced as part of process innovation should be addressed through legislation. Farmers should be encouraged to use environmentally friendly materials instead.

The use of urine and other materials as organic fertilizer and organic pesticides, which demonstrates a high demand and compliments water that is required by plants, should be scaled up through the formulation of appropriate policies by the government.

References

Anderson E (2015) Turning waste into value: using human urine to enrich soils for sustainable food production in Uganda. J Clean Prod 96:290–298

Berkes F (2004) Rethinking community-based conservation. Conserv Biol 18(3):621–630

Carrillo-Hermosilla J, Del Río P, Könnölä T (2010) Diversity of eco-innovations: Reflections from selected case studies. J Clean Prod 18(10):1073–1083

Cheng CC, Shiu EC (2012) Validation of a proposed instrument for measuring eco-innovation: An implementation perspective. Technovation 32(6):329–344

Christensen Clayton M (1997) The innovator's dilemma: when new technologies cause great firms to fail. University of Illinois at Urbana-Champaign's Academy for Entrepreneurial Leadership Historical Research Reference in Entrepreneurship. Available at SSRN: https://ssrn.com/abstract=1496206

Cuerva MC, Triguero-Cano Á, Córcoles D (2014) Drivers of green and non-green innovation: empirical evidence in Low-Tech SMEs. J Clean Prod 68:104–113

Dumanski J (1997) Criteria and indicators for land quality and sustainable land management. ITC J 3(4):216–222

European Commission (2000) Community innovation survey. https://circabc.europa.eu/ui/group/47133480-29c1-4c23-9199-72a631f4fd96/library/bfcf3592-83a3-4066-ab70-f9a5cf492253

Ford J (2000) Traditional ecological knowledge, ecosystem science, and environmental management. Ecol Appl 10(5):1249–1250

Hamann R, Smith J, Tashman P, Marshall RS (2015) Why do SMEs go green? An analysis of wine firms in South Africa. Bus Soc Doi: 10.1177/0007650315575106

Horbach J, Rammer C, Rennings K (2012) Determinants of eco-innovations by type of environmental impact–The role of regulatory push/pull, technology push and market pull. Ecol Econ 78:112–122

Huarng KH, Yu THK (2011) Entrepreneurship, process innovation and value creation by a non-profit SME. Manag Decis 49(2):284–296

Jacobides MG, Knudsen T, Augier M (2006) Benefiting from innovation: Value creation, value appropriation and the role of industry architectures. Res Policy 35(8):1200–1221
Keijzers G (2002) The transition to the sustainable enterprise. J Clean Prod 10(4):349–359
Kemp R, Pearson P (2007) Final report MEI project about measuring eco-innovation. UM Merit, Maastricht 32(3):121–124
Kemp R, Pearson P (2008) MEI project about measuring eco-innovation, final report. Maastricht. In.
Klewitz J, Zeyen A, Hansen EG (2012) Intermediaries driving eco-innovation in SMEs: a qualitative investigation. Eur J Innov Manag 15(4):442–467
Lewis K, Cassells S (2010) Barriers and drivers for environmental practice uptake in SMEs: A New Zealand perspective. Int J Bus Stud 18(1):7
Loader K (2010) Is local authority procurement "lean"? An exploration to determine if 'lean'can provide a useful explanation of practice. J Purch Supply Manag 16(1):41–50
Morand F (2008) Developing eco-innovation: opportunities for education and policy integration. Ecoinnovation network
Muhanguzi S, Kyobe M (2014) Aligning work practices, mobile technology and strategy for performance improvement: the case of SMEs in Uganda. Electron J Inf Syst Developing Countries 60(1):1–22
Nill J, Kemp R (2009) Evolutionary approaches for sustainable innovation policies: From niche to paradigm? Res Policy 38(4):668–680
OECD (2009) Sustainable manufacturing and eco-innovation. OECD, Paris
Oltra V, Saint Jean M (2009) Sectoral systems of environmental innovation: an application to the French automotive industry. Technol Forecast Soc Chang 76(4):567–583
Oncioiu I (2015) Eco–innovation in European SMEs: between limitation and possibilities. EIRP Proc 10
Oxborrow L, Brindley C (2013) Adoption of "eco-advantage" by SMEs: emerging opportunities and constraints. Eur J Innov Manag 16(3):355–375
Pinget A, Bocquet R, Mothe C (2015) Barriers to environmental innovation in SMEs: Empirical evidence from French firms. Management 18(2):132–155
Schumpeter JA (1934) The theory of economic development: an inquiry into profits, capital credit, interest and the business cycle (1912/1934). Trans Redvers Opie. Harvard University Press, Cambridge
Smith A, Stirling A, Berkhout F (2005) The governance of sustainable socio-technical transitions. Res Policy 34(10):1491–1510
Stebbins RA (2001) Exploratory research in the social sciences, vol 48. SAGE Publishing, California
Tilley F (1999) Small-firm environmental strategy. Greener Manage Int 25
Walker H, Preuss L (2008) Fostering sustainability through sourcing from small businesses: public sector perspectives. J Clean Prod 16(15):1600–1609
Warren DM (1996) Indigenous knowledge, biodiversity conservation and development. In: Sustainable development in third world countries: applied and theoretical perspectives, pp 81–88
Yin RK (2003) Case study research design and methods. Applied social research methods series, vol 5, 3rd edn

Chapter 3
An Analysis of Factors Influencing Green Entrepreneurship Activities in South Africa

Chipo Mukonza

3.1 Introduction

Entrepreneurship has been touted as a catalyst for economic development in Africa. In that regard, governments have put in place supporting measures in the form of loan guarantees, tax incentives and research credit designed to boost innovation or systems to encourage self-employment (Organization for Economic Development–OECD 2011). The advent of green economy/growth has even widened entrepreneurship activities. While global inequality and rising unemployment pose major challenges to policy makers, green entrepreneurship is seen as the driving force for the establishment of a holistic and sustainable economic, environmental and social system. The point of departure of the study is the realization that economic development is dependent on dynamic capabilities of a society to generate goods and services that, at prevailing factor prices, are of higher quality and lower costs than those that were previously available. Accordingly, the International Labor Organization (ILO) (2015) states that "green entrepreneurs provide the hotbed for starting and sustaining a green economy by providing green products and services, introducing greener production techniques, boosting demand for green products and services and creating green jobs." OECD (2010) posits that small and medium-sized (SMEs) enterprises are essential for green growth as key drivers of entrepreneurship and key players in emerging green industries. Green entrepreneurship is an increasingly relevant phenomenon from a development perspective, but it is still largely under-researched (Farinelli et al. 2011). However, much of the burgeoning literature

This chapter is extracted from a working paper by Mukonza (2016) published in the UNU-INRA working paper series (for more information, follow the link http://collections.unu.edu/view/UNU:5926).

C. Mukonza (✉)
Institute for Corporate Citizenship: Exxaro Business and Climate Change, UNISA, Pretoria, South Africa
e-mail: chiponyam@gmail.com

C. Atewamba and D. Yong Ngondjeb (eds.), *Inclusive Green Growth*, Advances in African Economic, Social and Political Development,
https://doi.org/10.1007/978-3-030-44180-7_3

on green entrepreneurship lacks substantial empirical analysis. Farinelli et al. state that a green economy is a bottom-up approach. That is, it needs to be driven by entrepreneurs, who respond to policy incentives through innovation in management and technology.

South Africa in recent years has promulgated policies and strategies in a bid to transition to a green economy. Consequently, the country has witnessed numerous green business initiatives emerging in the country. Bibri (2013) aptly pointed out that green entrepreneurship has been socially constructed as having a catalytic role in reshaping the socio-technical landscape of the economy and engendering cultural changes and institutional developments associated with ecological modernization. The maintenance of an ecological balance has been addressed by the King 111 Report, which emphasizes sustainability in the outcome, social and environmental spheres (Carroll and Buchholtz 2010). Against this background, the study seeks to investigate and critically analyze the factors affecting green entrepreneurship activities in South Africa. The remainder of the study is structured as follows: The next section provides the problem statement, research objectives and significance of the study. This is followed by Sect. 3.2, with a discussion on definitions, conceptual and theoretical issues and followed by Sect. 3.3, which examines the methodology, while Sect. 3.4 discusses and analyzes the results and, lastly, Sect. 3.5 proffers recommendations and draws conclusion for the study.

3.1.1 Problems Statement

Policy instruments that have been crafted to facilitate green growth are aimed at identifying the technological innovations capable of mitigating the human impact on the environment and addressing global environmental issues, such as climate change, land degradation and loss of biodiversity. It is, therefore, imperative to enhance green entrepreneurship. However, green entrepreneurship in South Africa is still very unsatisfactory and faces challenges, such as access to finance and lack of a strong entrepreneurial culture (Booyens 2011; Ayyagari et al. 2003). Identifying the factors affecting/influencing green entrepreneurship activities will therefore inform policy in order to boost green entrepreneurship in South Africa.

Chiloane-Tsoka et al. (2014) are of the view that green economy by nature requires high levels of intelligence and skill. This poses challenges for business entrepreneurs to take advantages of opportunities created by a green economy without having to leave a trail of heavy carbon footprints in their businesses. Lekhanya (2014) discovered that in as much as South African SMEs are aware of green marketing and its managerial implications, the South African Environmental Act and Consumer Protection Act are additional factors that influence their business operations. Against these revelations, the present study sought to do a comprehensive analysis of factors influencing green entrepreneurship activities in the country.

3.1.2 *Research Objectives*

The major objectives of this study are to:

- Identify some of the challenges, which green entrepreneurs face in South Africa.
- Determine factors influencing green entrepreneurship activities in South Africa.
- Provide recommendations to policy makers on what can be done to make green entrepreneurship activities flourish.

3.1.3 *Significance of the Study*

This is a relevant and well-timed study as it examines green entrepreneurship activities and transition to a green economy in South Africa. For a long time, SMEs have been considered to be the vehicle for economic development, helping in alleviating unemployment. The study adds to the growing literature on entrepreneurship, with a special emphasis on activities that are environmentally and economically sustainable. South Africa has chosen green economy as a sustainable path to development that addresses environmental, economic and social equity. Though the concept of green economy is fairly new, green entrepreneurship activities are being promoted as part of green growth strategies. Examining the manner in which these activities are being implemented and the challenges they pose is fundamental. The study sought to provide lessons that can be learnt from other countries, especially in Africa. The study also aims to make an original, empirically grounded and theoretically sound contribution to examining the experiences, reflections and perspectives of green entrepreneurs in the country.

3.2 Literature Review

The aftermath of global financial and economic crisis in 2008 compelled many countries to realize the central role of entrepreneurship in boosting the economic activity (OECD 2011). Since then "green economy" has become a buzz word, and it has ushered in new opportunities for small business and entrepreneurs. A green economy has been defined as one that results in improved human well-being and social equity while significantly reducing environmental risks and ecological scarcities (UNEP 2011). With the advent of a green economy, industries and not-for-profit organizations are seen as having shifted from a position of denial about their impacts on the environment to one, where they have the potential to mitigate the detrimental effects. This perspective is essentially one of ecological modernization, at the heart of which is a belief in technology, innovation and progress to solve environmental problems (Ndubisi and Nair 2009). Thus, a green economy, with a combination of new technologies and changing institutions, is increasingly becoming a source of policy responses and initiatives in the developed world (Parrish and Foxon 2009). Nonetheless, there is a paucity of literature when it comes to defining green entrepreneurship. Therefore, it has become difficult to separate green and non-green entrepreneurship activities. Several terms have been adopted to mean green entrepreneurship. OECD (2011) states that green entrepreneurship could be defined regarding the technology used for production in any sector of the economy or regarding the sectors that firms are active in. Nikolaou et al. (2011) opine that scholars have not yet agreed on the meanings and terms of the concept of green entrepreneurship. Thus, the literature that is available offers a range of terms with different meanings for the concepts of green entrepreneurship, such as green, environmental, ecological, sustainable entrepreneurship, eco-entrepreneurship and eco-premiership (Nikolaou et al. 2011). There are schools of thought that prefer to use the term social entrepreneurship to refer to green entrepreneurship as it is said to be all-encompassing (Abu-Saifan 2012; Groot and Dankbaar 2014). Social entrepreneurship is a term that has gained popularity, and it also means different things to different people. According to Shepherd and Patzelt (2011), social entrepreneurship is the creation of viable socio-economic structures, relations, institutions, organizations and practices that yield and sustain social benefits. On the other hand, defines social entrepreneurship as the art of simultaneously obtaining both a financial and social return on investment.

3.2.1 Benefits of Green Entrepreneurship

The benefits of green entrepreneurship include the following:

- Corporate social responsibility
- Environmental respect
- Easy implementation of environmental legislation
- Better insurance terms
- Attainment of green investments
- Acquirement of loans with better terms
- New trade opportunities
- Reduction of operational costs.

In this study, green entrepreneurship activities are defined as those that recognize environmental issues and whose business ventures are in the environmental market place. Such entrepreneurs pursue real opportunities that show good profit prospects.

3.2.2 Defining What Constitutes a Green Entrepreneur

Attempts to profile "green" or environmentally responsible entrepreneurs are a relatively under-researched field and lack substantial empirical analysis (OECD 2011).

Subsequently, it has been problematic to define a "green entrepreneur." Other stakeholders, such as lobby groups, non-governmental organizations (NGOs), venture capitalists, industry associations and local communities can also influence the context in which green entrepreneurs operate. Mwakambirwa (2013) posits that entrepreneurs are now making changes in their ways of doing business in order to address the society's new concerns on environmentalism and other ecological problems as they affect their enterprises. The transition to green economies has placed a significant responsibility on green entrepreneurs. The concept of a green economy was popularized around 2003 when the European Commission presented the Green Paper on entrepreneurship in Europe. The paper placed more emphasis on Europe becoming a competitive and dynamic knowledge-based economy. The European Commission proposed that Europe needs to foster entrepreneurial drives more efficiently. A green entrepreneur is defined as an entrepreneur, who provides ecological awareness to the firm. A green entrepreneur has a positive attitude toward environmental protection and restoration. International Labor Organization (ILO) (2015) defines green entrepreneurs as those who provide the hotbed for starting and sustaining a green economy by providing green products and services, introducing greener production techniques, boosting demand for green products and services and creating green jobs. ILO (2015) further states that green entrepreneurship can be defined from two perspectives related to the output (products and services), as well as the process (or production) of an economic activity. Entrepreneurs can enter into an overtly "green" business sector, providing green and environmentally friendly products and services (e.g., waste management). Alternately, green entrepreneurs can provide their products or services through an environmentally friendly process or with the help of clean technologies (e.g., ecotourism). Numerous terms have been used to describe a green entrepreneur (i.e., ecopreneurship, environmental entrepreneurship, entrepreneurship and sustainable entrepreneurship). The present study will use green entrepreneurs, based on the form in which green entrepreneurship can take place as described by ILO. The study also interviewed non-governmental organizations that are providing green and environmentally products and services.

3.2.3 Challenges and Issues Associated with Green Entrepreneurship Activities

OECD (2010) contends that for green entrepreneurs to fully participate in the transition toward sustainable economic patterns and seize the opportunities arising, it is essential that the main barriers to green growth and eco-innovation are identified. It further states that the willingness and capability of SMEs to adopt sustainable strategies or seize green business opportunities meet with size-related resources constraints, skills deficits and knowledge limitations, not to mention the need for enterprises to survive in the aftermath of a global crisis. Market failure has been suggested as one of the triggers of green entrepreneurship (Nikolaou et al. 2011). This comes from the environmental economics school of thought, which alludes to the fact that market failures are the main factor responsible for contemporary environmental problems and may have positive consequences for entrepreneurial development.

A study was done in Pakistan to identify enabling factors of green entrepreneurship. This was a case study of organic agriculture produce. A household survey was carried out from the green and potential green entrepreneurs. The results of the study showed the significance of reliable organic markets, price premiums and improved production conditions as key enablers of green entrepreneurship (Pandrani and Ferguson 2014). In another study, a SWOT analysis was done to evaluate the prospects of green entrepreneurship development in Greece (Nikolaou et al. 2011). The results indicate that institutional, structural, social and economic factors can play a critical role in whether entrepreneurs invest in new business ventures with natural resources. Brand et al. (2007) conducted a study that looked at factors influencing the sustainability of informal SMMEs in South Africa and suggested a method that can be used to transform them into more formal businesses that are integrated into the formal economy. Herrington et al. (2009) also tracked entrepreneurship activities in South Africa. Their study details South African comparative international standing and concluded that, as far as total entrepreneurial activity (TEA) rate is concerned, it is very low and has been dropping, compared to other middle income, efficiency-driven economies.

Research that is available in the public domain tends to focus on the motives of business to adopt environmental management practices on a voluntary basis, the benefits or barriers for new entrepreneurs to invest in the environmental sector, how to measure green entrepreneurship (OECD series) and the development of sustainable entrepreneurs. The present study acknowledges the development of green entrepreneurs in the country, as well as the policies and the institutions that have been put in place. However, most of the green entrepreneurs are not progressing well or dying in their infancy stage. Therefore, there is need to critically examine and investigate the factors influencing green entrepreneurship activities in the country is imperative.

3.2.4 Theoretical Considerations

The dynamic capabilities theory has evolved over the years, with its roots emanating from the resource-based view (Teece et al. 1997). Dynamic capabilities theory espouses that, for a firm to encompass competitive advantage, it must not be static but have dynamic capabilities; i.e., the capacity of an organization to purposefully create, extend or modify its resource base. (Teece et al. (1997) defined dynamic capabilities, thus: "the capacity to renew competencies so as to achieve congruence with the changing business environment" by "adopting, integrating, and reconfiguring internal and external organizational skills, resources, and functional competencies."

The importance of the dynamic capabilities theory is that it builds on the fundamental understanding that a firm needs to have resources for it grows. Furthermore, it questions how firms first develop firm-specific resource and capability bundles and how they renew their resource and capability configurations in order to respond to shifts in their environment (Teece et al. 1997). Ambrosini et al. proposed that there are three levels which exist that are related to managers' perceptions of environmental dynamism. The first level consists of the incremental, dynamic capabilities, which is concerned with the continuous improvement of the firm's resource base. At the second level are renewing dynamic capabilities, those that refresh, adapt and augment the resource base. At the third level is dynamic regenerative capacity, which impacts not on the firm's resource base but on its current set of dynamic capabilities, i.e., these transform the way the firm changes its resource base.

The study therefore adopts this definition to decipher green entrepreneurial activities and resources in South Africa. The thinking and notion behind this are that dynamic capabilities of green entrepreneurs will allow the organization to grow, generate profit and achieve competitive advantage under conditions of technological and market change (Teece et al. 1997). Thus, the framework attempts to explain why some organizations succeed and others fail (Arend and Bromiley 2009). Green entrepreneurship plays a catalytic role in the economic transition to a sustainable economy (Parrish and Foxon 2009).

3.3 Research Methodology

The study employed a mixed method approach. That is both the use of qualitative and quantitative methods. A five-pronged approach was used and is revealed as follows: comprehensive literature review, administering of questionnaires (pilot and reconnaissance visits included), in-depth interview, observation and focus group discussions with case study respondent and analysis of data and write-up. A questionnaire was administered, consisting of range-based questions, scale questions, YES/NO and open-ended questions, which would make it easy to use statistical techniques. The researcher got permission to use some databases that already exist mainly from SEED (United Nations Environmental Programme-UNEP), Association Network Development Entrepreneurs (ANDE) and the links that were provided by a knowledgeable person from the Department of Environmental Affairs (DEA). The SEED (UNEP) has 28 South African recipients, who have so far received the SEED Award from 2005 to 2015. A questionnaire was sent to all of them initially by email, and then, follow-ups were done telephonically, and this proved to be more efficient and yielded positive responses as most respondents preferred to talk over the phone. The majority of the respondents indicated that they did not have time to fill in the questionnaire. At the end, 25 out of the 28 individuals responded to the questionnaire. Snowball sampling, using knowledgeable people from the department of Environmental Affairs and ANDE, was then used to administer to the other entrepreneurs. In addition, to chain referral sampling, the researcher had the opportunity to attend conferences and workshops, where some of the green entrepreneurs were showcasing their businesses. For instance, UNISA, under the guidance of the

Exxaro Chair in Business and Climate change, hosted the First International Conference on Innovation for Sustainability under Climate Change and Green Economy/Growth (May 2015), where some of the green entrepreneurs showcased their work. In August 2015, the researcher also had an opportunity to attend the Women in Environment Conference, where some green entrepreneurs across the country showcased their business. Thus, the researcher had an opportunity to interview, observe and note some of the emerging technologies, products that are being developed by emerging green entrepreneurs. Finally, between May 2015 and December 2015, the researcher administered 103 questionnaires to green entrepreneurs fully in business and those that have partially adopted the concept. It is important to note that some green entrepreneurs that were SEED Award winners were also working with ANDE or with the DEA, and these were not interviewed twice.

Data collected was captured using SPSS Version 22 cleaned and analyzed, and the outputs were presented in form of frequency tables. Mean and standard deviation from Likert scale statements were used to interpret the data. Thematic analysis was used to analyze the data. In-depth interviews, observations and key informants helped to acquire a deeper understanding of the entrepreneurship environment from the aspect of green entrepreneurs and understand motives, challenges and obstacles they face in developing their businesses. The study draws on some of the vantage points presented by the dynamic capability theories and sustainability transition. The paper is epistemologically founded in interpretivism. Indeed, interpretivism, in contrast to positivism, poses multiple realities that validate internal research design through the subjectivism that will permeate the analytical assessment.

3.4 Statements of Results Findings and Discussion

The aim of the study was to investigate and critically analyze the factors affecting green entrepreneurship activities in South Africa. A Cronbach test was done to determine the reliability of data, and the following results were established (Table 3.1).

The results indicate that the data is reliable as it is above 50%. Thus, if the study can be replicated somewhere, there are almost 80% chances that it will yield the same results.

Table 3.1 Reliability statistics

Cronbach is alpha	N of items
0.879	4

Source Survey data (2015)

3.4.1 *Characteristics of Green Entrepreneurs*

One of the aims of the study was to establish the gender of the green entrepreneurs interviewed, and the following results were established. Figure 3.1 indicates gender proportions of the respondents. 43% of the people interviewed were female, and 57% were male. The majority of the respondents were 21–40 years (Fig. 3.1). This might be an indication that the young ones are the one who are willing and quick to adapt to changes in the environment. The older ones take time to adjust and to make decisions.

In trying to understand the factors affecting green entrepreneurship activities in South Africa, it is important to understand the age and the gender of the green entrepreneurs and the potential green entrepreneurs. The assumption is that gender and age play a critical role in determining green entrepreneurship activities. Figure 3.2 reveals the number of years in business of the green entrepreneurs interviewed.

The majority of the respondents were still at infancy stage, i.e., 2–4 years. There was one exception, where one of the respondents indicated that she had been in

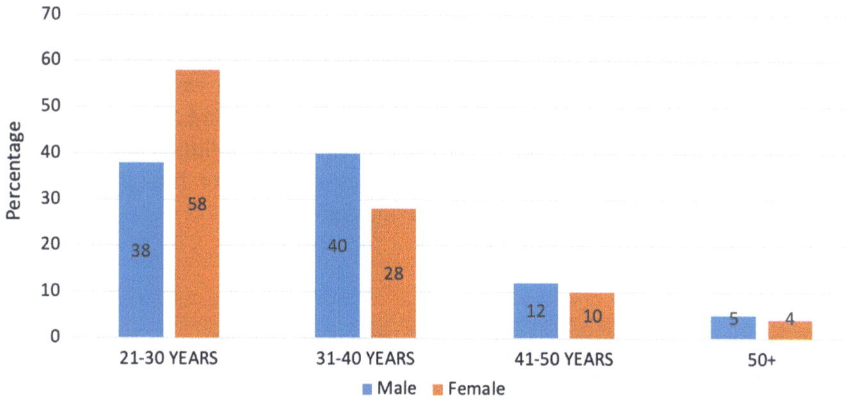

Fig. 3.1 Determining gender and age of entrepreneurs. *Source* Survey data (2015)

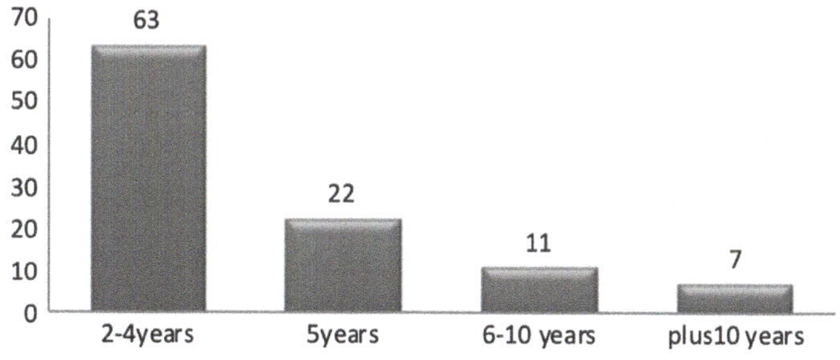

Fig. 3.2 Determining years in business. *Source* Survey data (2015)

the business for the past twenty-five years, only that then it was not called green entrepreneurship then, but rather acknowledged as sustainable practices. The reason most of the green activities were still in nascent stages may be attributed to the lack of policies and strategies, as well as start-up funding. National Small Business Chamber (NSBC) is dedicated to the success of small business in South Africa. A follow-up question was included in the questionnaire to distinguish those that have adopted green entrepreneurship activities from inception and those that adopted green entrepreneurship activities. About 73 of the respondents revealed that they had started during inception, courtesy of funding opportunities made available in the last four years, whereas 21 respondents said they had adopted it along the way. About nine of the respondents said they were partially green, have other activities or programmes in their organizations, which they were running. An example was one non-governmental organization, which used to deal with peace and conflict issues. The result corroborates findings from an SME Survey, where they interviewed about 1400 SMEs, and about 86% of them agreed on the importance of sustainability (SME Survey 2015). The analysis of the results proves that more and more SMEs in the country were becoming environmentally conscious or moving from sustainable practices to sustainable entrepreneurship. Figure 3.3 shows the type of activities the entrepreneurs are involved in.

When the questionnaire was being designed, the Department of Trade and Industries classification was followed. Fetola[1] organization, which worked with SMEs in South Africa, classified that business operating in the green economy included those that employed clean technology (reducing emissions, pollution and improved material efficiency), provided resource efficiency solutions or brought about an environmental benefit. Examples of such businesses include

- Recycling and/or waste,

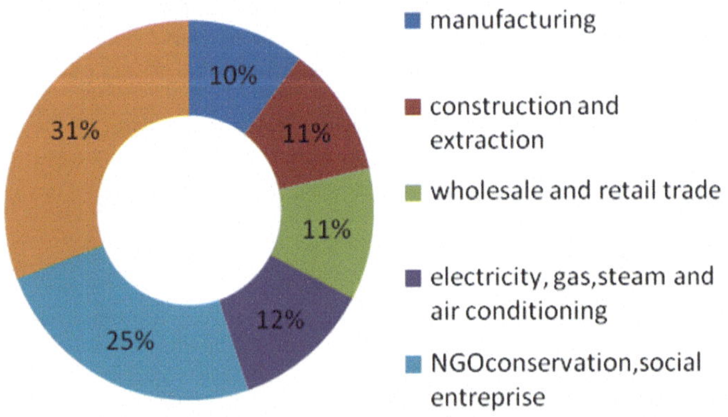

Fig. 3.3 Determining business field occupations. *Source* Survey data (2015)

[1] Fetola are enterprise and supplier development specialists with 30 years' experience in stimulating entrepreneurial success, http://www.fetola.co.za/.

- Management services,
- Clean energy generation,
- Energy efficiency solutions,
- Water management solutions,
- Sustainable agribusiness,
- Manufacturing businesses,
- Employing improved or cleaner technology,
- Green building initiatives,
- Green transport solutions.

Respondents were asked to indicate area clusters of their business and what they do in their business. The results revealed that 31% of the respondents were into waste and sanitation, while 25% were the non-governmental organizations or cooperative trusts that were into conservation, social enterprises organic farming and environmental awareness/advocacy. According to the definition provided earlier by ILO (2015), there are basically two firms, which can be taken by green entrepreneurs, i.e., entrepreneurs can enter into an overtly "green" business sector, providing green and environmentally friendly products and services (e.g., waste management). Alternately, green entrepreneurs can provide their products or services through an environmentally friendly process or with the help of clean technologies (e.g., ecotourism). Thornton defines entrepreneurship as the creation of new organizations, which implies a certain degree of innovation and size. This creation occurs as a context-dependent, social and economic process. On green entrepreneurship, involving the introduction of new technologies as well as new products aimed at greening the economy, the study also reveals that there were some entrepreneurs who had ventured into stoves, using waste to minimize the amount of carbon dioxide CO_2 emitted into the air. Thus, interesting innovations have been unleashed as entrepreneurs take advantage of the opportunities posed by the transition to a structural green economy. A report produced by SEED (UNEP) highlighted that SMMEs struggled in selecting a legal entity (e.g., cooperative, trust, PTY) under which they are able to conduct simultaneously for-profit and non-profit activities. Figure 3.4 shows whether the green entrepreneurs had received green entrepreneurship education or not.

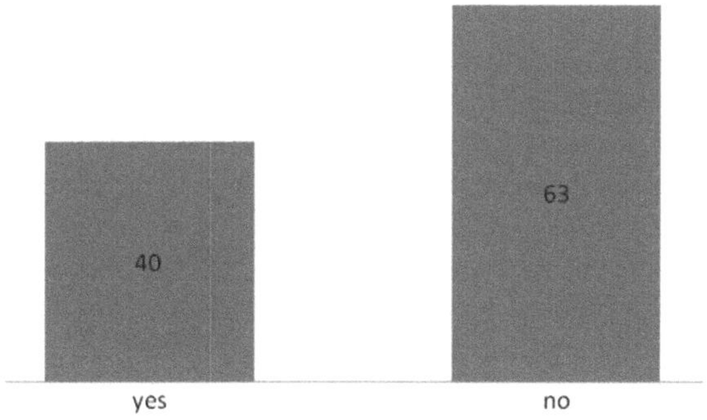

Fig. 3.4 Determining participation in green entrepreneurship education. *Source* Survey data (2015)

3.4.2 Development of Green Entrepreneurship

The majority of the respondents had not received an education. Most of the respondents said that they felt the need to be an environmentally responsible citizen or the need to be social entrepreneurs. A chi-square test was conducted in relation to the impact of green education and green entrepreneurship activities. A chi-square goodness of fit tests showed this to be significant (df 102, $x\, P = 0.000$). The concept of green entrepreneurship education has gained prominence with business schools, which now include it their curriculum. Consequently, studies are being done to determine the green culture in education, teaching through modeling, inculcating green aesthetics and how it contributes to both sustainable education. Green entrepreneurship education plays a significant role in developing green culture, green aspirations and promotes the development of green entrepreneurs in the country. There are organizations in South Africa, which are working hard to promote green entrepreneurship activities and complement government efforts as the country embraces green economy transition. Organizations, such Fetola,[2] have been offering courses through their #JustAddGreenprogramme, which was open to all, who have been in operation for at least six months (Fetola 2016: Online). This growing initiative aims to support a range of value-adding SMEs in that space.

Table 3.2 is a summary of the opinions of green entrepreneurs interviewed regarding their activities.

Table 3.2 shows that the existence of support for green product innovation, by SMEs, on green procurement is growing. Fetola organization, which works with the majority of the SMEs, reckons that access to green technology is one of the principal indicators of the prevalence of green entrepreneurship activities (Fetola 2016). As previously indicated, the number of green entrepreneurs has been growing in South Africa, with innovations being introduced every day. This is predicted to grow, with microecopreneurs exploiting opportunities in new niche markets that emerge (Holt 2011). The growth in green entrepreneurs is also influenced by changes in lifestyles to focus on organic food production, recycling initiatives and the like. Despite the fact that the demand for sustainability-related practices in South Africa is increasing, it is still too small to be sufficient for most companies to undertake these activities. Support in the start-up phase, i.e., the total early-stage entrepreneurial activity (TEA) that is also crucial for sustainability is still small (SME Survey 2015). However, the majority, if not all of the respondents, expressed the opinion that green activities gave their business a competitive advantage. The concept of green procurement has not been practiced although some of the respondents highlighted that they had always been environmentally conscious in buying their stationery or equipment. Those that expressed that expenditure on green procurement was growing are those that were introducing solar energy, going off grid, etc. Access to green technology seems not to be a problem among the green entrepreneurs interviewed. Figure 3.5 highlights public support for green initiative activities in the country.

[2]Fetola are enterprise and supplier development specialists with 30 years' experience in stimulating entrepreneurial success, http://www.fetola.co.za/.

Table 3.2 Determining the extent to which green entrepreneurship activities are taking place

	Strongly disagree	Disagree	Neutral	Agree	Strongly agree	Total	Mean	Standard deviation
Expenditure on green procurement is growing stronger	11	7	71	13	1	103	2.17	0.991
Expenditure on green procurement is reducing	1	14	63	17	8	103	2.64	1.237
The number of repeat buyers for green products is increasing	–	–	83	12	8	103	3.21	1.300
I have access to green technology	–	9	47	33	14	103	2.61	1.237
Green activities give my business a competitive advantage	–	–	5	22	76	103	3.47	1.347

Source Survey data (2015)

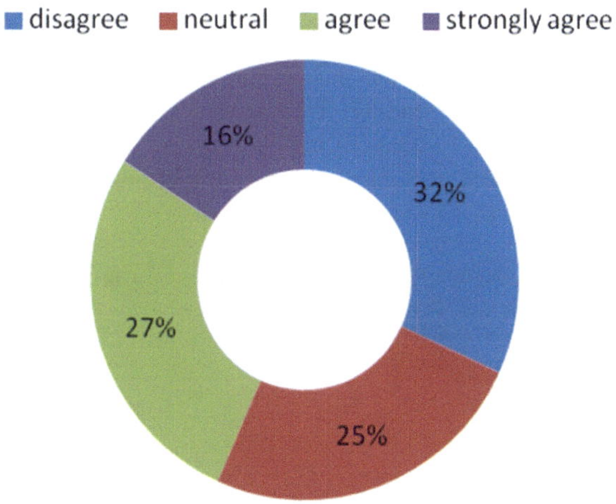

Fig. 3.5 Perceptions about public policies support for green initiatives. *Source* Survey data (2015)

Green entrepreneurs are optimistic and driven by the desire to see a positive effect on the natural environment and economic sustainability and also consciously aim at ensuring a more sustainable future (Gibbs and O'Neill 2015). The researchers further state that green entrepreneurs often struggle to survive, due to an unstable commitment from the public sector, whose support is easily withdrawn in response to frequent changes in politics and the efforts of lobbyist groups. The study found out that 43% of the respondents were neutral regarding their knowledge of public policies' support for green initiatives. Thus, majority of the respondents preferred to be neutral. In South Africa, there are a number of existing policy documents, such as the New Growth Path, the Industrial Policy Action Plan (IPAP), the green paper on a climate change response strategy, the long-term mitigation strategy, the Integrated Resource Plan (IRP) 2010, the Medium-Term Strategic Framework (MTSF) and the deliberations of the green economy summit, which identified the need for flagship programmes to demonstrate green economic activity. In addition, the Department of Environmental Affairs and the Development Bank of South Africa have funded some green projects since 2011. Therefore, the strategic intent to support green entrepreneurs by the government is evident. The issue might be the implementation of the policies and the strategies. Also, the small number of green entrepreneurs, who have been able to receive the funding, may account for the hesitation by the significant proportion of respondents preferring to be neutral. The Development Bank of Southern Africa (DBSA) contends that substantial policy reform is required if South Africa is to realize its vision of a green economy (Development Bank of Southern Africa 2011). This leads the researcher to probe further by investigating whether there was access to information. Figure 3.6 highlights the extent of access to information by green entrepreneurs.

Again, the majority of respondents were neutral, with 26% disagreeing and 18% agreeing. Meaning and understanding, as well as how one thinks about phenomena, are essential as these influence human actions. DBSA (2011) emphasizes that greening the economy requires a certain level of innovation and risk taking. However, decision makers, especially at local level, are often conservative in their approach to policy and spending. Therefore, for green entrepreneurs to take risks, there is a need to have access to information and examine their options. Studies on green entrepreneurs available in the public domain often focus on individual entrepreneurs,

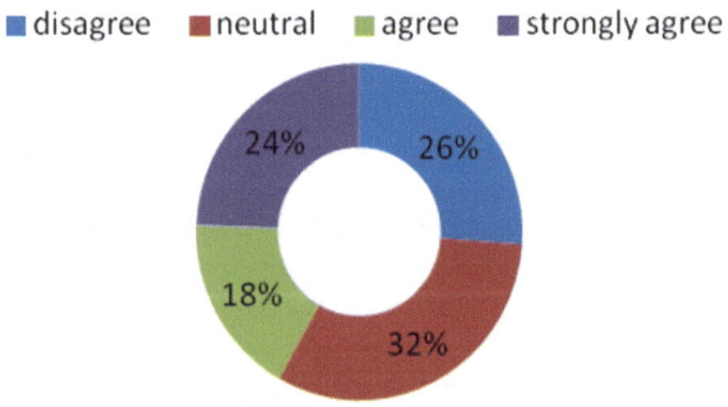

Fig. 3.6 Determining access to information on green entrepreneurship activities. *Source* Survey data (2015)

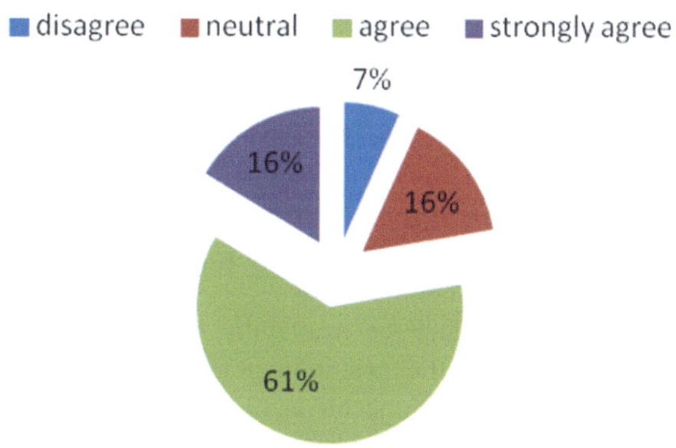

Fig. 3.7 Determining the extent of taking advantage of the opportunities. *Source* Survey data (2015)

neglecting wider economic and social contexts within which they operate. Sustainability transitions theories offer a potentially valuable means of understanding the role of businesses in engendering a green economy (Gibbs and O'Neill 2014). Figure 3.7 reveals the extent to which green entrepreneurs can recognize the opportunity.

Green is taken to mean green either by the nature of the product (e.g, renewable energy) or substantially green policies and practices within their business (Kirkwood and Walton 2015). 16% of the green entrepreneurs strongly agreed that they can easily recognize the green business opportunity while 61% agree. An interesting observation from the study was of not-for-profit organization that had been promoting peace initiatives all along but in the last two years has introduced another component to their projects, i.e., recycling and environmental awareness, thus taking advantage of the funding opportunities available in this area. Interesting and innovative ideas abound, but the challenge with budding entrepreneurs is to bring innovations to societies. Green entrepreneurs attempt new business opportunities and undertake ventures, which usually involve a very high risk. The outcome of these business ventures is often unpredictable (Farinelli et al. 2011).

3.4.3 Factors Affecting Green Entrepreneurship Activities in South Africa

While green entrepreneurship activities are being promoted in this era as South Africa transits toward a green economy, the greatest challenge is in identifying traits of green entrepreneurship. Numerous factors were raised by the green entrepreneurs interviewed for the study. Below is an extract of some of the activities and the issue raised by one organization interviewed: "They sell organic vegetables. They received training from development banks on waste management. In 2013, they were trained in Cuba in organic farming. The company works with local collectors but the challenge is that there is no space of garbage as Municipalities are not delivering. They work with community's households about 1050. There is no access to information.

There is a need to raise awareness. Corruption in government tenders. Lack of information among the communities," Duncan Village. Another organization interviewed revealed that green entrepreneurship is a subjective term. This particular organization had been in business for a long time, engaged in conservation and vegetation surveys. However, their concern was that support was not available unless if you have someone supporting you. Green is becoming popular, but it has interesting dynamics coming up. Table 3.3 indicates some of the influential factors that are influencing green entrepreneurship activities.

Table 3.3 indicated that availability of opportunities to go green with a mean rank of 4.01 was one of the most influencing factors on green entrepreneurship activities. This was followed by incentives (3.83), then public policy (3.60), access to finance (3.20), green education (2.87) and lastly institutions (2.43). The reason why 61% of the respondents ticked opportunities as the reason why they were going green or implementing sustainable entrepreneurship practices could be attributed to the number of programs and initiatives that have been unveiled in the country in the last five years. Initiatives encouraging entrepreneurs who address both socio-economic and environmental challenges in SA are becoming more common place. The most recent of these is a massive R300-million funding by risk financier Business Partners Limited (PBL) in January 2016. A SEED–UNEP report established that some of the challenges that were hindering green entrepreneurship activities in South Africa are few incentives for following green production and procurement practices and also training and learning of green practices manifest predominantly in-house, while community learning is a spillover effect. Mwakambirwa (2013), in his study, established that entrepreneur motivation had the greatest influence on green entrepreneurship practices in Kenya. According to surveys done by local financial institutions regarding access to finance, the percentage of SMEs seeking bank finance now stands at 91% and only 7% targeted private investors (Mungadze 2015). The Nedbank Small Business Index (NSBI) for the last quarter of 2014 has shown an improvement in the ease of obtaining finance, which it said was at a record high (Mungadze 2015). The next section discusses some of the observations regarding the types of entrepreneurs that were interviewed.

Table 3.3 Means ranking of factors influencing green entrepreneurship activities in South Africa $N = (103)$

Factor	Percentage	Mean ranking	Standard deviation
Opportunities	61	4.01	1.170
Incentives	15	3.83	1.160
Public policy and/or strategies	11	3.60	0.865
Access to finance	7	3.20	1.200
Institutions	4	2.43	1.205
Green education/training	2	2.87	0.90

Source Survey data (2015)

3.4.3.1 Survivalist or Opportunists?

Availability of opportunities was found to be one of the major factors influencing green entrepreneurship activities in the country. This was further crystallized with the in-depth interviews, where respondents stressed that they had taken advantage of the opportunities that were being unveiled. This raised the question then whether these organizations were survivalists or opportunists, which have taken advantages of the opportunities of going green. Walley and Taylor (2002) proposed four typologies of entrepreneurs, the four "ideal types" of green entrepreneurs: innovative opportunists, visionary champions, ethical mavericks and ad hoc enviropreneur. The "visionary champion" type is consistent with Thompson's "sea-change" paradigm, in that, this type of green entrepreneur is one who embraces a transformative, sustainability orientation. Innovasit-opportunist is one who has been influenced mainly by hard structural drivers, such as regulation. The "ethical maverick" type of green entrepreneur is characterized by a sustainability orientation and soft structural influences. In other words, the most significant influences on the business formation have been friends, networks and experiences, rather than visions of changing the world. With sustainability, value-driven motivation, these people might tend to set up alternative style businesses on the fringes of society. The "ad hoc enviropreneur" is a kind of accidental green entrepreneur. The motivation of such people is financial, not value-driven, and they are influenced most by personal networks, family and friends. The next section discusses some of the limitations of green entrepreneurship activities in the country (Fig. 3.8).

3.4.4 *Emergent Themes/Constructs*

3.4.4.1 How Green Is Green?

When it comes to green entrepreneurship activities, the most pressing issue is how to distinguish unambiguously relevant activities within "green" sectors of activities occurring in the rest of the economy (OECD 2011). From the study, one of the respondents expressed the opinion that there was nothing green about his technology, except that the reduction in transport costs and the traveling. This was quite intriguing to the researcher, prompting the question: How green is green? How do you measure green entrepreneurship activities? Alternatively, what are the traits of green entrepreneurs? Kirkwood and Walton conducted a study to outline the motivations for starting the business, the key green aspects of the product or service that the green entrepreneurs produce or sell, as well the degree of greening the organizations. Their study indicated that motivations for green entrepreneurs were diverse though a green entrepreneur is one with determination to make the world a better place. In this instance, this desire often exceeds the profit motive. The expression "green" is used in both a relative and absolute sense of the word. Therefore, a green business could refer to one that has been set up on a green basis or one that has become

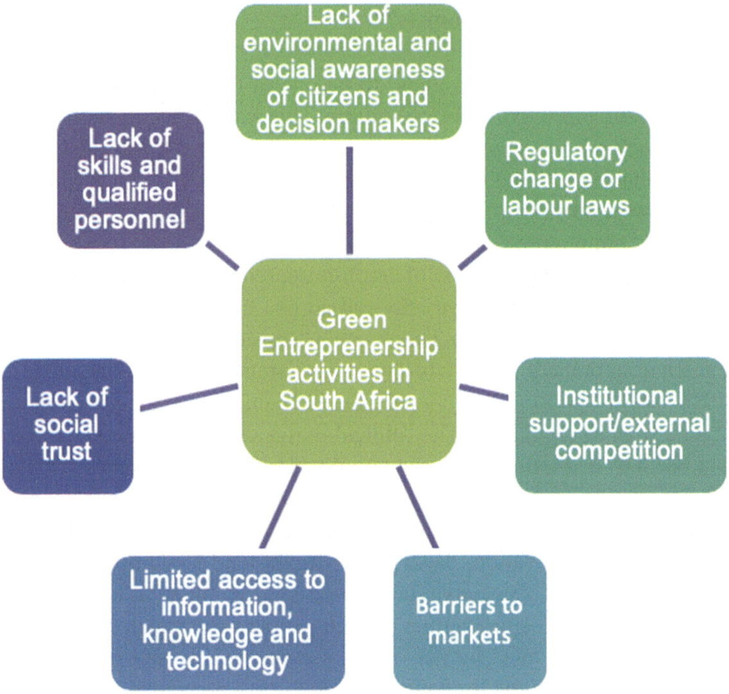

Fig. 3.8 Limitations of green entrepreneurship activities in South Africa. *Source* Author

relatively green. Greenness can also refer either to the product or to the process. The term "greening" is used as a kind of symbol for "moving toward environmental or ecological sustainability" (Gibbs and O'Neill 2015).

From the study, it emerged that there are at least two types of green entrepreneurs: One that did not start out with environmental concerns but became innovative along the way for either marketing, funding advantage or because of ethical concerns. On the other hand, there are those that embraced green activities from their inception, with a vision to transform the market in which it participates.

3.4.4.2 Green Entrepreneurship/Social Entrepreneurship

One of the respondents interviewed argued that green entrepreneurship is more of a vague term or a buzz word, which is complex and means many things to different people. The respondent said she believed in social entrepreneurship, as it was more sustainable: "business needs to understand that we need to have a positive impact to the society we live in. Green Entrepreneurship is great, but we need to understand people and the plan." She had been positively conscious of procurement process even before the concept became famous. Her business was expanding and selling online. She had agents in Kenya and on the ground in townships areas.

Çigdem and Arun (2016) opined that the dynamics of a budding green entrepreneurship can be different in the emerging markets because the nature of the society and relations of business are different. As a result of these differences, green entrepreneurship has also distinguished dimensions and specific motivations. Therefore, the definition of green entrepreneurs should also be distinguished from green social entrepreneurs' community. A green entrepreneur can be either making a business "green" or simply entering a "green business." In other words, green entrepreneurship seeks for ecological sustainability, while green social entrepreneurs' community, in their definition, considers the relationship between economy, ecology and society toward a sustainable approach. Two main roles are suggested for green social entrepreneurs' community: first, as a part of economy, which changes the structure of economy to more sustainable manner, and another, as norm creators in society. The second role acts as a holder for development (Zahedi and Otterpohl 2015). These roles were more pronounced among the SEED Award winners, as the social aspect was more distinctive, when compared to the other two facets of a green economy, which are environmental and economic.

3.5 Conclusion and Policy Recommendation

South Africa has put some policies and strategies to support and promote green entrepreneurship activities. Entrepreneurs have touted these as the strong signals toward the economic development. The concept of green entrepreneurship has been defined as the creation of new products, services or organizations to meet market opportunities. South Africa, in the past five years, has witnessed green entrepreneurs mushrooming across the country. However, some of the factors are affecting the continued development of green entrepreneurship activities in the country. The study sought to investigate and critically analyze the factors influencing green entrepreneurship activities in South Africa. It established that much of the green entrepreneurship activities have been started. Significantly, two types of green entrepreneurs exist: the ones that were started as a green business from inception and those that have adopted innovations along the way, taking advantage of funding and marketing opportunities. The term green entrepreneurship proved to be subjective and more of buzz words to some, with some preferring to call themselves social entrepreneurs. Factors, such as lack of finance, institutional barriers, external competition and lack of environmental and social awareness among citizen and decision makers, proved to be some of the issues affecting green entrepreneurship activities in the country. The study's findings are limited by its exploratory, quantitative nature and small nature. It is, therefore, recommended that the results are treated with care, and further research is carried out with a larger sample. The paper recommends that there is need to increase green entrepreneurs' awareness and to connect them to knowledge. The local ecosystem of firms, institutions and organizations constitutes key factors for the execution of strategies that better link the different stakeholders at the local level.

References

Abu-Saifan S (2012) Social entrepreneurship: definitions and boundaries. Technol Innov Manag Rev. Available online: http://timreview.ca/sites/default/files/article_PDF/GrootDankbaar_TIMReview_December2014.pdf. Accessed 22 June 2016

Arend R, Bromiley P (2009) Assessing the dynamic capabilities view: spare change, everyone? Strateg Organ 7(1):75–90

Ayyagari M, Beck T, Demirguc-Kunt A (2003) Small and medium enterprises across the globe: a new database. World Bank policy research working paper 3127. World Bank, Washington

Bibri SE (2013) The potential catalytic role of green entrepreneurship technological eco-innovations and ecopreneurs acts in the structural transformation of a low-carbon or green economy. A Foucaudian discursive approach. Thesis submitted for completion of Master of Science in Entrepreneurship, New Venture Creation, Department of Business Administration, School of Economics and Management, Lund University, Sweden

Booyens I (2011) Are small, medium- and micro-sized enterprises engines of innovation? The reality in South Africa. Sci Public Policy 38(1):67–78

Carroll AB, Buchholtz AK (2010) Business and society, ethics and stakeholder management. South Western Publishing, Cincinnati, Ohio

Chiloane-Tsoka E, Mabiza-ma-Mabiza J, Mbohwa C (2014). Green economy and SMEs in South Africa. How green are South African SMEs. Acad J Interdiscip Stud 3(6). MCSER Publishing, Italy

Çigdem V, Arun K (2016) What color is the green entrepreneurship in Turkey? J Entrep Emerg Econ 8(1):25–44. https://doi.org/10.1108/JEEE-07-2015-0042

Development Bank of Southern Africa (2011) Programmes in support of transitioning South Africa to a green economy. Development planning division working paper series no 24. Development Bank of Southern Africa Limited

Farinelli F, Bottini M, Akkoyunlu S, Aerni P (2013) Green entrepreneurship: the missing link towards a greener economy. ATDF J 8(3/4):42–48

Gibbs D, O'Neill K (2014) Rethinking sociotechnical transitions and green entrepreneurship: the potential for transformative change in the green-building sector. Environ Plan A 46(5):1088–1107

Gibbs D, O'Neill K (2015) Green entrepreneurship: building a green economy? Evidence from the UK. In: Social and sustainable enterprise: changing the nature of business. Contemp Issues Entrep Res 2:75–96. Published online: 08 Mar 2015. Available online: https://doi.org/10.1108/S2040-7246(2012)0000002008. Accessed 10 July 2015

Groot A, Dankbaar B (2014) Does social innovation require social entrepreneurship? Technol Innov Manag Rev. Available online: http://timreview.ca/sites/default/files/article_PDF/GrootDankbaar_TIMReview_December2014.pdf. Accessed 22 June 2016

Fetola (2016) An overview and analysis of results from the 2013 Fetola. SME growth survey, Nov 2016. Available online: http://www.fetola.co.za/wpcontent/uploads/2015/07/JustAddGreen_Brochure_Update_Small.pdf. Accessed 17 Apr 2016

Herrington M, Kew J, Kew P (2009) Tracking entrepreneurship in South Africa: a GEM perspective. Global entrepreneurship monitor. Country report

Holt D (2011) Where are they now? Tracking the longitudinal evolution of environmental businesses from the 1990s. Bus Strateg Environ 20:238–250. Available online: http://www.bdlive.co.za/business/2015/05/03/smes-access-to-finance-improves-survey-finds. Accessed 27 Jan 2016

International Labour Organization–ILO (2015) Green entrepreneurship creating green jobs through sustainable development. ILO

Kirkwood J, Walton S (2015) How green is green? Ecopreneurs are balancing environmental concerns and business goals. Australas J Environ Manag 21(1):37–51. https://doi.org/10.1080/14486563.2014.880384

Lekhanya LM (2014) The level of awareness of green marketing and its managerial implications amongst selected South African manufacturing small, medium and micro enterprises (SMMEs) in KwaZulu–Natal. J Econ Behav Stud 6(8):625

Mukonza C (2016) Analysis of factors influencing green entrepreneurship in South Africa. UNU-INRA working paper no. 20. UNU-INRA, Accra, Ghana

Mungadze S (2015) SA's emerging entrepreneurs 'among the best in the world'. Available online: http://www.bdlive.co.za/business/2015/05/15/sas-emerging-entrepreneurs-among-the-best-in-the-world. Accessed 20 Mar 2016

Mwakambirwa B (2013) Green entrepreneurial practices among small and medium enterprises in Mombasa County, Kenya. A master thesis, The University of Nairobi

Ndubisi NO, Nair SR (2009) Green entrepreneurship (GE) and green value added (GVA): a conceptual framework. Int J Entrep Innov 13(1):21–24

Nikolaou EI, Ierapetritis D, Tsagarakis KP (2011) An evaluation of the prospects of green entrepreneurship development using a SWOT analysis. Int J Sustain Dev World Ecol 18(1):1–16

Organization for Economic Corporation Development–OECD (2010) SMEs and green growth: promoting sustainable manufacturing and eco-innovation in small firms. Lessons from the global crisis and the way forward to job creation and growth. OECD working party on SMEs and entrepreneurship (WPSMEE), issue paper 3, Paris, 17–18 Nov 2010

OECD (2011) Measuring green entrepreneurship. In: Entrepreneurship at a glance 2011. OECD Publishing. Available online at https://doi.org/10.1781/97892640977l1-4en

Pandrani MD, Ferguson D (2014) Enabling factors of green entrepreneurship: a case study of organic agriculture produce in the Sindh and Balochistan provinces of Pakistan. Entrepr Learn 181–191

Parrish BD, Foxon TJ (2009) Sustainability entrepreneurship and equitable transitions to a low-carbon economy. Greener Manag Int 55:47–62

Shepherd DA, Patzelt H (2011) The new field of sustainable entrepreneurship: studying entrepreneurial action linking "what is to be sustained" with "what is to be developed". Entrep Theory Pract 35(1):137–163

Teece D, Pisano G, Shuen A (1997) Dynamic capabilities and strategic management. Strateg Manag J 18(7):509–533

United Nations Environmental Programme-UNEP (2011) Towards a green economy. Pathways to sustainable development and poverty reduction. Available online at www.unep.org/greeneconomy. Accessed 11/03/2015. ISBN 978-92-801-3143-9

Walley EE, Taylor DW (2002) Opportunists, champions and mavericks? A typology of green entrepreneurs. Greenleaf Publishing

Zahedi A, Otterpohl R (2015) Towards sustainable development by creation of green social entrepreneur's communities. Procedia CIRP 26:196–201

Chapter 4
Analysis of Crop Yield Volatility Among Smallholder Farmers in Ghana

James Atta Peprah, Clifford Afoakwah, and Isaac Koomson

4.1 Introduction

Agriculture is one of the major sectors of the Ghanaian economy as it employs majority of the populace. As it stands, the agricultural sector employs 54.2% of the total population and this translates into 45.8% of households (Ghana Statistical Service–GSS 2013). This means that the contribution of the agricultural sector to the economy of Ghana is overarching. In explaining further, it is evident that majority (73.5%) of these agricultural households are located in rural areas and this confirms the generally accepted fact that majority of agricultural households in Sub-Saharan Africa live in rural areas, where poverty and deprivation are most severe (Diao et al. 2007). These households live in rural areas and have relatively small farm sizes, thus making them smallholder farmers.

According to Chamberlin (2007), Ghanaian agriculture is highly dominated by smallholder farmers (mainly located in rural areas) who produce many commodities.

This chapter is extracted from a working paper by Peprah et al. (2016) published in the UNU-INRA working paper series (for more information follows the link http://collections.unu.edu/view/UNU:5923#viewAttachments).

J. A. Peprah (✉)
Department of Applied Economics, School of Economics, University of Cape Coast, Cape Coast, Ghana
e-mail: jpeprah@ucc.edu.gh

C. Afoakwah
Centre for Applied Health Economics (CAHE), Griffith University, Brisbane, QLD, Australia
e-mail: c.afoakwah@griffith.edu.au

I. Koomson
Faculty of Science, Agriculture, Business and Law, UNE Business School, University of New England, Armidale, NSW, Australia
e-mail: ikoomso2@une.edu.au

Network for Socioeconomic Research and Advancement (NESRA), Accra, Ghana

C. Atewamba and D. Yong Ngondjeb (eds.), *Inclusive Green Growth*, Advances in African Economic, Social and Political Development,
https://doi.org/10.1007/978-3-030-44180-7_4

More than 70% of farmers in Ghana cultivate about three hectares (ha) of land or even smaller. The small nature of these farms notwithstanding, they still play a vital role in the economy. A study by Irz et al. (2001) found a significant positive relationship between crop yield and the number of people living in poverty, implying that increased yield improves livelihoods. The corresponding elasticity coefficients estimated for Africa were higher than that of Asia, which suggests that compared to Asia, Africa could experience a larger reduction in the poverty if policies that increase crop yields are adopted.

In spite of the important role smallholder farmers play in the agricultural sector of the Ghanaian economy, these farmers lack access to credit (Chamberlin 2007; Christen and Anderson 2013; Peprah and Koomson 2015) and as a result are not able to adopt modern (appropriate) technology in their operations. Their problems in accessing credit are also compounded by their farm sizes and locations (Koomson et al. 2016) and are sometimes compelled to resort to personal savings (Peprah et al. 2015). This is because security of land tenure and the land size thereof are major determinants used by financial institutions to advance credit to loan applicants. Unfortunately, for those who would like to use their profits to finance the purchase of needed machines and adoption of new and improved technology, they are highly constrained due to smaller and unstable profits realized from their farming operations.

Unlike the usual expectation of many studies coming from this interesting area, there exist no or limited studies on the subject matter. This is the gap that this paper aims at filling and consequently contributes to knowledge. It is presumed that access to credit will enable farmers to embark on more green economy activities such as organic farming instead of traditional application of inorganic fertilizers. With access to credit, they will be able to incorporate improved production techniques which will lead to increased efficiency, higher yields, and improved product quality. This in turn means increased household food security and higher household income from more and better-quality produce. Especially in cases where farmers are able to save money through purchasing and using less fertilizer and pesticides, there will be positive environmental knock-on effects.

Owing to the problems confronting smallholder farmers and how these challenges affect their crop yields, this study confirms whether smallholder farmers in rural areas face higher volatility in crop yield compared to their counterparts in the urban areas. It also assesses the effect of access to credit on crop yield, as well as examining the effect of multiple inputs technology use on crop yield. In doing so, this study employs the stochastic dominance analysis to determine whether one can predict crop yield among rural farmers with higher levels of certainty compared to their urban[1] counterparts or vice versa (using outcomes of crop yield). We then proceed to use ordinary least squares (OLS) method to estimate the effect of credit and technology use on crop yield for the pooled model and rural–urban (subsample) models, to cater for heterogeneity among rural and urban farmers.

[1]This refers to farm households that have their farms located in the urban areas. Generally, urban farmers compared to rural farmers are relatively advantaged when it comes to access to credit and usage of technology for farming.

The rest of the paper is organized as follows: Section 4.2 looks at the literature review and considers smallholder farming in Ghana; the methodology is explained in Sect. 4.3; Section 4.4 presents the results and discussion and the final section concludes with policy recommendations.

4.2 Literature Review

4.2.1 Access to Credit and Crop Yield

Over the past two decades, Sub-Saharan Africa (SSA) has experienced impressive economic growth averaging about 4.5% with some non-oil-exporting countries reaching an average of more than eight percent (Chauvin et al. 2012). Despite this impressive economic performance, agricultural transformation has been slow and growth rather sluggish. Notably, agricultural productivity is still below yield potentials, agricultural mechanization is weak and declining, and the state of agricultural financing is still nascent. In this context, the worse affected group of farmers is smallholder farmers.

Globally, an estimated 500 million agricultural smallholders farm up to two hectares of land, with 2–2.5 billion people living in these smallholders' households worldwide (Christen and Anderson 2013; Hazell 2011). These farms feed a great number of the rural poor. According to the International Finance Corporation (2011), of the three quarters of the world's poor that live in rural areas, 80% depend on agriculture as their main source of income and employment. These smallholders also play a key role in increasing food supply, more so than large farms in poor countries, and increasingly supply large conglomerates and corporations with produce for their production processes (Carroll et al. 2012). Despite their socioeconomic importance, smallholder farmers tend to have little or no access to formal credit, which limits their capacity to invest in the technology and inputs they need in order to increase their yields and incomes and reduce hunger and poverty. The situation is not all that different from some Sub-Saharan African countries.

It is well known that the effectiveness of technology adoption has direct impact on yield. Even though yield in turn depends on land characteristics, such as soil quality, water availability and accessibility, and farmers' land tenure system, and other factors such as income levels, wealth and access to credit, these factors have also been identified as key determinants that influence technology adoption (Rosen et al. 2012). For instance, among the factors that contribute to low fertilizer usage is lack of access to credit. Smallholder farmers who recognize the need to adopt the right amount of fertilizer may be constrained by lack of credit facilities, as they cannot afford to purchase the fertilizer from their own means. Thus, lack of access to credit has contributed to low fertilizer applications among smallholder farmers in SSA.

Credit is an important element in agricultural production systems. It allows producers to satisfy the financial and resource needs induced by production cycle, which characterizes crucial activities such as land preparation, planting, cultivation, harvesting, storage, and transportation typically done over a period of several months. In the absence of credit markets, farmers would have to maintain cash reserves to facilitate the various activities along the production cycle. The availability of credit allows greater investment in inputs, implements, and infrastructure, which subsequently result in increases in farmers' yield.

Access to credit does not only influence land use but has also impacted on the efficiency and yields of agricultural households (Jorge 2004). Credit services are useful in overcoming the financial and resource constraints faced by agricultural households in their efforts to manage the risk associated with productivity growth (*risk-coping effects*). Households with access to credit and financial deposits are more inclined to invest in agricultural intensification, which typically offers higher returns.

The majority of smallholder farmers in SSA are credited constrained. It has been argued that the lack of access to credit adversely affects the productivity of farmers. Chisasa and Makina (2013) examined the supply of credit to smallholder farmers in South Africa and concluded that smallholder farmers are indeed credit constrained mainly due to their inability to provide collateral. Wynne and Lyne (2003) confirmed that lack of credit was hampering the development of smallholder farmers in South Africa but did not test the contribution of credit to farm performance. As some authors (e.g., Chisasa and Makina 2013) have verified that at a macro level, credit makes a positive and significant contribution to the overall agricultural output; by the same token, the lack of it has affected yield among agricultural households, especially rural smallholder farmers.

The effectiveness of credit as an input depends on the economic and financial policies that go with it. If well applied, credit should increase the size of farm operations, introduce innovations in farming, improve marketing efficiency, and enhance farmers' consumption (Nwaru 2004). Unfortunately, the moral hazard problem in the credit market may contaminate the application of credit thus producing undesired results. This might be one of the major reasons why smallholder farmers are excluded from the credit market, thus contributing to the decline in domestic credit to agriculture.

In Ghana, for example, domestic credit to the agricultural sector continues to decline. In 2003, the agricultural sector received 5.40% of domestic credit, and by 2009, the share had reduced to 4.74%. The sector saw an increase in 2010 (6.13%), however, since then credit to the sector has dwindled to 4.08% in 2013 (Institute of Statistical, Social and Economic Research 2014). Apart from this, the yield of some major food crops has also not been consistent. Other crops have recorded decreases in yields. This presumes that a possible relationship between access to credit and yield exists and that with credit being made available smallholder farmers' yield could increase.

4.2.2 Technology and Crop Yield

The main factors that influence the adoption of technology among smallholder farmers in SSA are assets, vulnerability, and institutions (Meinzen-Dick et al. 2002). Assets are requisite physical (material) and abstract possessions (e.g., education and finance) that are essential for technology adoption. For farmers to adopt more technology, the new technologies should require little use of assets, so that even the asset constrained smallholder farmers can adopt the technologies easily. Vulnerability factors relate to technologies that impact on the level of exposure of farmers to economic, biophysical, and social risks. In this regard, farmers are generally more attracted to technologies with lower risk. Finally, institutional factors like credit, insurance, and information dissemination are those that focus on the extent or degree to which smallholder farmers adopt technology. Other institutional factors relate to mechanisms and facilities that improve farmers' access to productive inputs and product markets (Meinzen-Dick 2004).

Nkonya et al. (2004) advised that farmers should be trained not only for them to adopt improved yield-raising technologies, such as improved seed varieties, but also for fertility-restoring and conservation technologies. In implementing programs that promote the adoption of technologies among smallholder farmers, synergies should be created among government departments, non-governmental organizations (NGOs), researchers, donors, and local communities. Effectively doing this will lead to increased agricultural productivity, reduced environmental degradation, and improved soil quality.

In SSA, agricultural technology development is an essential strategy for increasing agricultural productivity, achieving food self-sufficiency and reducing poverty among smallholder farmers. This strategy is particularly relevant for smallholder farmers in the subregion because they are disadvantaged in many ways, which make them a priority for development efforts. Usually, these farmers live and farm in areas where rainfall is low and erratic, and soils tend to be infertile. In addition, infrastructure such as irrigation and roads, and institutional support factors such as input and product markets, credit and extension services tend to be poorly developed. It is recommended that further research and rural development efforts should focus on the development of infrastructure and institutions in these areas (Muzari et al. 2012).

According to Muzari et al. (2012), the technologies that farmers adopt play significant roles in determining how fast agricultural productivity grows and how that growth affects the poor and the condition of natural resources. The development of agricultural technology for both food and non-food crops, improvement of the rural financial markets, the dissemination of assets and information, development of agricultural research and extension facilities targeted toward smallholder farmers, all work together to increase their agricultural productivity and improve their livelihoods. Overall, the experience and evidence from countries within and around the SSA region indicate that returns to agricultural technology development could be very high and far-reaching, not only in the smallholder sector but in the entire economy as well. However, improved technologies are of little value unless farmers perceive them to be appropriate and subsequently adopt them. It is therefore imperative not only to develop new agricultural technologies but also to promote their adoption by smallholder farmers (Meinzen-Dick 2004).

As highlighted before, prominent among the factors that affect technology adoption are: assets, vulnerability, and institutions. Lack of assets, such as land, education, or equipment, will limit technology adoption. This necessitates the need for further studies and development efforts to pay more attention to technologies that require fewer assets. Decision makers also need to recognize that technologies that build on assets that the poor farmers already have are more likely to be adopted (Meinzen-Dick et al. 2002). In order to encourage adoption of new technologies, pro-poor agricultural researchers must look beyond simply boosting productivity. They should emphasize certain variables which reduce the farmers' vulnerability to loss of income, bad health, natural disasters, and other factors. In addition, an understanding of local cultural practices and preferences is important if smallholder farmers are to benefit from agricultural technologies developed through research. All these form a potentially useful area of study for future research (ibid.).

In some countries, female-headed households are discriminated against within their local communities and/or by credit institutions. This is because male and female farmers have different access to many aspects including credit and land tenure systems, with female farmers generally having lower access. Besides the challenges faced by women, they are known to contribute a lot in terms of labor and overall production in the smallholding farming sector in SSA. There is clearly a case for improving current credit systems to ensure that a wider spectrum of smallholder farmers is able to access credit, especially female-headed households (Mkandawire and Matlosa 1993).

Other steps that may be taken to encourage the adoption of technologies that increase agricultural productivity and reduce land degradation include reducing the prices of inputs, offering credit, and waiving some of the taxes levied on input trading businesses. The promotion and strengthening of research-extension linkages will also improve technology adoption. Stronger partnerships between agricultural researchers and other agents of change, including local organizations, farmers, community leaders, NGOs, national policy makers, and donors, are also important in stimulating technology adoption for increased agricultural productivity (Muza et al. 1997).

Promotion of various smallholder income sources, such as off-farm employment, remittances, and livestock production, can lead to higher total household income to finance the purchase of inputs such as fertilizers, seed, and hired labor. Introducing technologies that require less labor is also likely to lead to their adoption because the smallholder farming sector in the subregion is beset with chronic shortages of labor during the agricultural season (Muzari et al. 2012).

It is noteworthy to highlight that there are some rational, positive aspects in certain traditional agricultural practices that are done by smallholders in SSA. Modern researchers should therefore seek to investigate the reasons why smallholder farmers do the things they do and attempt to improve on them. This is a more effective strategy than the prevailing approaches which seek to displace traditional technologies outright on the grounds that they are irrational, unscientific, primitive, and backward.

The use of improved technology has the potency of increasing technical efficiency and consequently agricultural production among smallholder farmers. In a study by Maurice et al. (2015) in the Adamawa State in Nigeria, the authors concluded that under existing technology, the use of *agrochemicals, inorganic fertilizers, and improved seeds contributed significantly to food crop output. However, these technologies are anti-green economy practices due to their repercussive effects; hence, the use of green economy practices such as organic fertilizer applications (e.g., composts) is believed to yield better and sustainable results.*

4.2.3 Smallholder Farming in Ghana

Smallholder, according to Chamberlin (2008), connotes limited land availability, and when the connotations are stretched, it is broadened to include smallholder farmers as those that are 'resource-poor' (e.g., those with limited capital including animals, fragmented holdings, limited access to capital). Because resource-poor households are a diverse group of people, Asuming-Brempong et al. (2004), in doing the Ghana's Poverty and Social Impact Analysis (PSIA), argued that different resources and risk conditions better define smallholders than simple measures of landholdings. A smallholder farmer in any part of Ghana is one who farms on fewer than five hectares (Ekboir et al. 2002). In another descriptive definition, the Ministry of Agriculture (MoFA 2011) stated that agriculture is predominantly on a smallholder basis in Ghana and that about 90% of farm holdings are fewer than two hectares in size. In Ghana, the breakdown of smallholder farmers into groups has largely been done to operate around agroecological zones; as pointed out by Asuming-Brempong et al. (2004), this has resulted in these farmers' inability to benefit from government targeted interventions such as provision of high-input and high-output technologies.

Smallholder farmers in general own small-based plots of land on which they grow subsistence crops and one or two cash crops relying almost exclusively on family labor. Some of the main characteristics of the production systems of smallholder farmers are: simple, outdated technologies; low returns; highly seasonal labor fluctuations; and women playing a vital role in production. Smallholder farmers differ in individual characteristics, farm size, resource distribution between food and cash crops, livestock and off-farm activities, their use of external inputs and hired labor, the proportion of food crops sold and household expenditure patterns.

4.3 Methodology

4.3.1 Data Type and Source

Data for the study was sourced from the Ghana Living Standard Survey (GLSS 6) collected in 2012/2013. The GLSS 6 contains information on smallholder farmers, their input combination, and yield. It also contains information on farm levels, household-level characteristics, and socio-demographic characteristics. The needs of the Savannah Accelerated Development Authority (SADA) areas were adequately catered for in this survey by increasing the number of primary sampling units (PSUs) and households from 580 and 8700 to 1200 and 18,000 respectively. This indicates an expansion in coverage of about 107% when compared to the previous round of the survey (GLSS 5). The sampling approach adopted for the GLSS 6 was the multi-stage approach, where 1200 enumeration areas (EAs) were selected to form the primary sampling units (PSUs). The PSUs were allocated into the ten regions using probability proportional to population size (PPS), and the EAs were further divided into urban and rural localities of residence. The secondary sampling units (SSUs) were generated from a complete list from the selected PSUs. At the second stage, 15 households from each PSU were selected systematically. This yielded a total sample size of 18,000 households nationwide (GSS 2014).

4.3.2 Description of Methods

In order to achieve the objectives of this paper, two main methods are used. First, we test the hypotheses that access to credit and the use of technological equipment improves crop yield among rural and urban smallholder farmers, using an ordinary least squares (OLS) technique. Three separate models were estimated. The first is a pooled model for all rural and urban farmers, with rural farming as a dummy. The two other models are subsamples for only rural farmers and urban farmers independently. The Chow/F-statistics is used to determine whether the separate models or pooled model better fits the data. The other reason for the separate models for rural and urban farmers is because of the differences in yield that might be accounted for by rural and urban farming and also because rural dwellers have been documented as facing more problems in accessing credit. By so doing, we are able to tell how access to credit and technology use differently affect rural and urban farmers, which is important for policy making. The use of OLS estimation technique is appropriate as the asymptotic properties of crop yield suggest a normal distribution having logged it to control for outliers. As shown in Fig. 4.1, crop yield among rural and urban farmers is normally distributed, thus, symmetrical, hence the use of OLS is fit.

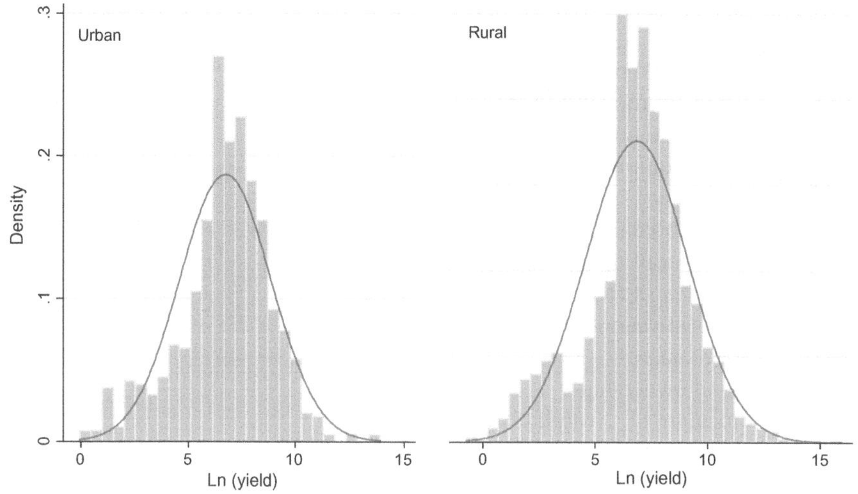

Fig. 4.1 Distribution of crop yield among rural and urban farmers. *Source* Graphed by authors using the GLSS 6 data

4.3.3 *Empirical Models*

The following econometric models are therefore specified and estimated using OLS method. Model one is the pooled model that regresses the yield per acre on access to credit, technology use, farm-specific, household-level, and individual-level characteristics regardless of farming location. Model two is specifically for smallholder farms in the rural areas while model three is for smallholder farms in urban areas.

Model 1: Pooled Model

$$\begin{aligned}\text{cropyldpa} = {} & \delta_0 + \delta_1\text{crop}_i + \delta_2\text{credit}_i + \delta_3\text{feqpt}_i + \delta_4\text{landsize}_i + \delta_5\text{hhsize}_i \\ & + \delta_6\text{malefms}_i + \delta_7\text{femfms}_i + \delta_8\text{age}_i + \delta_9\text{age}_i^2 + \delta_{10}\text{landdown}_i \\ & + \delta_{11}\text{rural}_i + \delta_{12}\text{female}_i + \delta_{13}\text{ezone}_i + \varepsilon_i\end{aligned}$$

Model 2: Rural Model

$$\begin{aligned}\text{cropyldpa}_{\text{rural}} = {} & \kappa_0 + \kappa_1\text{crop}_i + \kappa_2\text{credit}_i + \kappa_3\text{feqpt}_i + \kappa_4\text{landsize}_i + \kappa_5\text{hhsize}_i \\ & + \kappa_6\text{malefms}_i + \kappa_7\text{femfms}_i + \kappa_8\text{age}_i + \kappa_9\text{age}_i^2 \\ & + \kappa_{10}\text{landdown}_i + \kappa_{11}\text{rural}_i + \kappa_{12}\text{female}_i + \kappa_{13}\text{ezone}_i + \eta_i\end{aligned}$$

Model 3: Urban Model

$$\begin{aligned}\text{cropyldpa}_{\text{urban}} = {} & \lambda_0 + \lambda_1\text{crop}_i + \lambda_2\text{credit}_i + \lambda_3\text{feqpt}_i + \lambda_4\text{landsize}_i + \lambda_5\text{hhsize}_i \\ & + \lambda_6\text{malefms}_i + \lambda_7\text{femfms}_i + \lambda_8\text{age}_i + \lambda_9\text{age}_i^2 + \lambda_{10}\text{landdown}_i \\ & + \lambda_{11}\text{rural}_i + \lambda_{12}\text{female}_i + \lambda_{13}\text{ezone}_i + \gamma_i\end{aligned}$$

where

Cropyldpa	crop yield per acre
Crop	type of crop cultivated
Credit	access to credit (1 = yes; 0 = no)
Feqpt	farm equipment (1 = yes; 0 = no)
Land size	land size
Hhsize	household size
Malefms	number of male farmers
Femfms	number of female farmers
Age	age (in years)
Age^2	age square
Landown	land ownership
Rural	location (1 = rural; 0 = urban)
Female	female-headed households
Eczone	ecological zones

A Priori Expectations
The a priori signs are indicated for parameters of the pooled but these also apply to variables in the rural and urban models since the variables are the same in all models

$$\delta_9 = \delta_{12} < 0 \text{ and}$$
$$\delta_1 = \delta_{10} = \delta_{11} = \delta_{13} = \pm\delta_2$$
$$= \delta_3 = \delta_4 = \delta_5 = \delta_6 = \delta_7 = \delta_8 > 0$$

Stochastic Dominance Analysis
The second method which ascertains crop yield volatility among rural and urban smallholder farmers is the stochastic dominance analysis (SDA). The SDA is used to refer to a set of relations that may exist between a pair of distributions. Commonly used in the analysis of income distributions and income inequality, SDA has also been applied in many other domains including agricultural economics. In this paper, the SDA is used to determine crop yield variability among smallholder rural and urban farmers in Ghana. To determine whether a relation of stochastic dominance holds between two categories of farmers (rural and urban), the distributions are first characterized by their cumulative distribution functions (CDFs). For a given set of yields, the value of the CDF at yield y is the proportion of yields in the set that are no greater than y. In the context of a random variable Y, the value of the CDF of the distribution of Y at y is the probability that Y should be no greater than y.

We consider two distributions of farmers, rural (A) and urban (B), characterized, respectively, by CDFs F_A and F_B. According to Davidson and Duclos (2013), the distribution B dominates distribution A stochastically at first order if, for any argument y, $F_A(y) \geq F_B(y)$. If y is measured as crop yield, then the volatility or disparity means that the proportion of farmers in distribution A with yields no greater than y is no smaller than the proportion of such farmers in B. Thus, there exits at least as high a proportion of low yield farmers in A as in B, if low yield means a yield smaller than y. If B dominates A at first order, then whatever maximum yield we choose as efficient, there is always more low yield in A than in B, which is why A is defined as a dominated distribution.

4.4 Results and Discussion

4.4.1 Descriptive Statistics

The descriptive statistics are presented in Table 4.1. In general, the average yield per acre for all crops for all farmers is about 4173.025 kg per acre. When disaggregated, the averages are 4452.468 kg and 3006.637 kg per acre for rural and urban farmers, respectively. These values have very high standard deviation indicating the presence of outliers. Figure 4.2 shows that such high yield is driven by yield among farmers who cultivate industrial crops (crops grown to produce goods to be used in the production sector, rather than food for consumption such as cocoa, coffee, and cotton production). The average cereal yield per acre among urban and rural farmers is 2986.12 kg and 2664.7 kg, respectively. Farmers in rural areas experience lower yield in vegetable production than any other type of crop.

About 11.3% of rural farmers accessed credit as compared to 14.6% of urban farmers. This suggests that relatively more farmers in the urban areas access credit compared to those in the rural areas. This is possibly due to the fact that many of the financial institutions are normally located in the urban settings as noted by Amoa–Mensah et al. (2013). More than 50% of farmers in the rural areas adopted at least one input technology on their farms while less than 50% of farmers in the urban areas adopted multiple input technologies.

The findings revealed that about 81% of smallholder farmlands in Ghana are personally owned either through purchase or distributed by families, and in other cases, used by a farmer for free, though owned by another person. The descriptive statistics suggest that the average number of people within a farm household is about four, while the average number of male and female laborers employed on a farm is about two each.

4.4.2 Regression Analysis of Crop Yield Among the Smallholder Farmers in Ghana

The analysis and discussion are done for three models: pooled, urban, and rural. For access to credit and technology, all models are analyzed and discussed to ascertain the relative effect of these variables in these locations and to also inform policy directions. With regard to the rest of the variables, the model of focus is the pooled model that covers all smallholder farmers in Ghana, regardless of whether they are located

Table 4.1 Descriptive statistics

	Variable description/definition	Pooled		Rural		Urban	
		Mean	SD	Mean	SD	Mean	SD
Crop yield	Crop yield per acre; continuous variable	4173.025	56,354	4452.468	61,794.89	3006.637	22,201.55
Dummies for crop type harvested							
Cereals	Cereals or not (1 = yes/0 = no)	0.655	0.476	0.642	0.480	0.708	0.455
Industrial	Industrial or not (1 = yes/0 = no)	0.216	0.411	0.222	0.416	0.188	0.391
Legumes	Legumes or not (1 = yes; 0 = no)	0.091	0.287	0.097	0.296	0.062	0.242
Fruits	Fruits or not (1 = yes/0 = no)	0.023	0.152	0.022	0.146	0.031	0.0172
Vegetables	Vegetables or not (1 = yes/0 = no)	0.001	0.038	0.002	0.039	0.001	0.033
Other crops	Other crops (1 = yes/0 = no)	0.014	0.118	0.015	0.122	0.010	0.099
Credit	Dummy variable for whether the household has accessed credit or not (1 = yes/0 = no)	0.119	0.0324	0.113	0.316	0.146	0.353
Farm equipment use	Dummy variable for whether the household uses equipment for farming or not (1 = yes/0 = no)	0.526	0.499	0.533	0.499	0.489	0.499
Land	Total land size cultivates in acres	0.552	0.471	0.544	0.472	0.586	0.461
Household size	Total household members	4.904	2.701	4.946	2.746	4.728	2.498
Male farmers	Number of males engaged on farm	1.739	2.177	1.751	2.260	1.688	1.782
Female farmers	Number of females engaged on farm	1.694	2.531	1.727	2.518	1.553	2.581
Age	Age of the household head	44.410	14.145	44.251	14.352	45.081	13.223
Age square	The square of the age of the individual	2172.319	1367.201	2164.10	1388.63	2206	1273.071
Land ownership							
Personal ownership	Dummy variable for whether farmland is personally owned or not (1 = yes/0 = no)	0.808	0.394	0.807	0.394	0.809	0.392

(continued)

Table 4.1 (continued)

	Variable description/definition	Pooled		Rural		Urban	
		Mean	SD	Mean	SD	Mean	SD
Rented	Dummy variable for whether farmland is rented or not (1 = yes/0 = no)	0.062	0.241	0.059	0.236	0.073	0.261
Shared cropping	Dummy variable for whether farmland is shared cropping or not (1 = yes/0 = no)	0.130	0.337	0.133	0.340	0.117	0.322
Rural	Dummy variable rural farming (1 = yes/0 = no)	0.809	0.393	–	–	–	–
Female	Dummy variable for female-headed households (1 = yes/0 = no)	0.245	0.429	0.241	0.427	0.260	0.439
Ecological zone							
Coastal	Coastal or not (1 = yes/0 = no)	0.082	0.272	0.089	0.284	0.053	0.205
Forest	Forest or not (1 = yes/0 = no)	0.553	0.497	0.548	0.498	0.574	0.495
Savannah	Savannah or not (1 = yes/0 = no)	0.365	0.481	0.363	0.481	0.373	0.483

Source Computed from GLSS 6

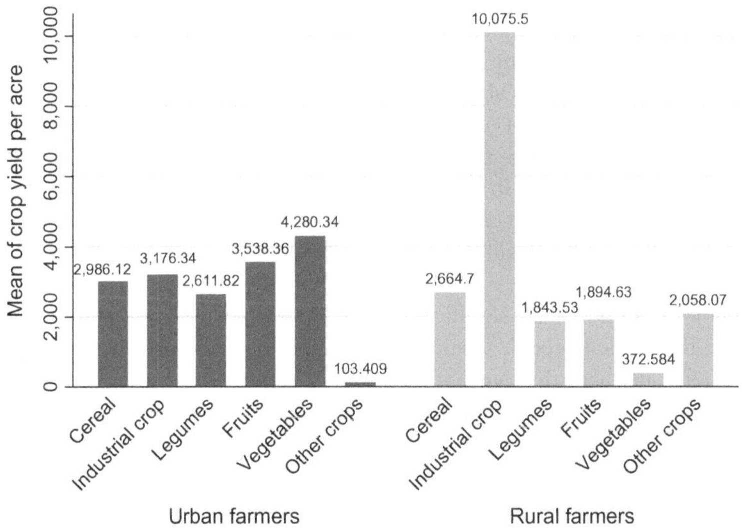

Fig. 4.2 Distribution of crop yield per acre among rural and urban farmers. *Source* Graphed by authors using the GLSS 6 data

in urban or rural areas. The chow test shows an F-statistic, which is significant at 1% level and this leads to the rejection of the null hypothesis that the coefficients in the urban and rural models are equal. This suggests that the differences in the coefficients in both models can be attributed to differences in locational characteristics of smallholder farmers in rural and urban areas of the country. Thus, running the subsampled models gives preferable estimates compared to the pooled model (i.e., the separate models best fit our data). The set of explanatory variables included in the models explains about 17% of the variability of crop yield among rural and urban farmers, respectively. These R-squares (although relatively small) are good enough for a cross-sectional analysis of this nature (Wooldridge 2010). In addition, the Chi-square statistics in the two models indicate that the regression lines are of good fit at 1% significance level.

The results of the regression analysis (Table 4.2) revealed that, in general, access to credit has a significant effect on the productivity of smallholder farmers. The pooled model showed that smallholder farmers that have access to credit obtain yields per acre that are 35.5% more than their counterparts that do not have access to credit at a significance level of one percent, holding all other variables constant. This significant effect is driven by increased yield among rural farmers as a result of accessing credit. Rural farmers stand a chance of increasing their produce by about 38.5% if they access credit. On the contrary, access to credit has no significant effect on yield among farmers in the urban areas. It suggests a relatively high opportunity cost for use of credit on farm activities among urban farmers. In the urban areas, farming is normally not the primary occupation among households; hence, farmers will always have to sacrifice a significant part of their returns to invest in accessing credit for use in their farming activities. For the rural folks, farming is seen as a primary economic activity as about 80.9% of farmers used in this study are in a rural setting. This finding does not only support similar studies conducted in developing economies (e.g., Nwaru 2004; Chisasa and Makina 2013) but also demonstrates the significant role credit plays in improving the yields of rural smallholder farmers. This also rehashes the several calls that have been made to, increasingly, make credit available to rural farmers to aid them in increasing their productivity in terms of the yield per acre.

Technology adoption also significantly affects crop yield of the smallholder farmers. From the pooled model, smallholder farmers that adopted technology experienced yields per acre that are 65.7% more than farmers who did not adopt any farming technology, at a one percent significance level, holding all other variables constant. Among urban smallholder farmers, those who adopted technology had about 70.30% extra yield compared to farmers who did not. In the rural areas, farmers who adopted technological equipment had yields that were 61.90% more than their counterparts who did not adopt technology in their farming operations. To this end, it can be inferred that adopting technology and the mix of it is very beneficial in increasing yield. This warrants the call on smallholder farmers to embrace the use of technology. They also need to be given more credit to enhance their ability to acquire the technology necessary for increasing yields. The technological effect is much felt among urban farmers than among rural farmers. Nkonya et al. (2004)

Table 4.2 Determinants of crop yield among smallholder farmers in Ghana

Log (crop yield per acre)	Pooled	Rural	Urban
Explanatory variables	Coefficients	Coefficients	Coefficients
Dummies for crop type harvested (Base = Industrial crop)			
Cereals	−0.868 (0.083)***	−0.858 (0.093)***	−0.844 (0.180)***
Legumes	−0.790 (0.150)***	−0.817 (0.167)***	−0.243 (0.340)
Fruits	−0.423 (0.212)**	−0.331 (0.233)	−0.699 (0.479)
Vegetables	−0.740 (0.980)	−1.712 (920)*	0.831 (0.232)***
Other crops	−0.904 (0.357)**	−0.682 (0.38)*	−2.543 (0.628)***
Credit	0.355 (0.102)***	0.385 (0.119)***	0.155 (0.186)
Technology use	0.657 (0.068)***	0.619 (0.076)***	0.703 (0.147)***
Land size	−1.626 (0.072)***	−1.702 (0.079)***	−1.274 (0.161)***
Household size	0.025 (0.012)**	0.037 (0.014)***	−0.020 (0.029)
Male farmers	0.019 (0.016)	0.013 (0.017)	0.098 (0.046)**
Female farmers	0.057 (0.012)***	0.063 (0.013)***	0.021 (0.023)
Age	0.029 (0.013)**	0.027 (0.015)*	0.038 (0.032)
Age square	0.0003 (0.0001)**	0.0002 (0.0001)**	−0.0005 (0.0003)
Land ownership (Base = Rented)			
Personal ownership	−0.243 (0.123)**	−0.247 (0.140)*	−0.197 (0.251)
Shared cropping	−0.261 (0.146)*	−0.272 (0.164)*	−0.252 (0.316)
Rural	−0.009 (0.0818)	–	–
Female	−0.483 (0.084)***	−0.466 (0.096)***	−0.454 (0.171)***
Ecological zone (Base = Coastal)			
Forest	0.274 (0.154)*	0.315 (0.165)*	0.510 (0.443)
Savannah	0.577 (0.159)***	0.463 (0.172)***	1.347 (0.425)***
Constant	6.739 (0.425)***	6.729 (0.421)***	−8.886 (1.081)***
N	4002	3234	768
Prob > Chi^2	40.51***	36.88***	30.59***
R^2	0.156	0.165	0.165
Chow test (F = 2.186)	0.0021		

Source Authors' own computation using the GLSS 6
Robust standard errors are in parentheses. *, **, *** significant at 10%, 5%, and 1%, respectively

advise that when implementing programs that promote the adoption of technologies among smallholder farmers, there is need to create synergies among government departments, NGOs, researchers, donors, and local communities. This will not only lead to increased agricultural productivity among smallholder farmers but will also facilitate the reduction in environmental degradation and the improvement of soil quality.

The analysis controlled for own labor and hired labor input in the cultivation process. With respect to own labor, household size which was used as proxy showed a positive impact on yield. In general, inclusion of an additional person to the size of smallholder farmers' households leads to an increase in crop yield per acre by 2.50%, at a five percent significance level, holding all other variables constant. Thus, households with many adult members have higher yields compared to those with fewer adult members. Such positive effect is not realized among farmers in the urban settings as this variable does not significantly affect crop yield per acre. Consistent with the general findings, rural farmers who have larger household size are better off in terms of crop yield. This variable is statistically significant at one percent level. The reason for the differences in this effect between urban and rural areas is that in the urban areas, farm labor comes from outside the home while farm labor in the rural areas mainly comes from within the household. This is why an additional person to the household increases farm productivity in the rural areas but has no effect on the productivity of the urban smallholder farms.

In terms of hired labor, it was proxied with the number of female and male workers engaged on the farm. For this gender dynamics of farm labor, we observed that while male laborers were much productive in the urban areas, such significant effect was not the case in rural areas. Instead, increased number of female laborers leads to increased productivity among rural farmers. This finding is in line with other studies in the literature which indicate that, relative to men, women with little training in good farm management practices make marginal productivity very high (Applefield and Jun 2014). An increase in the number of females engaged in the smallholder farm by an additional person increases farm yield per acre by 6.3% at a one percent significance level, holding all other variables constant. This supports the fact that in Ghana, women dominate rural farming. In terms of decision-making, we noticed that female decision makers harvested lower yield. Thus, female smallholder farmers experienced about 48.30% lower yield compared to their male counterparts, all else being equal.

Regarding age and productivity, a nonlinear relationship was found among rural farmers. While farmers are very productive below age 45, their productive capacity diminishes beyond age 45. Surprisingly, no significant effect was found among urban farmers. Among the rural folks, farmers below age 45 have about 29% extra yield more than those beyond 45 years. The implication is that farming should not be left at the mercy of the aged as is witnessed in most developing countries today.

Smallholder farmers that cultivated cereals, legumes, fruits, and other crops all experienced crop yield per acre that were 86.80%, 79%, 42.30%, and 90.40%, respectively, lower than that experienced by farmers cultivating industrial crops. Cereals and legumes were significant at one percent while fruits and other crops were significant at five percent.

Smallholder farmers who farm on their own lands as well as those who practice shared cropping realized lower yield per acre compared to farmers who cultivate on rented plots. Output of farmers that owned farmland was 24.3% and that of farmers that share cropped was 26.1% lower than that of those who rented plots. The reason for these differences could have resulted from land-renting farmers' obligation to work and pay for rent and still reap some profit from their enterprise.

Using coastal zone as a reference category, ecological dummies turn out to significantly influence crop yield among farmers in both urban and rural settings. Among the rural farmers, those in the forest and savannah belts had higher yields than those in the coastal belt.

4.4.3 Stochastic Dominance Analysis of Crop Yield Volatility

In order to analyze crop yield volatility among the smallholder farmers, we controlled for the three main ecological zones in Ghana. This is because the volatility of farmers' yield also depends on the climatic conditions within which the farmer cultivates. According to Wood (2013), variation in precipitation and temperature is controlled by the movement and interaction of continental and maritime winds. The evergreen rain forest, deciduous rain forest, transition, and coastal savannah zones make up the southern half of the country. These agroecological zones have a bimodal equatorial rainfall pattern, allowing for minor and major growing seasons within the year. The Guinea and Sudan Savannah make up the northern half of Ghana. These agroecological zones have a unimodal tropical monsoon, allowing for only one growing season (major season). The single growing season in the north is bound by the harmattan period, which begins in December and ends in March. Generally, annual precipitation in Ghana ranges from 600 to 2800 mm (ibid.). Annual precipitation generally decreases from the hot and humid southwest coast, north, to the relatively hot and dry savannah (average of 1000 mm), whereas the lowest annual precipitation typically occurs within the warm southeast coastal savannah zone (600–1200 mm) (Oppong-Anane 2001). Relative humidity also tends to decrease from south to north, creating a general increase in evapotranspiration potential in the north.

The three main ecological zones considered in this paper are the forest, coastal, and the savannah. In the forest zone, the analysis shows that rural farmers first order stochastically dominate urban farmers. Thus, one can predict yield with higher degrees of certainty (i.e., lower cumulative probability levels) for rural farmers compared to the urban farmers within the same climate zone. The analysis for farmers in the coastal zone mimics similar trends as those in the forest zone. Farmers who cultivate in the rural coastal areas are certain of higher crop yield compared to their fellow farmers in the urban savannah (Fig. 4.3). Thus, in spite of the favorable weather conditions enjoyed by farmers in the forest and coastal zones of the country, urban farmers in these areas seem not to benefit much from this as their crop yields tend to be relatively lower.

Contrary to the certainty of predicting yield among farmers in rural forest and coastal zones of Ghana, urban farmers in the savannah zone seem to perform better than rural farmers. Urban farmers stochastically dominate rural farmers in terms of crop yield. While urban farmers are certain of higher yield, their rural counterparts are not because the latter have higher cumulative probability levels in crop yield. This

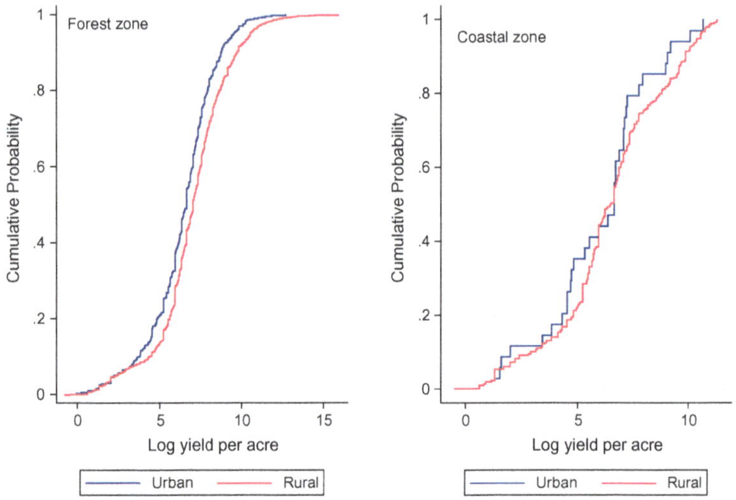

Fig. 4.3 SDA of crop yield volatility-forest and coastal zones. *Source* Graphed by authors using the GLSS 6 data

is an indication that although the savannah zone of the country has low rainfall and other unfavorable weather conditions, urban farmers within the zone seem to employ adaptive measures to reduce the negative impacts of such weather conditions on their crop yield (Figs. 4.3 and 4.4).

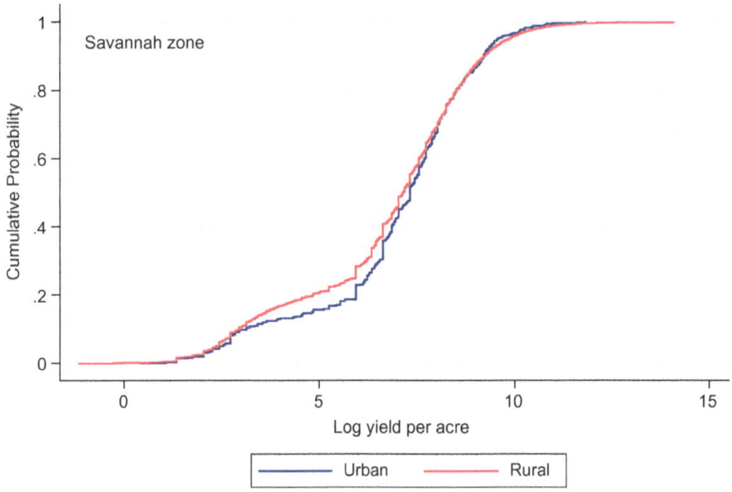

Fig. 4.4 SDA of crop yield volatility among smallholder farmers-savannah zone. *Source* Graphed by authors using the GLSS 6 data

4.5 Conclusions and Policy Recommendations

The study sought to explore yield volatility among rural and urban smallholder farmers and also find out how access to credit and technological usage affect crop yields among smallholder farmers in Ghana. The OLS technique and the stochastic dominance analysis were employed on the sixth round of the GLSS 6 data.

The findings showed that in the forest and coastal zones, rural farmer's first order stochastically dominates urban farmers, while in the savannah zone, it is vice versa. This implies that rural farmers in the forest and coastal areas are more certain of higher crop yields compared to their urban counterparts, while in the savannah zone, urban farmers seem to be more certain and perform better than rural farmers. In terms of access to credit, the findings showed that, across the country, smallholder farmers that have access to credit obtain yields per acre that are 35.5% more than their counterparts that do not have access to credit. This effect is significantly experienced by rural farmers, where they stand a chance of increasing their produce by about 38.5% if they have access to credit. On the contrary, access to credit did not have any significant effect on yield among farmers in the urban areas. Smallholder farmers that adopted farming technology experienced yields per acre that are 65.7% more than farmers who did not adopt any farming technology. This technology-adoption effect was more pronounced in the urban areas than in the rural areas (the effect was 8.4% more in the urban than rural areas). Based on these findings, we call for policies that increasingly make credit available to smallholder farmers, especially those in rural areas, so that they can increase their crop yields significantly. Such policies should ensure that cost of credit is reduced to promote access among rural farmers. The availability of credit will help smallholder farmers to acquire the needed technology that will increase their productivity. There is also the need for education on the use of technology among smallholder farmers in both rural and urban farmers.

References

Amoa–Mensah JJ, Aubyn A, Amoah I, Opoku C, Amponsah P, Osei Baah F (2013) Challenges faced by farmers in accessing credit from rural banks. Unpubl. PhD thesis, Christian University College, Kumasi, Ghana

Applefield A, Jun J (2014) Working with women: an essential component of global food security and agricultural development. Fletcher Forum World Aff 38(2):185–192

Asuming-Brempong S, Al-Hassan R, Sarpong DB, Kwadzo GTM, Akoena SKK, Sakyi-Dawson O, Mensah-Bonsu A, Amegashie DPK, Egyir I, Ashley S (2004) Poverty and social impact analysis (PSIA) studies in Ghana: economic transformation of the agricultural sector. Final report submitted to National Planning Development Commission, Ministry of Food and Agriculture and DFID, Accra, Ghana

Carroll T, Steyn A, Zook D, Funes R, Rastegar A, Lien Y (2012) Catalyzing smallholder subsistence agricultural finance. Dalberg Global Development Advisors

Chamberlin J (2007) Defining smallholder agriculture in Ghana: who are smallholders, what do they do and how are they linked with markets? GSSP background paper 6. International Food Policy Research Institute, Washington, DC

Chamberlin J (2008) It's small world after all: defining smallholder agriculture in Ghana. Discussion paper 00823. International Food Policy Research Institute, Washington, DC
Chauvin ND, Mulangu F, Porto G (2012) Food production and consumption trends in Sub-Saharan Africa: prospects for the transformation of the agricultural sector. Working paper, 2012-11. UNDP
Chisasa J, Makina D (2013) Bank credit and agricultural output in South Africa: a Cobb-Douglas empirical analysis. Int Bus Econ Res J 12(4):387–398
Christen RP, Anderson J (2013) Segmentation of smallholder households: meeting the range of financial needs in agricultural families. Focus note 85. CGAP, Washington, DC
Davidson R, Duclos J-Y (2013) Testing for restricted stochastic dominance. Econom Rev 32:84–125
Diao X, Hazell PB, Resnick D, Thurlow J (2007) The role of agriculture in development: implications for Sub-Saharan Africa. International Food Policy Research Institute, Washington, DC
Ekboir J, Boa K, Dankyi AA (2002) Impact of no-till technologies in Ghana. CIMMYT, Mexico
GSS (2013) 2010 population & housing census of Ghana: national analytical report. Ghana Statistical Service, Accra
GSS (2014) Ghana living standards survey round 6 (GLSS 6): child labour report. Ghana Statistical Service, Accra
Hazell P (2011) Five big questions about five hundred million small farms. Paper presented at the IFAD conference on new directions for smallholder agriculture, 24–25 Jan 2011. International Fund for Agricultural Development, Rome
International Finance Corporation (2011) Scaling up access to finance for agricultural SMEs: policy review and recommendations. International Finance Corporation, Washington, DC
Institute of Statistical, Social and Economic Research (2014) The state of the Ghanaian economy in 2014. Institute of Statistical, Social and Economic Research, College of Humanities, University of Ghana, Legon
Irz X, Lin L, Thirtle C, Wiggins S (2001) Agricultural productivity growth and poverty alleviation. Dev Policy Rev 19:449–466
Jorge MH (2004) Relationships among poverty, financial services, human capital, risk coping, and natural resources: evidence from El Salvador and Bolivia. Electronic thesis (https://etd.ohiolink.edu/), Ohio State University
Koomson I, Annim SK, Peprah JA (2016) Loan refusal, household income and savings in Ghana: a dominance analysis approach. Afr J Econ Sustain Dev 5(2):172–191
Maurice DC, Joseph M, Garba A (2015) Analysis of technical inefficiency in food crop production systems among small-scale farmers in some selected local government areas of Adamawa State, Nigeria. ATBU J Sci Technol Educ 3(1):1–14
Meinzen-Dick R (2004) Science and poverty: an interdisciplinary assessment of the impact of agricultural research. International Food Policy Research Institute, Washington, DC
Meinzen-Dick R, Knox A, Place F, Swallow BM (2002) Innovation in natural resource management: the role of property rights and collective action in developing countries. International Food Policy Research Institute, Washington, DC
Mkandawire R, Matlosa K (1993) Agrarian change and food security among smallholder farmers in Malawi. In: Food policy and agriculture in Southern Africa, pp 63–84
MoFA (2011) Agriculture in Ghana: facts and figures (2010). Statistics, Research and Information Directorate (SRID), Ministry of Food and Agriculture, Accra
Muza L, Dhliwayo HH, Twomlow SJ (1997) Dryland maize response to different combinations of tillage and weeding methods. In: Proceedings of the Eastern and Southern Africa regional maize conference 5, Arusha (Tanzania), 3–7 June 1996. CIMMYT
Muzari W, Gatsi W, Muvhunzi S (2012) The impacts of technology adoption on smallholder agricultural productivity in Sub-Saharan Africa: a review. J Sustain Dev 5(8):69–77
Nkonya E, Pender J, Jagger P, Sserunkuuma D, Kaizzi C, Ssali H (2004) Strategies for sustainable land management and poverty reduction in Uganda. International Food Policy Research Institute, Washington, DC

Nwaru JC (2004) Rural credit markets and resource use in arable crop production in Imo State of Nigeria. Unpubl. PhD thesis, Micheal Opkara University of Agriculture, Umudike Abia State, Nigeria

Oppong-Anane K (2001) Country pasture/forage resource profiles: Ghana. Food and Agriculture Organization of the United Nations (FAO), Rome

Peprah JA, Koomson I (2015) Capital and credit sources and household non-farm income in Ghana. J Arts Soc Sci 3(1):74–97

Peprah JA, Afoakwah C, Koomson I (2015) Savings, entrepreneurial trait and self-employment: evidence from selected Ghanaian Universities. J Glob Entrep Res 5(1):1–17

Peprah JA, Afoakwah C, Koomson I (2016) Crop yield volatility among smallholder farmers in Ghana. UNU-INRA working paper no. 15. UNU-INRA, Accra, Ghana

Rosen S, Meade B, Shapouri S, D'Souza A, Rada N (2012) International food security assessment, 2012–22/GFA-23. A report from the economic research service. United States Department of Agriculture

Wood TN (2013) Agricultural development in the northern savannah of Ghana. Doctoral documents from doctor of plant health program, paper 1, University of Nebraska-Lincoln

Wooldridge JM (2010) Econometric analysis of cross section and panel data. MIT Press, Cambridge

Wynne AT, Lyne MC (2003) An empirical analysis of factors affecting the growth of small-scale poultry enterprises in KwaZulu-Natal. Dev South Afr 20(5):563–578

Chapter 5
Prospects for and Constraints to Investment in Environmentally Friendly Infrastructure in Western Nigeria

Chioma Ifeyinwa Madueke and Mmaduabuchukwu Mkpado

5.1 Introduction

Our environment is under threat by climate change. The climate change is exacerbated by energy consumption pattern. Energy is invaluable, as it is used to drive the world's economic and social life; but the current trend of energy consumption cannot support sustainable development in the emergent era of climate change. The quest for good environment, cleaner and sustainable energy resources and usage has become reoccurring issues in world development. The achievement of the desired goal of using cleaner energy sources and good environment will essentially require increased investment and change of attitude, with respect to environmentally friendly infrastructure, which are green technologies and cleaner energy sources (Mkpado 2013a). EFIs include solar energy resources, such as solar panels and inverters, low-energy bulbs (LED), as well as rechargeable appliances and wind energy systems. Green technologies include, among other things, change in attitude regarding the use of resources, as such the idea of reduce, reuse, and recycle, which have become innovations in resource management.

Sustainable development cannot occur in a vacuum; it can be achieved through a good mix of financial and human resources. Human nature can be considered a complex phenomenon that needs to be handled in a unique way to avoid stifling initiatives. The level of awareness, skills, and attitude of the people is to be taken into consideration, when making investments that will be driven by the people. Thus, sustainable and lucrative investments are determined by market demand; consumers with their

C. I. Madueke (✉)
Department of Metallurgical and Materials Engineering, Federal University Oye-Ekiti, Ekiti, Nigeria
e-mail: chioma.madueke@fuoye.edu.ng

M. Mkpado
Department of Agricultural Economic and Extension, Federal University Oye-Ekiti, Ekiti, Nigeria
e-mail: mmaduabuchukwu.mkpado@fuoye.edu.ng

C. Atewamba and D. Yong Ngondjeb (eds.), *Inclusive Green Growth*, Advances in African Economic, Social and Political Development,
https://doi.org/10.1007/978-3-030-44180-7_5

taste thus direct the nature and magnitude of investment. Innovative approaches used make for efficiency in the economy. Effective climate change adaptation and mitigation mainly depend on the nature of innovations, investment, and practices on green economy (Igbokwe and Mkpado 2011; Mkpado 2013b).

The paper seeks to examine the factors affecting the demand for EFI products in Nigeria. EFI products, like energy-saving bulbs, solar panels and inverters, energy-storage devices and bicycles, which are often in competitive markets, as the products are of different brands, standards, prices, efficiency, and durability. What factors affect the demand and use of the aforementioned products in Nigeria?

Next are reduce, reuse, and recycling issues, which are important practices and attitudes for green economy because these practices involve a reduction in the carbon footprint. These approaches are particularly recommended for the use of water (domestic activities), cellophanes/plastics, as well as tins and others, which are very helpful in waste management. What is the level of awareness and practices of these innovations?

Objectives

The objectives of the study are:

1. To examine the effects of age, income, educational qualification, and gender on the use of EFI, such as energy-saving bulbs, solar panels and inverters, as well as assess the awareness of EFI technologies in rural Nigeria;
2. To investigate the constraints to the use of EFI in Nigeria.

The main research questions were to identify the factors affecting the use of EFI innovations and the opportunities for environmentally friendly investments in rural areas, as well as the constraints to the use of EFI in Nigeria.

The null hypotheses guiding the research include:

1. Income does not affect the use of EFI technologies
2. Educational qualification does not affect the use of EFI innovations
3. There is no difference with respect to the use of EFI technologies for private and commercial purposes.

Significance of the Study

Africa is experiencing adverse effects of climate change, and the consequences can increase over the years, as rising temperatures are predictable with weather variables. Changes to new innovations with respect to EFI, as well as behavioral changes with respect to reduce, reuse, and recycle, are inevitable for African societies. Sustainable development tends to start with the people at their present level and push them upward, with good motivation for participation. The study has given insight into the peoples' capacity, behavioral changes, and application of innovations in EFI. It has shown the line of actions that will increase use and investments in EFI. In the course of conducting the research, it stimulated interests of respondents on innovations in EFI. Policy makers have been provided facts on factors requiring actions for EFI practices in rural Africa, especially in Nigeria. The study has provided other scholars with literature for expansion of the frontier of knowledge in the area of EFI innovations.

5.2 Literature Review

Environmentally friendly infrastructures are eco-friendly and are a subset of green technology. Environmentally friendly investments in infrastructure are those investments that take cognizance of the carbon footprint and do not increase it, do not reduce the soil fertility, and do not involve non-degradable materials. Environmentally friendly investments in infrastructure include the use of solar panels; energy-saving devices such as energy-saving bulbs; and energy-storage devices, such as inverters and bicycles. This is also complemented by fundamental practices of reduce, reuse, repair, refurbish, and recycle to assist in keeping the environment healthy. Development in environmentally friendly infrastructure can lead to better technological innovations, which are likely to advance the economy of African continent. The objectives of investing in environmentally friendly physical infrastructure are multifaceted. Such facilities contribute to economic growth and are therefore closely co-related with development (Kessides 1993). Topping the list is the use of solar energy via solar panels. Owing to very poor electricity supply in Nigeria, several dominant players in industrial sectors, such as Michelin tyres and Dunlop tyres, have fled the country, with serious implications for employment and fiscal revenues, which should have contributed to the well-being of the Nigerian people. The power generation with solar cells system has received great attention in research because it appears to be one of the possible solutions to the environmental problems (Wai et al. 2008). Climate change, largely driven by global warming, mainly refers to the persistent increase in atmospheric temperature, often associated with dryness. Climate change results from the emission of carbon (IV) oxide and other greenhouse gases from fossil fuels, bush burning, and other factors; it is driven by anthropogenic activities, as mankind engages in development activities and better life facilities. However, it is now apparent that mankind may witness disastrous consequences, if the current trends associated with fossil fuel consumption continue unabated. The quest for greener environment and secured future has reinforced and motivated the search for green technologies and renewable energy resources (IPCC 2007; Igbokwe and Mkpado 2010).

Africa and the rest of the world are developing environmentally friendly infrastructural products, which are green technology products and advocating their uses. Green and renewable energy would provide good opportunity to drastically reduce greenhouse gas emissions, pursue economic growth, and enhance energy security for achieving sustainable development. Africa and other less developed economies are suffering the adverse effects of climate change, although they account for the least carbon (iv) oxide emissions because of low capacity to invest in greener technologies and adopt them (Igbokwe and Mkpado 2010). And this trend may likely continue due to the poverty trap that exists in Africa.

Green energy technologies are components of environmentally friendly infrastructure, which take cognizance of the carbon footprint; they do not reduce the soil fertility and do not give rise to non-degradable materials. Environmentally friendly infrastructure includes the use of solar panels; energy-saving devices, such as energy-saving bulbs; and energy-storage devices, such as inverters and bicycles. For optimal results, this is often complemented by fundamental practices of reduce, reuse, repair, refurbish, and recycle to assist in keeping the environment healthy (Kessides 1993). Green technologies and innovation can open new opportunities for skill acquisition, job, and wealth creation, as human resources will be required to manage the new developing sub-sector. It can lead to increased production, as the energy resources, such as solar energy, can be more stable and efficiently used in tropical Africa (Wai et al. 2008; Kessides 1993).

5.2.1 The Poor Situation of Power Sources and Solar Energy Projects in Nigeria

Agbongiarhuoyi (2015) reported Dr. Simon Bradshaw, as saying, "Four out of five people without electricity live in rural areas that are often not connected to a centralized energy grid, so local, renewable energy solutions offer a much more affordable, practical, and healthy solution. The country has a target in 2007 to produce 7% of its 2025 energy needs from renewable with solar and hydro as the major priority." According to a report by Charles Opara-Ndudu, in THISDAY Newspaper of 15th March 2015, Nigeria has the potential to exploit its abundant solar energy resources, considering its geographic location around the equatorial sunbelt. The country receives abundant sunshine throughout the year, ranging from 6.70 kwh/m^2/day in Borno State to roughly 4.06 and 5.86 kw h/m^2/d in such locations as Calabar, in Cross River State.

Suleiman (2010) noted that in 2005, the federal government constituted a presidential committee to design a 25-year power development plan. The report projected electricity demand profile for Nigeria as 15,000 MW; 30,000 MW; and 19,000 MW in the short, medium, and long terms, respectively, providing the basis to project a 10% economic growth scenario over the period in view. Based on this projection, the Energy Commission of Nigeria conducted a study, which indicated that renewable energy electricity is expected to contribute about 14% in the short term; 23% in the medium term; and 36% in the long term of the total energy and electricity supply, as dictated by the National Energy Policy. The lapsing period of the short, medium, and long terms is 2008, 2015, and 2030, respectively.

The Bank of Industry (BoI) Chief Executive was reported saying that Nigeria's current electricity situation had been unsatisfactory and put the total electricity supply on the National Grid at less than 4000 MW, relative to the electricity demand, conservatively estimated at 40,000 MW (Opara 2015). For an African economy like Nigeria, with a population of more than 170 million people, this is grossly inadequate. Many Nigerians and Nigerian businesses that can afford it have resorted to electric generators, at a great expense. Kerosene lanterns, used largely by rural dwellers, adversely affect the environment and are dangerous, untidy, and dim, besides high cost of petroleum products. The unsteady supply of petroleum products needed to power these generators or lanterns has exposed the unreliability of using generators or lanterns, as a long-term alternative source of electricity, not to mention the environmental/noise pollution associated with several generators in the city's neighborhoods and factories (Opara 2015).

Nigeria has recorded some milestones in the area of solar energy. The Bank of Industry (BoI), in collaboration with United Nations, has established solar-powered home systems under the BoI/United Nations Development Programme, which was expected to last for a minimum of 25 years and started in 2013/2014 (IEEE 2015). In 2013, the bank implemented a six-kilowatt PV solar-powered rural mini-grid at Egbeke, Rivers State, and is currently implementing an 18 kW PV solar mini-grid at the same location. It also implemented stand-alone PV solar-powered systems in communities in Lagos and Ogun states. The BoI further made a provision of long-term financing for the installation of off-grid solar-powered home systems in six communities, with an average of 200 homes each in a pilot phase, as part of its renewable energy partnership with the UNDP. The communities selected are located in Anambra, Delta, Gombe, Kaduna, Niger, and Osun states.

The facility will enable "each home to power three LED lightbulbs, one electric fan, one radio/TV set, and, of course, mobile phone charging. These are the basic energy needs of the average rural family (Opara 2005). The BoI wanted rural communities to take control of their energy generation and pay only for the energy used. The BoI medium-term vision of the scheme is to have 100,000 homes installed with solar-powered systems in the next five years, through a combination of micro-grid and stand-alone solar-powered home systems (Opara 2015). This is essentially a program aimed at poverty alleviation and rural economic development.

RUWES (2015) noted that the Rural Energy Access Project (REAP) represents the lighting component of RUWES, designed to address the need to source for and deploy alternative and sustainable, clean and renewable energy sources for lighting and heating purposes, which will impact positively on income, health, and environment, as well as create sustainable livelihoods for the rural poor and those most affected by the shortage of electricity and energy supply in the country. REAP is striving for the reduction in our National Emission Factor (by using clean, energy-efficient LED bulbs) and introducing household stand-alone solar-powered kits to replace incandescent bulbs, single-wick kerosene, and oil lanterns, as well as small generators in rural areas.

IEEE Smart Village program is to support electricity projects in off-grid communities of Cameroon, Haiti, India, Kenya, Nigeria, South Sudan, and Zambia, among others. IEEE (2015) noted that Nigeria, Green Village Electricity (GVE) Projects Limited, did receive a US$675,655.40 equity and debt investment from the country's Bank of Industry, and to further support the investment; IEEE Smart Village will provide GVE with US$65,535.20 of matching funds, as part of a seed funding, program-supporting grant, bringing the total investment to US$741,190.60 (IEEE 2015).

Eleri et al. (2013) noted that the SUNGAS project, which is funded by the European Union, is being implemented by the International Institute for Environment and Development (IIED); the Niger Delta Wetlands Centre (NDWC); and the Living Earth Foundation (LEF), with the aim to catalyze the development of Nigeria's natural gas and renewable energy markets through innovation, demonstrations, policy dialogue, and advocacy. Small demonstration projects for both renewable and gas-to-power will show that community-based energy facilities are technically viable, financially sustainable, and can ensure better access to modern energy services for rural communities.

Investments in clean energy facilities in Nigeria have been in large and small hydroelectricity generation, solar PV, biomass, biofuel, and wind energy. These can be used for domestic, agro-industrial purposes, and health facilities. The agro-processing endeavors—such as flour grinding, oil expelling/extraction, crop drying, and threshing—and small-scale industry—such as sawmilling, wool and cotton processing, stone crushing—have been noted. Dayo (2008), Usman (2012), and Huzi (2014) observed that there were a number of projects to improve the life of the rural people, using green energy technologies in western Nigeria. Some of these projects include streetlighting; improving education and health services; and using environmentally friendly infrastructure.

5.2.2 Possible Uses of Environmentally Friendly Infrastructure in Nigeria

Usman (2012) reported that villagers use solar energy for cooking, water heating, refrigeration, house lighting, domestic water pumping distribution, ironing, television, and radio powering. Green infrastructure is particularly useful for the agricultural economy, providing solar technology for drying, processing and storage of agricultural products, especially the perishable ones; water pumping and distribution for irrigation; flour grinding; threshing; oil expelling; crop drying; and operation of various agricultural implements. Green infrastructure is also useful for commercial purposes. These include usage of solar energy in businesses, such as shops lightening, community banks, restaurants, stone crushing, sawmilling, wool/cotton processing and local bakeries, especially mobile cellphones battery charging business. The list can include cinema houses, powered with solar systems for watching sports events like football. Domestic usage includes solar cookers, solar water heaters, lightening and refrigeration, electricity for refrigerators and deep freezers, in which vaccines and drugs can be safely stored, without losing their potencies, especially during immunization in remote areas. Solar stills are designed to produce distilled water from brackish water and will be useful for hospitals, industry, and laboratories. When sized appropriately, they can provide for the needs of comprehensive health centers of semi-urban areas. Solar water heaters based on flat-plate collectors, with appropriate storage units, can produce water at temperatures of up to 800 °C, which will find applications in hospitals, hotels, industry, and private residences and can significantly reduce electricity bills. Water pumping can be demonstrated by the use of photovoltaic solar modules for pumping water from wells and boreholes, especially in rural areas for providing the water requirements of entire communities. Photovoltaic powered pumps can also be used for irrigation purposes Usman (2012).

5.2.3 *Some Constraints to EFI Development Finance, Poverty, and High Initial Capital Outlay*

Financial problems constitute a huge constraint to the development of EFI/green economy. In developing nations with very limited financial resources like Africa, green energy facilities, like solar panels, inverters, and energy-saving bulbs, are more expensive than the traditional electricity generators and incandescent lamps. Due to higher initial cost of solar panel installation, even though in the long run it becomes cheaper, majority of the people cannot afford to install it (Arun and Gopalkrishnan 2012). With respect to the higher cost, King and Lenox (2002) noted that energy Star-rated compact fluorescent bulbs to replace 100-w incandescent may cost between $2 and $15 per bulb, depending on the manufacturer, while the first 100-w LED bulbs cost as much as $50.

Skills and knowledge: There is little knowledge among the masses on environmentally friendly infrastructure amid climate change. There is a crucial need for governments, non-government organizations, and the educational systems in Africa, especially Nigeria to create awareness and training on the types, uses, and advantages of green technology infrastructure. This will accelerate acquisition and use of environmentally friendly infrastructure. For instance, the use of inverters requires a general knowledge on the size of the inverter that will be appropriate to the appliances to be used during power outages, including the voltage, which could vary from 110 to 120 V. Some of the appliances could have voltages much higher than the above, in which case a special purpose inverter should be used. The electric current usually applicable is 6 A.

Power = current × voltage; therefore if $V = 110$ and $I = 6$ A therefore $P = \mathrm{IV} = 110 \times 6 = 660$ W. If an inverter of 660 W is to be used, then the appliances should not exceed 660 W. Also, 1750 W is the commonest inverter. To use this inverter, the appliances must be worked out and should be lower than 1750 W. Furthermore, 12 V batteries connected in parallel are also required to work with the inverter. A back-up battery is essential should power failure last longer than expected. Majority of both urban and rural dwellers are not well acquainted or knowledgeable with these facts. Also, improved skills on how to design, manufacture, and market green products are seldom available. This only shows that most of the green technology devices are being imported, creating difficulty in availability of spare parts, repair and servicing, whereby in the case of major breakdown, expatriates are sought for, leading to enormous resources being expended.

Lack of awareness on EFI products available: A good proportion of African people, particularly Nigeria, is very much ignorant of the concept of green economy. There should be a focus on creating interactive environment for the experts in green economy to communicate with the masses. There should be credible platforms through which genuine change to green economy should be engendered. Public enlightenment program should be made available for the masses. The Head, Operations Integrity, OER, Mr. Kayode Boladale, said, "It is necessary to develop a communication and awareness action plan to raise awareness on the benefits of using renewable energy and energy efficiency against other forms" (Wai et al. 2008).

Low promotion and marketing: Marketing and promotion is highly critical to the growth of any business. EFI should be promoted and the products extensively marketed. There is an immediate need for green economy businesses to embark on e-commerce solutions in order to grow and improve productivity and boost economic growth. Seminars and workshops on the use of energy-saving bulb, inverters, and solar panel, where people will be gathered and tutored on the associated benefits, should be organized. Digital and physical marketing interaction is crucial at this stage. This will make a lot of meaningful impact on EFI across Africa. There is poor explosion/buildup in social media on environmentally friendly infrastructural products, such as the use of LED bulbs, inverters, and others. Their use and benefits are seldom advertised on social media, as well as on such mass media as newspapers, magazines, radio, and television. Also, they are hardly featured on local Nigeria television and radio platforms. Promotion and marketing must be stepped up to enhance growth.

Attitude to change: General African attitude to technological changes is highly discouraging. The average Nigerian is slow to embrace change and is of the old school; he is not easily taken by new technological innovations and would prefer to keep up with what he is already familiar with.

Lack of robust policies on EFI technology in Nigeria: Practical policies must be rolled out and put in place to encourage the use of EFI and also to enable investors to embrace the endeavor. A sound and profitable economy is realistic only when there are basic laws and order in the society. There is an imperative to develop legislative and regulatory frameworks to drive renewable energy use (Wai et al. 2008). Policy can stipulate percentage of financial involvement through government budgets, strategies for measurement and evaluation of objectives, tax incentive structures, and mass campaign and educational systems to improve green energy development (Mkpado 2013b). Muhammed (nd) aptly noted that low literacy level; lack of/low level of awareness of the mitigation technologies and their potential benefits; inefficient energy use; inappropriate energy pricing; financial constraints; low-income level; technological incapability; and pressures from population growth are constraints affecting green energy development. Abubakar et al. (2015) study identified weak institutional framework, poor policy implementation, inadequate financing, and lack of awareness of the socioeconomic, technological, and environmental merits of renewable energy technology, as the major barriers hindering its development and diffusion. Usman (2012) observed that in Nigeria, there is little national policy dedicated specifically to energy efficiency and renewable energy, or its applications to rural and remote areas. There are, however, other policy instruments on the environment and energy that may be relevant to supporting such applications. For example, the Nigerian National Policy on Environment provides that the goal of the policy on the environment is to achieve sustainable development in the country and, particularly, to conserve and use the environment and natural resources for the benefit of present and future generations.

Poor research capacity on EFI technology: Across Africa and Nigeria, in particular, researches carried out on green energy devices are quite few. Greater proportions of Nigerians do not know much about these devices and their benefits. These devices are yet to become common household commodities in Nigeria. Intensive effort is highly imperative to ensure that all African countries are actively included in collaborative research on EFI, as well as green technology. Teamwork is to get good encouragement, and brainstorming can produce better results than individual approaches. No indigenous and result-oriented research has been recorded on green technology products. Mkpado and Onuoha (2008) have demonstrated that technologies are better appreciated and adopted by the people, when local scientists have modified it with local content, thus indigenizing the process.

Low level of investment on EFI technology: For environmentally friendly infrastructure products to reach the next level, governmental and non-governmental organizations, instead of going back and forth akin to a simple pendulum, should arrange an injection of cash into it by first providing some of these products and services free or at subsidized prices to the market. A total of €24.5 m (N5.33bn) has been earmarked by the European Union and the German Federal Ministry of Economic Cooperation and Development to fund Nigeria's renewable energy, energy efficiency, and rural electrification programs. The amount, which has its beneficiary as the Federal Ministry of Power, will be used to promote the Nigerian Energy Support Programme for five years.

The availability of the funds was disclosed by the Head of Unit, Capacity Development, NESP, Mr. Felix Nitz, during the signing of a Memorandum of Understanding between the NESP and the Renewable Energy and Efficiency Programme with 11 Nigerian training institutions in Abuja. He explained that the MoU was part of efforts by the EU, the German government, and the United States government to support skills development for Nigeria's growing clean energy sector. Nitz noted that the objective of the MoU was to create a skilled workforce in Nigeria's clean energy sector by empowering local institutions to offer seven training courses on renewable energy and energy efficiency for engineers, architects, and technicians. He said the courses included off-grid renewable energy design for engineers, energy-efficient building design for architects, and solar photovoltaic installations for technicians (Punch newspaper; July 31, 2015, Okechukwu Nnodim).

5.2.4 *Prospects in Environmentally Friendly Infrastructure in Nigeria*

Improvement of quality of life for rural Africa: In the case of Nigeria, there is acute deficiency in electricity supply. Consequently, the populace do not have any option than to resort to the use of generators, thereby increasing exposure to the inhalation of carbon monoxide (CO), which is highly detrimental to health. The advent of solar energy is accompanied by the prospects to transform human lives. In the area of recycling, Nigeria is bereft of a robust recycling culture. Therefore, if adequate awareness is created on this issue, it will generate a lot of employment and transform the health conditions of Nigerians, in particular, and Africans, in general.

Increased life span: While energy-saving lightbulbs may cost more initially, these devices can pay for themselves through energy savings over a greatly increased life span. A typical incandescent bulb may last 1000–2000 h before burning out. Compact fluorescent lightbulbs have a life span approaching 10,000 h, while LED bulbs could potentially last 25,000–50,000 h or more before failing. The precise life span of a bulb depends heavily on usage patterns and environmental conditions, but consumers can expect to go considerably longer between bulb changes, once they switch to energy-saving lightbulbs.

Less monolithic economy: Shares in some of the world's biggest energy groups tumbled, as the collapse in oil prices battered the profits of ExxonMobil and Chevron, capping a week in which European rivals announced thousands of job losses and slashed spending by billions of dollars; the oil price has continued to decline (Nigerian Punch Newspaper; August 1, 2015, vol. 7290, no. 1790 www.punchng.com). Nigeria's monolithic economy is under threat, and the Nigerian Association of Chambers of Commerce, Industry, Mines and Agriculture (NACCIMA) has warned. It is expected that Nigerian government, in particular, would therefore look at opportunities in other sectors of the economy, such as green technology; according give it a priority by developing an action plan, akin to the oil and gas sector. This is aimed at diversifying the economy, which invariably brings about more opportunities, thereby generating more revenue and increasing the Gross National Product (GNP).

Creation of jobs: For Africans, the use of solar panels, in particular, is a novel technology. It calls for massive job creation, as people will surely demonstrate an interest in learning, as well as marketing the technology. Jobs are not created in a vacuum. The foundation of economic growth and job creation is the presence of "Startup communities," which come into existence when "Some People" bring, blend, and develop synergies between (1) "Energy," (2) "Aggregate of Socioeconomic Activities," and (3) Innovation in specific geographical areas (Nzamujo 2013). There are multiple windows of huge opportunities for green technology for those who understand and can leverage the terrain. There are still enormous opportunities for economic growth associated with green technology.

Poverty alleviation: Boladale, while speaking at the "Live Well Initiative 2015" Annual Multi-sectoral Grand Health Bazaar in Lagos, stated that the use of renewable energy would help alleviate poverty, especially as it could enhance technological capacity and economic development in the developing world.

Innovation: The use of solar panels, which is relatively new in Nigeria, calls for innovation in that angle. Nigerians, exercising prudence, will definitely work on a lot of innovative ways of installing, fixing, and general maintenance of the setup and bringing tremendous improvement, innovation, and growth on environmentally friendly infrastructural products.

Safety concern: On energy-saving design, the compact fluorescent has raised safety concerns because of the materials used in its design. Compact fluorescent lights (CFLs) contain a small amount of mercury, enough to present a health hazard if the bulb breaks. In addition, when a CFL reaches the end of its life span, the electronics at the base of the bulb implode, as a safety precaution. This process can produce a puff of smoke and even scorch the plastic housing.

5.3 Methodology

The study was conducted in South-West, Nigeria. Primary data were generated by conducting surveys, featuring both structured questionnaires and oral interviews. Two states were purposely selected, based on population density and level of business development, or active programs for climate change mitigation in the states concerned. One hundred and sixty (160) households were randomly selected from the study area, which means eighty respondents from each state.

Analytical techniques

Logistic regression model was used to analyze (i) objective and (ii) hypotheses

The implicit formula is:

$$Y = b_o + b_1X_1 + b_2X_2 + b_3X_3 + b_4X_4 + b_5X_5 + b_6X_6 + e$$

where

Y = a dummy variable, with one for use of green economy innovation/EFI; *otherwise*, zero.

An index was constructed for the use of energy-saving bulbs, solar panels, and inverters, as well as bicycles. Households that score up to 60% were judged as adopting green innovation and scored one; otherwise, zero. This is necessary to have a smilingly composite view of green technology adoption in the study area.

X_1 = gender: a dummy variable for gender, with a value of one for females and zero for males.

X_2 = income in naira

X_3 = educational qualification, measured as the number of years of formal education

X_4 = family size, number of people in a family.

X_5 = government equipment ownership status: This measures the state of ownership of climate change facilities in use by government or NGOs, with one for government-provided facilities; otherwise, zero.
X_6 = private equipment ownership status: This measures the state of ownership of climate change facilities in use by private entities, with one for privately provided facility, or zero otherwise.
X_7 = age range
X_8 = level of awareness of climate change: scaled variable measured unaware = 0, slightly aware = 1, very much aware = 2.
b_o = constant
b(s) = coefficients
e = sample error term.

The model above was also used in testing the hypotheses.

Descriptive statistics and Likert scale were used to analyze objectives (i), (ii), and (iii). The descriptive statistics and Likert scale were also useful in answering the research questions that have to do with the nature of institutional and government policies that will favor green economy and evaluating people's opinions on the issue, with respect to usage, opportunities, and constraints. The researchers developed the questions with options to enable good survey, coding, and analysis.

Likert scale: Five points Likert scales with the following format: strongly agree, agree, undecided, disagree, and strongly disagree. The mean (neutral or undecided) response was computed as: 5 + 4 + 3 + 2 + 1 = 15/3 = 3. In order to have three points of inference, take a deviation of 0.5, then any response less than 2.5 (3 – 0.5) is considered to be disagreeing, while any response up to, or greater than, 3.50 (3 + 0.5) is considered to be agreed upon.

The Chi-squared statistics were also used to test a hypothesis, and the implicit format is presented as:

$$\chi^2 = \sum \frac{(O - E)^2}{E}$$

$$\mathrm{df} = (c - 1)(r - 1)$$

χ^2 = Chi-squared statistics, O = observed frequency, E = expected frequency, df = degree of freedom, c = number of columns, r = number of rows.

Data source

Ekiti and Ondo states were selected because of availability of community solar-powered water supply systems and solar-powered ICT in rural schools, besides individuals' use of other climate change adaptation and mitigation innovations, such as energy-saving bulbs. In fact, a clean energy organization has reported the use of many facilities/activities in favor of green economic development in the states proposed for this study, especially in Ondo State (Huzi 2014). There are government initiatives to improve the peoples' welfare with environmentally friendly infrastructure in the proposed study area. The study concentrated on the communities, where these infrastructures exist.

5.4 Results and Discussions

5.4.1 Descriptive Analysis

Socioeconomic characteristics of respondents

The socioeconomic characteristics of respondents are presented in Tables 5.1 and 5.2. Table 5.1 starts with the age distribution of respondents. Six percent of the respondents were less than 25 years of age. This represented the minimum age range of the people surveyed. Actually, 23, 66, 38.5, and 9.9 years represented the minimum, maximum, mean, and standard deviation of the age, respectively. The frequency distribution of the age range is presented in Table 5.1. It may be heartwarming to note that the majority of the respondents are still within their useful and productive age. This age group can be a strong force in driving the concepts of green technology (Mkpado and Arene 2003).

Gender is another important socioeconomic variable presented. Majority of the respondents were male, accounting for 74% of the respondents, while 26% of respondents were female. The survey tried to include both genders, as excluding one gender can lead to conflict and stifling of efforts to transfer innovation. Related to gender is the marital status: Majority of the respondents (71.3%) were married, while 25% were single and 3.7% were experiencing widowhood.

The uses of environmentally friendly infrastructural products involve many people from different walks of life. This can be seen from the distribution of respondents according to their occupation. Table 5.1 showed that about 10% were farmers, while about 5% were artisans or skilled workers. Traders accounted for eleven percent, and the majority (66%) were civil servants. This development could be attributed to targeting, where the EFI technologies and facilities exist. Also, some people use such facilities in their business or place of work. Socioeconomic features are important, as Abubakar et al. (2015) noted, "weak institutional framework, poor policy implementation, inadequate financing, and lack of awareness of the socioeconomic, technological and environmental merits of Renewable Energy Technologies, as the major barriers hindering its development and diffusion."

Other socioeconomic characteristics, such as education status, household size, and income, are presented in Table 5.2. Only about three percent of the respondents did not possess any form of formal education. Table 5.2 also showed that 3.75% had secondary education, while 22.5 had National Certificate of Education, or National Diploma. Those possessing higher national diploma and first degree accounted for

Table 5.1 Socioeconomic characteristics of respondents

Characteristics	Description	Frequency	Percentage
Age in years	Less than 25	12	7.5
	26–30	30	18.75
	31–35	32	20
	36–40	30	18.75
	41–45	24	15
	46–50	14	8.75
	51–55	6	3.75
	56–60	10	6.25
	Above 60	2	1.25
	Total	160	100
Gender of household head	Sex		
	Male	120	75
	Female	40	25
	Total	160	100
Marital status	Marital status		
	Married	114	71.25
	Single	40	25
	Widowhood	6	3.75
	Total	160	100
Major occupation	Occupation		
	Farming	16	10
	Trading	18	11.25
	Salaried job	106	66.25
	Artisan/skilled worker	8	5
	Others	12	7.5
	Total	160	100

Source Authors' computation, using data from field survey, 2015

15.5 and 27.5%, respectively. The relatively higher levels of education possessed by respondents may be due to the fact that employees at the offices or organizations, where EFI technologies exist, were selected for the interview.

Household size can be a variable determining expenditure profile. The minimum, maximum, mean, and standard deviations of household size are 1, 12, 5, and 2, respectively. Large family size characterizes the people, as 13 and 38% have 3–4 and 5–6 members in their households, respectively (Table 5.2). Household income is a variable that can determine welfare and ability to try new technologies. Table 5.2 showed relatively higher levels of income for the rural people. This can be explained by the fact that most of the respondents possess higher educational certificates and

Table 5.2 Other socioeconomic characteristics

Characteristics	Description	Frequency	Percent
Educational level/status	No formal education	6	3.75
	Primary school	20	12.5
	Secondary school	6	3.75
	NCE/ND	36	22.5
	HND	28	17.5
	BSc.	44	27.5
	Others	20	12.5
	Total	160	100
Household size	1–2	8	5
	3–4	50	31.25
	5–6	62	38.75
	7–8	30	18.75
	9–10	6	3.75
	Greater than 10	4	2.5
	Total	160	100
Monthly income in Naira[a]	Less than 15,000	4	2.5
	15,000–25,000	2	1.25
	25,001–35,000	26	16.25
	35,001–45,000	38	23.75
	45,001–55,000	36	22.5
	Greater than 55,000	54	33.75
	Total	160	100

Source Authors' computation using field survey, 2015
[a]Naira exchange to Dollar = 200 official, parallel market = 225

are employed either by local, state, or federal government of Nigeria. Mkpado et al. (2012) reported that income is a major factor determining the choice of cleaner energy for cooking purposes. Also, Mkpado and Ugbaja (2008) noted that socioeconomic issues are to be taken into consideration for motivating certain groups and youth for purposeful actions.

Level of Awareness of EFI Technology in the Study area

The level of awareness of EFI technologies could be an important factor affecting the use of such technologies. Table 5.3 shows that over 63.75% of the respondents claimed to be very much aware of them, while nineteen (23.75) and ten (12) percent, respectively, claimed to be slightly aware and unaware of EFI technologies. This result is acceptable because some EFI technologies are used to provide public goods and services. There is need to create more awareness on the use of EFI technologies. This may not be limited to slight introduction but detailed discussion, including

Table 5.3 Indicators of awareness of LED bulbs and bicycles in the study area

Awareness of EFI technology	Level of awareness	Frequency	Percentage
	Very much aware	102	63.75
	Slightly aware	38	23.75
	Unaware	20	12.5
	Total	160	100
Need of assistance on the use of EFI technology?	I need assistance on it	128	80
	I do not need assistance	32	20
	Total	160	100
Use of LED bulbs?	Yes, the bulbs	38	23.75
	A few bulbs	78	48.75
	Not at all	46	28.75
	Total	160	100
Attitude to energy saving	It is good	122	76.25
	I don't care/or like it	18	11.25
	I am neutral	18	11.25
	It is for the white man	2	1.25
	Total	160	100
Any household member using bicycles?	Yes	60	37.50
	No	100	62.50
	Total	160	100

Source Authors' computation using data from field survey, 2015

merits and prospects of using these technologies. Also to be given emphasis is skills acquisition and training on these technologies. The need for this is reinforced by the fact that eight (80) percent of the respondents indicated that they needed some assistance on the use of EFI technology. The types of assistance specified by respondents included both financial and technical assistance. It may be easier to provide technical assistance by some type of training and mass enlightenment program.

Energy-efficient bulbs emit less heat and are climate change compliant. Households are yet to engage fully the use of energy-saving bulb innovation. Only about 23% of respondents fully engaged in this by making use of only energy-saving bulbs. About 48% were partially using energy-saving bulbs as they used both energy-saving bulbs and ordinary ones, concurrently. On the other hand, 28.75% of the respondents were yet to change to use of energy-saving bulbs (Table 5.3).

It may be heartwarming to note that the people's attitude toward the concept of saving energy is impressive, as the majority (76.25%) accepted the ideas as a good initiative (Table 5.3). It is often easier to accelerate the speed of people on the direction they voluntarily choose. This indicates that programs and projects on energy saving will be welcomed. This result is in consonance with the use of energy-saving bulbs, since the majority is either using such bulbs fully or partially.

The use of bicycle is one of the green transportation facilities. The results showed that only 37.50% of households have at least one person using bicycles to either school, market, or farm, while the majority (62.5%) do not use it (Table 5.3). There may be disdain on the use of bicycles. This is associated with the high risk prevalent on the roads, fueled by the absence of marked lanes for bicycle users. All vehicles, both big and small, as well as motorcycles and bicycles use the same roads, which are often in poor conditions. This tends to reduce the motivation of using bicycles in the area studied.

Types of EFI technologies in the study area

The types of EFI technologies in the study are presented in Table 5.4. The communities can have more than one type of EFI technology at a time. This is indicated by the multiple responses recorded in Table 5.4. Result indicates that only about 31% of the respondents had streetlight powered by solar technology. The low proportion experiencing it may be due to its high cost. Muhammed (nd) noted that the cost of solar rural electrification and water pumping project in three villages in Jigawa State, Nigeria, is $600,000.00. Rural electrification can be achieved with solar-powered technology. NESP (2014) noted opportunities and solar resources available in Nigeria for using green technology for rural electrification. The use of solar-powered water boreholes and solar-powered technology in schools and public library had good response with the use of green technology in the area, accounting for over 68 and 75% of the response, respectively (Table 5.4).

Educational program has become the largest recipient of this technology. This helps with online library, registration of external examinations for certificate classes online, as well as easy teaching of computer basic education, including information and communication technology (ICT) skills in primary and secondary schools in the study areas. Next to these is the use of solar-powered technology in hospitals and health centers. Usman (2012) aptly noted that photovoltaic power components have also been shown to adequately provide the electricity for refrigerators and deep freezers, in which vaccines and drugs can be safely stored without losing their

Table 5.4 Types of EFI technologies for public usage in the study area

Green energy use in the community	Frequency	Percentage
ICT in schools and public library	110	68.75
Hospitals and community health centers	96	60
Streetlight	50	31.25
Water boreholes	120	75
Total	376	Multiple responses recorded

Source Authors' computation using data from field survey, 2015

potencies, especially during immunization in remote areas. Access to drinkable water is critical for the adoption of solar energy (Table 5.4). This may have been cushioned by the impetus given by the Millennium Development Goals (MDGs), with respect to improved health and education, which have been accorded priority by the MDGs, as they have a common trend with water access.

Building and maintenance of public EFI

The provision and maintenance of EFI technology facilities are critical to the usage. Table 5.5 indicates that government Ministries, Departments and Agencies (MDAs) constitute the major providers and also care for the maintenance. Politicians are also involved with the provision and maintenance of such facilities. This may be one of their campaign strategies; non-government organizations and philanthropists also have a share in this regard (Table 5.5). It is pretty significant to note that the community using such technology has a share in its maintenance. This information gives insight into the usage of the facility, as well as its sustainability. In a similar vein, 55 and 37% of the respondents view EFI technology facilities in their locality as very good and good, respectively (Table 5.5). Such an appreciation can serve as encouragement to motivate increased investment in the trend. It is good to sustain the culture of investment and maintenance with democratic processes, which will endear the project to the people (Mkpado and Arene 2007).

Table 5.5 Building and maintenance of public EFI

Question	Answers/options	Frequency	Percentage
Who installed solar panels in your community?	Government Ministry/Agency	80	50
	NGOs/philanthropists	30	18.75
	Politicians/political leaders	38	23.75
	I don't know	12	7.5
	Total	160	100
Who maintains it in your community?	The provider	74	46.25
	Community using it	32	20
	Both community and providers	38	23.75
	I don't know	14	8.75
	Total	160	100
How do you view use of green technology for public purposes?	very good	88	55
	Good	60	37.5
	Bad	0	0
	I don't know	12	7.5
	Total	160	100

Source Authors' computation using data from field survey 2015
NGO Non-government organization

Domestic EFI technology use in the study area

The domestic or household EFI technology use in the study areas included bulbs for lighting, refrigeration, water pumping, communications, radio and television (TV), but no one indicated using it for cooking or ironing. The commercial EFI energy usage was limited to lighting systems of the business facilities. The rural populace is yet to expand the use to agro-processing activities, such as flour grinding, oil expelling/extraction, crop drying, and threshing. Small-scale industry, such as sawmilling, wool and cotton processing, and stone crushing, has not seen the light of green technology in the study area. Dayo (2008) aptly noted that investments in clean energy facilities in Nigeria have been in large and small hydroelectricity generation, solar PV, biomass, biofuel, and wind energy.

The use of EFI technologies and innovations can be for domestic and commercial purposes. Energy-saving bulbs appear to be the most popular in the study area. About 35 and 40% of people use it for domestic and commercial purposes, respectively. Next to this are the rechargeable lanterns, which the respondents use for domestic and commercial purposes were 38.8% and 31.7%, respectively. However, the use of solar panels was more for commercial than domestic purposes. The Chi-squared test had a coefficient of 69.21, which is significant at 1% probability level. The significant likelihood ratio indicates that the trend may not be easily changed. Incidentally, Table 5.6 shows a similar trend for the use of the technologies for domestic and commercial purposes, as the Pearson's correlation coefficient (R) is 0.783, which is significant at one percent probability level. The positive sign of R indicates a positive trend with respect to the use.

However, the magnitude of the use differs. This is because the Chi-squared test is significant at one percent probability level, with a coefficient of 69.21, signifying that the number of EFI technology appliances/innovations/used for commercial and domestic purposes differs. It is reasonable because, as can be seen from Table 5.6, the number of units of each facility used differed.

Table 5.6 Use of EFI technologies for domestic and commercial purposes

Some EFI technologies	For household usage		For commercial purposes	
	Frequency[a]	Percentage	Frequency[a]	Percentage
Energy-saving bulbs	80	40.81632653	42	35
Energy-saving appliances	24	12.24489796	20	16.66667
Rechargeable lanterns	76	38.7755102	38	31.66667
Solar panels	6	3.06122449	20	16.66667
Inverters	10	5.102040816	2	1.6
	196	100	120	100

Source Authors' computation using data from Field survey, 2015

[a]Multiple responses recorded; Chi-squared test = 69.21, df 6, sig level 0.000 (sig at 1% probability level); likelihood ratio = 67.007 df 6, sig level 0.000; Pearson's correlation coefficient R = 0.783, sig level 0.000

Table 5.7 Recycling of materials and wastes

Items	Frequency	Percentage
Iron	26	16.25
Aluminium	12	7.50
Rubber/plastic	24	15.00
Others	8	5.00
Use of organic manure	32	20.00
Total	102	63.75

Source Authors' computation using data from field survey, 2015

Recycling

Recycling of materials is one of the desired behavioral changes in green economy. Although the materials that can be recycled are not mutually exclusive, which gives indication of multiple response, only 63.75% of respondents indicated that they have been involved in it (Table 5.7). Twenty percent of the respondents have used organic manure, which is an indication of recycling organic waste. The organic manure consists of animal droppings and/or crop residues used as fertilizers to enrich the soil, a common practice with farmers. Iron and aluminum are very important metals for building different structures, but iron is recycled more than aluminium (Table 5.7). Rubber and plastic materials are also involved in the recycling processes. Other things people recycle include bottles. However, many households may not see the need to recycle items due to very small money they can bring and the fact that people who buy these items are not always available. They can throw items that can be recycled away, which poorer people will scavenge on. This may account for the low response with respect to recycling items.

EFI technology, especially solar-powered energy systems, serves useful purpose in providing very clean water. The solar-powered borehole water, being more potable than water from other sources, is mainly used for domestic purposes, especially drinking and cooking food (see Table 5.8); some households use it especially for baby food. This usage is contrasted with the use of well water for laundry and flushing toilets. Thus, it is apparent that rural households understand the significance of good water source, and EFI technology serves very well in this context. The menace of health risk posed by poor water sources is well reported in literature (Yanda et al. 2005).

Factors affecting the use of EFI Technology

High costs and poverty are factors that pose problems when both occur together, and the rural poverty level in less developed economies, like Nigeria's, is usually more severe than urban ones (Mkpado and Arene 2010). This may account for the reason why over 72% claim that cost and financial burden constitute major problems to the use of EFI technologies. There are also the problems of how to use the technology, as about 13.75% claim that they do not have skills to operate them; therefore, while they have the finance, the lack of awareness and skills will limit the usage. Standardization of green technology products is very important, as consumers are aware of

Table 5.8 Sources and uses of water

Sources	Uses	Frequency	Percentage
Solar-powered borehole	Drinking and cooking food	38	23.75
	Washing	4	2.5
	Flushing toilet	8	5
	All of the above	70	43.75
	All having the water	120	75
Well water	Yes, flushing toilet	126	78.75
	No	26	16.25
	Total	150	93.75

Source Authors' computation using data from field survey, 2015

the presence of low or fake EFI products, such as LED bulbs. Fake products enrich their manufacturers, but deter the rate of development and attainment of desired adoption level. In Table 5.9, about 13.75% claim that the presence of fake products affects them. This, again, shows the need to regulate the quality of the products in the country.

Table 5.9 illustrates the major factors affecting the use of EFI technology by the households. Costs and financial problems were the most limiting factors; about 72% claimed that they were constrained by finance with respect to access and use of EFI technologies. The result is in consonance with Mkpado et al. (2012) report. Awareness and skills were also noted; to be aware of a product and know how to use it can motivate one to key-in and embrace the product. Some EFI technologies require the use of electricity, which is associated with the risk of electrocution. There is also the need of understanding the capacity of the facilities, such as those of inverters and solar panels to avoid overload and damaging of the facility. The very limited knowledge of the rural people with respect to capacity of inverters and solar panels can cause lack of interest and adoption of the innovations/technology. In a similar manner, RUWES (2015) advocated establishment of training schools and workshops on renewable energy in Nigeria. This can spur local content and indigenization of the processes that will increase acceptance and usage (Mkpado and Onuoha 2008).

Table 5.9 Major factors affecting the use of EFI technology

Factors	Frequency	Percentage
Cost/finance	116	72.50
Awareness/skills	22	13.75
Low-quality/fake products	22	13.75
Total	160	100

Source Authors' computation using data from field survey, 2015

The presence of low-quality and fake products in the market constitutes another problem. Unscrupulous businesspeople can use low-quality materials to build their products. This will create uncertainties in the market, as consumers may not be able to differentiate between the genuine and fake products at the point of purchase. Soon, the fake products will either become disused, or perform with very low efficiency, and the consumer will be disappointed and count such purchases as a waste. The Nigerian Customs Service and the Standards Organization of Nigeria can help with quality maintenance for clean energy technologies (RUWES 2015; Muhammed, nd). Suleiman (2010) aptly noted that lack of capacity and standard quality control, as well as quality assurance, limits consumer confidence in the new and growing market of renewable energy. This is also accompanied by the general low level of public awareness of the renewable energy technologies in Nigeria.

Some challenges to the use of EFI Technology in the study area

The qualitative assessment of some other challenges with respect to the use of EFI technology is presented in Table 5.10. The availability of fake or inferior rechargeable bulbs is reducing peoples' ability to use them; family size and income are also major factors determining how much is spent on solar-powered facilities. The needs of training on the use of solar-powered facilities are noted, as Power Holding Company

Table 5.10 Some challenges to the use of EFI technology/innovation

Some challenges to use of EFI technology/innovation	Total respondents	Total score	Mean sore	Remarks
The availability of fake or inferior rechargeable bulbs is reducing peoples' ability to use them	160	690	4.259259	Agree
Family size and income are major factors determining how much we spend on solar-powered facilities	160	662	4.08642	Agree
I need training on the use of solar-powered facilities	160	672	4.148148	Agree
Power Holding Company of Nigeria (PHCN) is spoiling many rechargeable bulbs due to poor voltage	160	650	4.012346	Agree
Used rechargeable bulbs are very poisonous to dispose	160	580	3.580247	Undecided
Increasing cost of materials and water affects recycling and household water usage	160	584	3.604938	Undecided

Source Authors' computation using data from field survey, 2015

Table 5.11 Classification of respondents by use of EFI technology in the rural areas

Index score	Frequency	Percentage	Remarks
Less than 60% = 0	82	51.25	Non-adopters, score 0
60% and above = 1	78	48.75	Adopters, score 1
Total	160	100	

Source Authors' computation using data from field survey, 2015

of Nigeria (PHCN) is damaging many rechargeable bulbs due to poor voltage and epileptic power supply. It is undecided that energy-saving bulbs and rechargeable lanterns are very poisonous to dispose and that cost of materials and water can affect their recycling (Table 5.10). The electricity suppliers in Nigeria need to overhaul and upgrade their facilities to serve the people efficiently, as their services can greatly affect the use of EFI technology appliances.

Use of EFI technology in the rural areas

An index was constructed for the use of energy-saving bulbs, solar panels, and inverters, as well as bicycles. Households that score up to 60% were adjudged as using EFI innovation and scored one; otherwise, zero (Table 5.11). This is a gross index for using EFI technology, both at domestic/private level and public places. The majority represented by 51% of the respondents are yet to be adjudged as using EFI innovation, while the minority represented by 48.75% of the respondent were adjudged to be using EFI technology to a reasonable level. Mkpado (2013c) aptly noted that future livelihood of rural people will be improved with innovations and technologies.

5.4.2 *Regression Analysis*

The overall estimate of determinants in the use of EFI technologies is acceptable because of its significant log-likelihood ratio (35.6) and relatively high Cox & Snell R Square (0.451); Nagelkerke R Square (0.664). The variables that were significant include income government ownership of facilities, private ownership of facilities, age, and level of awareness of climate change (Table 5.12).

Age has positively and significantly affected the use of EFI technologies. Its coefficient is 0.252, while its Wald statistics is 2.986, which is significant at five percent probability level. It means that older people are more likely to adopt the technologies.

Level of awareness of climate change had a positive and significant relationship with the use of EFI technologies. Its coefficient is 0.543 and its Wald statistics is 3.140, which is significant at ten percent probability level. It means that those who are very much aware of climate change are more likely to adopt the technologies.

Table 5.12 Logistic regression estimation of determinants of use of EFI technologies

Variables	Coefficient	Standard error	Wald statistics	Remarks	Exp (B)
Gender of household head	−1.091	0.894	1.491	Ns	0.336
Educational status	0.107	0.189	0.320	Ns	1.113
Income	0.671	0.357	3.530	**	1.956
Government ownership of facility	0.266	0.145	3.369	**	1.304
Private ownership of facility	0.639	0.324	3.902	**	1.895
Household size	−0.010	0.156	0.004	Ns	0.990
Age	0.252	0.146	2.986	**	1.287
Very much aware of climate change	0.543	0.306	3.140	*	1.721
Constant	−1.693	2.091	0.655	Ns	0.184

Source Authors' computation using data from field survey, 2015
Note **, * = significant at 0.5% and 10% probability levels, respectively; *Ns* Not significant

Private ownership of facility has positive and significant relationship with the use of EFI technologies. Its coefficient is 0.639 and its Wald statistics is 3.902, which is significant at 5% probability level. It means that increasing private investment/ownership of green technologies will bring more people to adopt it; it is reasonable to believe that those who invest in such technologies own and use them.

Income had a positive and significant effect on the adoption of green technology. Its coefficient is 0.671, while its Wald statistics is 3.53, which is significant at 5% probability level. It means that rising personal income will increase the likelihood of the use of the innovation/technology. It is possible because the use of such technologies will be at a cost to the household.

Government ownership of facilities had a positive and significant effect on the use of EFI technologies. Its coefficient is 0.266, while its Wald statistics is 3.369, which is significant at 5% probability level. This can be as a result of government investment on EFI technologies, like solar-powered water boreholes, solar-powered streetlights, solar-powered hospital facilities, and solar-powered ICT facilities in public schools and libraries. Thus, it is easy to identify people who use and benefit from these facilities. Increased investment by government is desirable and can be achieved by setting the right policy and priority. For example, African countries have been advised by NEPAD to invest at least 10% of their annual budget to agriculture (CSAC 2008; Mkpado 2013b). A similar pattern can be followed by African countries setting a goal to invest certain percentage of their budget for investment in environmentally friendly infrastructure.

The model indicates that gender, educational status, and household size all have not significantly affected the use of EFI technologies/innovations; but the level of awareness of climate change and ability to make investments counts.

5.5 Recommendations and Conclusion

It is recommended that adults need to encourage younger people to key-in an embrace EFI technologies, as age is positively correlated to the use of EFI. Government can improve regulation and standardization of EFI technology products sales in Nigeria; this will reduce the presence of substandard EFI facilities, which tend to reduce motivation for adoption of EFI facilities. Training and workshops on EFI technologies are important to motivate the rural people toward the desired goal; as knowledge and level of awareness of EFI have been identified as constraints. Increased investment in environmentally friendly technology by both public and private sectors should be encouraged as low income and associated poverty problems abound in rural areas. There is the need for development partners and African governments to work out modalities to provide finance for the needed transition to green economy. Adequate policy framework is required to spur the increases of government expenditure on environmentally friendly infrastructure and set targets, as well as regulate the general business conduct of EFI technologies in Nigeria.

The study reveals that EFIs are unattractive to local people. The present situation can be described as that of early users. The study further reveals that there is need to create more awareness on the use of EFI technologies, as well as awareness on climate change. This may not be limited to slight introduction but detailed discussions, including merits and prospects of using EFI technologies, public campaign on EFI technologies via electronics and mass media is imperative. The types of assistance specified by respondents include both financial and technical support. It may be easier to provide technical assistance by some type of training and mass enlightenment program. There is the need for development partners and African governments to work out modalities to provide finance, or subsidize costs associated with the transition to EFI technology. Government can improve regulation and standardization of EFI technology products, thus reducing fake products to the barest minimum in Nigeria.

References

Arun S, Gopalkrishnan R (2012) Resource-constrained product development: implications for green marketing and green supply chains. Elsevier

Charles Opara-Ndudu in this day newspaper of 15th Mar 2015

CSAC (2008) (Civil Society input unto the African Commission on effective development cooperation with Africa) Agriculture, growth and employment in Africa–need for institutions, infrastructure, information, innovation, investments and integration discussion paper number 2 Apr 2008 http://www.friisbach.dk/fileadmin/cfb/images/Africa/Agriculture-final.pdf

Dayo FB (2008) Clean energy investment in Nigeria—the domestic context. International Institute for Sustainable Development, Canada. https://www.iisd.orgpdf2008cei_nigeria.pfd

Eleri EO, Onuvae P, Ugwu O (2013) Low-carbon energy development in Nigeria: challenges and opportunities. International Institute for Environment and Development 2013 http://www.ppubs.iied.org.pdfs/G03555

Huzi IM (2014) Energy access for all: the role of clean energy in alleviating energy poverty. http://www.ng.boell.org/sites/default/files/uploads/2013/10/clean_energy-green_deal_nigeria_study.pdf

IEEE (2015) IEEE smart village: empowering off-grid communities http://ieee-smart-village.org/ieee-smart-village-celebrates-bank-of-industry-nigerias-investment-in-gve-projects-ltd-with-additional-funding-to-further-support-electricity-projects-in-off-grid-communities/

Igbokwe EM, Mkpado M (2011) Socioeconomic impact of climate change in Nigeria. Agro-Science 10(1):12–21. www.agrosciencejournal.com/article-222.html

Intergovernmental Panel on Climate Change (IPCC) (2007) Impacts, adaptations, and vulnerability, fourth assessment report. Cambridge University Press, Cambridge

Kessides C (1993) The contributions of infrastructure to economic development: a review of experience and policy implications. World Bank Discussion Paper Number 213. Washington

King A, Lenox M (2002) Does it really pay to be green? J Ind Ecol 5:105–117

Mkpado M (2013a) Green technologies for African Agriculture and energy and possible strategies for incubation and development. J Agric Sustain 2(1):43–65. http://infinitypress.info/index.php/jas/article/view/41

Mkpado M (2013b) The status of African agriculture and capacity development challenges for sustainable resilience from global economic shocks. J Sustain Dev Stud 3(1):45–79

Mkpado M (2013c) Prospective of Agribusiness for the future Sub-Saharan Africa. Int J Rural Stud 20(2):10–15

Mkpado M, Arene CJ (2003) Effects of cropping system and selected socioeconomic factors on the sustainability of arable crops production: the case of cassava in Okigwe Imo State, Nigeria. J Agro-Technol Extension 3(1 & 2):74–81

Mkpado M, Arene CJ (2007) Effects of democratization of group administration on the sustainability of agricultural micro-credit groups in Nigeria. Int J Rural Stud 12(2):1–9. www.vri-online.org.uk/ijrs/Oct2007/Democratization%20

Mkpado M, Arene CJ (2010) Does group design affect savings mobilization rural micro credit groups: evidence from Nigeria. Econ Aff 55:145–157

Mkpado M, Onuoha C (2008) Indigenous knowledge as sources of low input agricultural technologies agricultural in sub-Sahara Africa: Nigerian experience. Int J Rural Stud 15(2):2–8. www.vri-online.org.uk/ijrs/index.php

Mkpado M, Ugbaja MJ (2008) Socioeconomic issues and strategies for youth's involvement in sustainable agriculture amidst stabilizing democratic experience in Nigeria. Int J Econ Dev Issues 7(1):166–172

Mkpado M, Joel N, Nweze, Igbokwe EM (2012) Income and energy sources among Agrarian households in Nigeria: implications for low carbon energy development in less developed countries. Russ J Agric Socio Econ Sci 7(7):32–40. www.rjoas.com/issue-2012-07.html

Muhammed TL (nd) Policy, and regulatory challenges in promoting efficient and renewable energy for sustainable development and climate change mitigation in Nigeria. Legislative https://www.google.com/search?q=green+energy+%2C+climate+change+and+rural+nigeria&ie=utf-8&oe=utf-8

Muhammed AY, Abdullahi MG, Mohammed NY (2015) Critical factors affecting the development and diffusion of renewable energy technologies (RETS) in Nigeria. J Multi Eng Sci Technol (JMEST) 2(8)

Nigerian Energy Support Programme (NESP) (2014) The Nigerian energy sector–an overview with a special emphasis on renewable energy. Energy efficiency and rural electrification. https://www.giz.deendownloadsgiz/2014-en-nigerian-energy-sector.pdf

Nzamujo G (2013) Dynamics of youth-led rural growth: the Songhai Model for developing young agricultural entrepreneurs. How the redeemed Christian Church of God could harness this possibility set in its Ministry

Opara S (2015) Punch newspapers, 9 July

Rural Women Energy Security (RUWES) (2015) http://www.4.unfccc.intsubmissionsListsOSP SubmissionUpload39_122_130688171405397115-Nigeria%20gender.pdf

Suleiman LN (2010) Renewable energy as a solution to Nigerian energy crisis business. A thesis in economics and tourism of Vaasa University of applied sciences for degree program of international business

Usman M (2012) Rural solar electrification in Nigeria: renewable energy potentials and distribution for rural development. http://www.sases.conference-services.netresources/2522859pdf, SOLAR2012_0232_full%20paper.pdf electric rural

Wai RJ, Wang WH, Lin C-Y (2008) High-performance stand-alone photovoltaic generation system. Proc IEEE Trans Ind Electron 55(1)

Yanda PZ, Kangalawe RYM, Sigalla RJ (2005) Climatic and socioeconomic influences on Malaria and Cholera Risks in 3 the Lake Victoria Region of Tanzania. http://www.aiaccproject.org/AIACC. Working paper no. 12

Chapter 6
Sustainable Development and Small- and Medium-Sized Enterprises: How to Meet the Challenges of Mobilizing Ivorian SMEs?

Koffi Paul Assandé, Assémian Faustin Naouré, and Yann Cédric Armel Vangah

6.1 Introduction

The observation of the adverse effects of industrial development on the environment and, incidentally, on the living conditions of the people led the United Nations to take, at the Stockholm Conference[1] in 1972, resolutions stressing the need for states to consider the environmental and social constraints in growth and development projects. This commitment reiterated at successive conferences (Rio in 1992 and Johannesburg in 2002) was materialized through an action plan aimed at implementing the sustainable development program in each state.

To date, the research has clarified the main concepts, sharing a set of findings, particularly about the "barriers to engagement" in CSR and the positive spin-offs associated with CSR (Labelle et al. 2010). In large companies, research on social responsibility (CSR) attempts to indicate how to integrate economic, social, and environmental constraints into their business model (Jenkins 2009). However, in developing countries, corporate ownership of CSR remains a major concern for governments, researchers, and business leaders alike despite the existence of a legislative framework (in most countries) and the gradual adoption of SD strategies by the various UN Member States.

[1] For the first time, this conference placed environmental issues on the international agenda.

K. P. Assandé (✉) · A. F. Naouré (✉)
Université Félix Houphouet Boigny D'Abidjan, Abidjan, Côte d'Ivoire
e-mail: assande2000@yahoo.fr

A. F. Naouré
e-mail: naf_assemian@yahoo.fr

Y. C. A. Vangah
Université Lorignon Guédé de Daloa, Daloa, Côte d'Ivoire
e-mail: vangahyann@gmail.com

C. Atewamba and D. Yong Ngondjeb (eds.), *Inclusive Green Growth*, Advances in African Economic, Social and Political Development,
https://doi.org/10.1007/978-3-030-44180-7_6

In Côte d'Ivoire, the government is stepping up efforts to promote the widespread ownership of CSR practices.[2] Indeed, after identifying the key players in sustainable development, the government's strategy now seems clear with recent meetings with the private sector[3] to raise awareness and mobilize support for SD. This commitment aims to raise awareness among Ivorian business leaders that SD is now a safe alternative to harmonious development in view of the opportunities and strategic options that flow from it. And this is for the benefit of all components of society, namely the state, businesses, and populations.

This political will clearly displayed by the government could not guarantee the adherence of business leaders to SD without the contribution of the scientific community through results of studies highlighting the significant relationship between performance firms and CSR. In this regard, much remains to be done in Côte d'Ivoire, both at the level of large companies and SMEs. In a singular way, considering their weight[4] in the Ivorian national economy (98% of formal national enterprises), their capacity to innovate and their contribution to GDP, SMEs must be, not only at the heart of the government, but also and above all considered as a fertile field of investigation for researchers. Unfortunately, to date, very little work has been done on the impact of sustainable development strategies on the performance of small- and medium-sized enterprises (SMEs) as well as on rural economies in Côte d'Ivoire.

Work on the appropriation of CSR practices in other countries shows that SME managers who adopt a CSR-friendly posture benefit their companies in terms of reducing costs, increasing employee loyalty, better relations with territorial institutions, acquiring new knowledge and, incidentally, a better reputation (Bonneveux and Saulquin 2009). If, despite these positive repercussions, the mobilization of SME leaders in SD remains problematical, Berger-Douce (2008) believes that it is because of the lack of time, but especially of knowledge and information about CSR. What about the Ivorian context?

In other words, what are the determinants of the commitment of the leaders of Ivorian SMEs to adopt a posture favorable to SD? This is the question that this research will attempt to answer.

In this perspective, the general objective of this research is to identify the determinants of the commitment of Ivorian SMEs in SD.

---

[2]According to the 2012 country report of the Ministry of the Environment in the perspective of RIO + 20, more than 40 international texts relating to the environment and SD have been ratified; a ministry in charge of the environment exists; the law of orientation n° 2014-390 of June 20, 2014 on the sustainable development was adopted, environmental impact studies are required to validate any project of activity, a national commission of the durable development was even created, and so on.

[3]September 10, 2015 and September 11, 2015 were the foundations for sustainable development in Côte d'Ivoire. These foundations led to the signing of a partnership agreement between the state through the ministry in charge of SD and the private sector represented by the General Confederation of Enterprises of Côte d'Ivoire (CGECI).

[4]INS Report 2012–2013.

Specifically, the research aims to:

- analyze the perception of the concept of SD by Ivorian SMEs;
- identify the "responsible" practices that prevail in Ivorian SMEs;
- analyze the motivational factors and obstacles to SME engagement in SD.

To achieve these objectives, we used a methodology that initially consisted of analyzing the existing work on SD issues in order to identify the main themes that made it possible to develop the collection tools. In a second step, the data were collected from SME managers and analyzed in order to understand the factors that could be levers as well as the obstacles to their adherence to SD.

The plan for this work has four sections. Section 1 deals with the theoretical foundations and concepts of SD and CSR. Section 2 focuses on the specificities of the SME and the determinants of the commitment of the leaders of this type of enterprise in SD. The methodological approach and the results of the research are presented in Sects. 3 and 4, respectively.

6.2 The Conceptual and Theoretical Foundations of SD and CSR

6.2.1 The Concepts of SD and CSR

Debates on SD are growing at the level of policy makers, business leaders, and the research community. They have gradually begun to clarify the contours of the concept even if it remains somewhat complex to define. Thus, SD is defined as "development that meets the needs of the present without compromising the ability of future generations to meet their own needs" (Brundtland 1987).

Translated in terms of allocations of factors of production or growth, the goal of SD is to bequeath to future generations the amount of factors enabling them to reach a standard of living at least equivalent to that of current generations (Sassenou 2006). We are talking about intergenerational solidarity.

The appropriation of SD values by companies in the development and implementation of their development strategy is known as CSR, i.e., corporate social responsibility, business or social and environmental responsibility. It is an extension of corporate responsibility, and in other words, the transition from a vision that is primarily concerned solely with the interests of shareholders to a vision that is concerned with the satisfaction of all stakeholders (Assandé 2014).

In this perspective, CSR "implies that profitability is certainly the main objective in order to ensure the long-term survival of the firm, but this profitability cannot be acquired at any price. Respect for the environment, society and people is equally important and the company pursuing a SD strategy must achieve the most harmonious balance possible between these three dimensions" (Spence et al. 2007).

For Abdelouahab, Ahmed, Samiha (2012), the notion of responsibility is graduated from the inevitable as the regulatory and economic obligations to that which gives the choice as the ethical and discretionary responsibility. Taking SD into account led Carroll (1979) to propose that the company's performance be understood from the point of view of the social performance that results from all the obligations that the company with respect to the company and encompassing the economic, legal, ethical, and discretionary categories. For the author, this type of performance is based on the company's relations with all the stakeholders concerned by its activity.

Despite the existence of several approaches to SD, many authors agree that there is the search for a new model of socioeconomic development. However, differences arise when it comes to identifying the types of activities to be carried out by the company to achieve sustainable development (Spence et al. 2007). In an attempt to address this concern, Starik and Rands (1995) identified five categories of activities to be carried out by a company in the context of sustainable development. These are activities related to economic prosperity, social justice, protection of the environment, culture of society, and organizational culture.

6.2.2 The Theoretical Foundations of SD

The vision of SD shared by the majority of authors retains three dimensions:

- the economic dimension that recommends a reasonable use of resources;
- the social dimension which implies the fair treatment of all;
- the environmental dimension which suggests the protection of the environment and its resources by all members of the community.

The consideration of these dimensions by companies in their development implies a greater investment in stakeholder relations, in human capital, and in the protection of the environment. While some authors agree that the desire to conceive of development differently is growing strongly within large companies, the fact remains that the issue of mastery of SD goals and evaluation criteria sustainable performance is always posited according to Lauriol (2004). The main reason for differences over the purposes of SD is that there is a plurality and variety of stakeholders in the organization with often conflicting perceptions of performance. A number of theories are used to explain the basis of these discrepancies and to analyze companies' SD behavior. Among them, stakeholder theory is the one that comes up most often in most works.

For Mercier (1999), the stakeholders are made up of all the agents for whom the development and the good health of the company represent important stakes. From this perspective, the resulting theory views the firm as a constellation of cooperative and competing interests (Donaldson and Preston 1995). Mercier (2001) finds that the essential characteristic of stakeholder theory is to be, at the same time, a managerial theory and an ethical theory. It seeks to replace the traditional view of the company, which assumes that executives have a fiduciary duty to act exclusively in the interests of their shareholders.

From the perspective of SD, beyond the variety of stakeholders, the challenge is to integrate their interests, sometimes antagonistic, but legitimate, in the strategy of the company so as to create lasting shared value. This is why, with Gendron (2000), we see that the company, whatever its size, today appears as a social institution that is legitimately accountable to society and as a social and political actor that must deal with CSR as strategic issues in the service of improving its performance.

In the same logic, Bonneveux and Saulquin (2009) identify in the literature three streams of stakeholder theory: the descriptive stream who considers that the organization is at the center of cooperations and competitions each having their intrinsic value; the instrumental stream which is based on the idea that companies that manage their stakeholders are more profitable in terms of profitability, stability, and growth; and the normative ethical current that insists on the ethical obligations of the firm and on the ways to assume them, without denying its interest in its economic performance objectives. This vision of the company is fully in line with CSR and clearly has the strategic objective of contributing to increasing the profitability of companies through their investments in the future. Thus, the CSR actions of the company are assimilated to signals sent to their various stakeholders as pledges of goodwill (Berger-Douce 2008). Moreover, this theory, which is part of the family of contractual theories, remains limited when it comes to conceptualizing the governance of SD (Lauriol 2004). The author justifies his position by saying that "structures and systems of governance are, for the moment, mainly designed for internal (or primary) stakeholders (Lauriol 2004)". Second, because the external stakeholders (or secondaries) do not necessarily feel the need to contract their relationship with the company. In this case, how to integrate parties that do not "want to take sides"?

Alongside this contractualist vision that has been the theoretical underpinning of many works on SD, other authors have mobilized the neo-institutional sociological theories which are based on "a symbolic representation of the decisions of the organization, embedded in a social network and seeking a legitimacy, a suitable image despite conflicting and contradictory expectations of the actors of the organizational field" (Spence et al. 2007). While these theories have been mobilized as part of the work on SD at the level of large enterprises, they are no less applicable at the level of SMEs. Indeed, despite their specificity and constraints, SMEs not only maintain relations with several actors who may suffer or benefit from the effects of their activities, but also they are seeking legitimacy for a better positioning and incidentally better performance.

In addition, Barney (1991) has developed a resource-based approach with the objective of evaluating companies' commitment to the path of SD and benefiting from it.

6.3 SMEs and Sustainable Development

As mentioned above, while there is a lot of debate about SD in large companies, the situation is different for SMEs, despite the exploratory work carried out by some researchers in the field. In this section, we highlight the elements that make the specificity of SMEs and the determinants of their commitment to SD.

6.3.1 *The Specificities of SMEs*

The problem of SD in SMEs can only be well understood if this particular economic entity is well perceived. For Torres (1999), the SME is perceived as a company with many virtues: adaptability, flexibility, creativity, and working atmosphere. For some authors, the SME differs from the big company according to various aspects: the ownership structure, the financing structure, the corporate culture, the sources of vulnerability, and so on.

6.3.2 *The Ownership Structure of SMEs*

The capital of the SME does not have the same configuration as that of a large company. It is most often owned by the owner-manager. Thus, the manager is both the manager and the contributor of capital. In SMEs where there are associates, the latter are either in the family sphere of the manager or are related to him in a certain way. In such a situation, the governance model as advocated in large companies is inadequate. It is the manager's perception of the environment and his beliefs that will determine his strategic choices and therefore the commitment or not of his company in SD.

6.3.2.1 The SME Financing Structure

Unlike large companies, whose assets often offer sufficient guarantees to benefit from bank financing, SMEs are obliged to resort to self-financing or at best to funding from the owner and/or his family and friends. This situation is mainly due to the absence or insufficiency in the assets of SMEs of transferable assets that can be used as a guarantee for indebtedness and constitutes, moreover, one of the obstacles to SME engagement in SD as noted in the 2014 report of the CSR platform in France. According to this report, the issue of financing CSR initiatives or at least their initialization is often put forward by the SMEs as a barrier to their commitment to SD.

6.3.2.2 Management and Corporate Culture in SMEs

Labor relations in an SME are very often embedded in relationships of dependence or allegiance to the boss who is an elder, a villager, and an uncle to whom the employee feels indebted. In these conditions, the management of human resources takes a particular form with a too strong presence of the leader that Torres (2008) qualifies as effect of egotrophy. The author defines this effect as the high concentration of the management of the SME on the leader of whom all depends. All that forces him to take only short- and medium-term decisions. The geotropic effect is negatively correlated with the size of the company. In fact, the smaller the size of the company, the more the weight and the role of the manager increases and the more the company is extremely dependent on the manager to the point of becoming indispensable or irreplaceable. This situation is uncomfortable for managers who need freedom to take initiatives, which is why there are very few executives in small companies. The combination of all these factors related to the human resources of SMEs, especially the leading role of the leader, the lack of managers, and the small staff can hinder the SME in its decision to engage in SD. In relation to the corporate culture, the development and sustainability of the SME reside in its ability to reconcile social and cultural values with economic efficiency insofar as it evolves in a system of social relations of belonging to the community.

6.3.2.3 Determinants of SME Engagement in SD

- The specificities of the SMEs identified above are for the most part obstacles to their commitment to SD. Despite these weaknesses, some SMEs have committed to the SD approach and derive significant benefits from the literature review on the issue. Most authors who have been concerned about the characteristics of SMEs engaged in SD have limited themselves to an exploratory approach and have identified as levers and/or brakes the following elements:
- organizational characteristics;
- the financial position of the company;
- the characteristics of the company's personnel;
- the relational networks of the company.

6.3.2.4 The Organizational Characteristics of SMEs

The Size Effect

Temri and Fort (2009) have identified in the literature a number of works in the literature that show the existence of a size effect in issues of sustainable development. Indeed, according to these authors, several studies reveal that SMEs do not have the same behavior as large companies and are not sensitive to the same pressures because their characteristics are different.

Thus, it appears that in SMEs, there is a big difference between attitudes, rather positive in relation to sustainable development issues, particularly the environmental dimension, and actual behaviors. According to Temri and Fort (2009), small firms do not see any particular benefits or constraints that would push them to take action beyond regulatory compliance, particularly in the environmental field. They do not also see the impact of their activities on the environment.

As for the European Commission (2002), it argues that size influences firms' SD practices, but it is also crucial in the perception of companies to be engaged. In the same logic, Pogutz et al. (2004) believe that SMEs, in view of their particular characteristics, should only collect "data that is important to stakeholders with a relationship of power, legitimacy or urgency."

Spence et al. (2007) subscribe to the existence of a size effect on SME engagement in SD. They concluded from their research that SMEs that are more likely to engage heavily in SD are of average size.

In addition to size, the presence of SMEs in certain sectors of activity more predisposed to engaging in SD, including potentially polluting sectors and/or exposing employees; we are talking about a business sector effect. Gimenez et al. (2003) came to this conclusion from their work by observing that companies located in sectors such as the chemical industry more easily adopted the EMAS standardized European environmental system.

The Ability to Innovate

The ability of SMEs to innovate is another determining factor of his commitment to SD as revealed by the European Commission (2002). For her, only the most innovative companies have the capacity to be part of a logic of sustainable development. The influence of innovation on the involvement of SMEs in SD finds its explanation in the definition of this concept which suggests a new way of understanding the economic development of companies. Indeed, according to some authors, including Spence et al. (2007), if we consider CSR and sustainable development, as a different and innovative way to apprehend the company, the contribution of research on innovation and diffusion technological innovations seem to be able to be mobilized to explain the mechanisms of appropriation of these concepts by SMEs.

However, we agree with these authors that "SMEs often do not have the knowledge to make a full assessment of available technologies and to fully capture the benefits." In addition, moving from a reactive culture to a proactive culture of sustainable development requires vision and managerial skills that small business leaders still do not have.

The Size of the Business Market

Unlike large companies whose markets often cross-national borders, most small- and medium-sized enterprises are content with a national, regional, or simply local market. In these circumstances, in the absence of a regulatory framework that can "force" them to engage in SD, there is no incentive for SMEs operating on a local or even national market to adhere to it. On the other hand, SMEs that have relationships with external partners or sellers internationally are often quick to adhere to the SD approach to meet the demands of these markets. Thus, we postulate that SMEs willing to engage intensely in SD are those that have foreign buyers or sell on international markets. In the same logic, Julien et al. (1999) suggest that the lack of information from SMEs on the differences in customer needs on distribution networks does not promote their commitment to SD.

The Financial Situation of the Company and CSR

From the summary of the literature on the link between company profitability and CSR practices carried out by Berger-Douce (2008), it follows that the financial situation of the company is an essential element in terms of environmental commitment in the extent to which the lack of financial resources is often considered as the major obstacle perceived by the managers of SMEs. For the author, adapting or developing a new environmental technology is a costly proposition whose economic benefits are uncertain and long term. Thus, only profitable firms would have the means to avail themselves of it. In these circumstances, it is clear that a sound financial situation is a prerequisite for the implementation of a CSR policy, especially in SMEs.

Characteristics of Company Employees and Commitment to SD

The skill of an employee is seen as a base of strategic resources through which reproduction can be painful for competitors, Barney (1991). From this perspective, the fact that an employee is competent is an engine for sustainable development. More specifically, employees' skills in technology help improve the environmental performance of SMEs by reducing the negative impact of production and consumption and also promote growth (Adeoti 2000). While following this vision, other authors consider training and information of employees as essential to facilitate the adhesion of SMEs to SD. Thus, for Friedman and Miles (2002), "the more employees are informed of the possible gains from responsible behavior, the easier it will be for SMEs to go down this path." Spence et al. (2007) will not say the opposite when they arrive at the conclusion of their work, to the conclusion that SMEs that engage intensively in SD are those that have employees trained and informed on the SD approach.

The Values of the Leader and the Commitment of the Company in SD

In most of the work on SME engagement in SD, the leader invariably returns as an important lever. Thus, according to Spence et al. (2007), business leaders who have a positive view of SD and perceive external pressures as opportunities and internal resources as easily mobilized forces engage their business more intensely in SD. In the same vein, Quairel and Auberger (2005) quoted in the editorial of the international SME magazine, note that "the objectives and the commitment of the entrepreneur are a key factor for the adoption of CSR strategies in companies SMEs and therefore to go beyond strict compliance and transform customer CSR requirements into opportunities."

From the foregoing, it is clear that personal characteristics of the leader of the SME, especially psychological or cognitive, have an influence on the behavior of companies in terms of sustainable development (Temri and Fort 2009). Thus, for these authors, the manager's motivations and personal beliefs are an important motivating factor for the company's commitment to environmentally friendly actions. Indeed, for Paradas (2006, 2007), the lack of resources is not very decisive in the commitment of SMEs in the SD toward the convictions of the leader. For the author, "when the leader's beliefs relate to responsible behavior and the awareness and lucidity are important, social responsibility can exist without always having to demonstrate what it brings within the limits the good health of the company. On the other hand, if the beliefs and the pleasures of the leaders do not go in the direction of responsible behaviors, it will be to be much more convincing as for the return on investment of the CSR."

Membership in Relational Networks as a Lever for SD Membership

In addition to the above-mentioned levers, the SME's membership in a relational network can be a motivating factor for SD engagement. Indeed, empirical work on the problem of SME engagement in SD shows that in the field of the environment, the relational networks of the enterprise are considered as a fundamental and unique source insofar as they facilitate the implementation of strategies for sustainable development. As for Friedman and Miles (2002), they postulate that SMEs belonging to professional networks benefit from the experience of other companies, experts, government institutions, etc., which not only provide them with practical advice, but also encouragement, a social group for the exchange of ideas, values, and experiences.

The results of the work of Spence et al. (2007) confirm this position by concluding that SMEs more committed to SD belong to networks relationships that facilitate the implementation of SD strategies or principles.

6.4 Methodological Approach to Research

Studies are certainly conducted on the issue of SD in SMEs, but ownership of the concept by the latter still raises concerns especially in developing countries such as Côte d'Ivoire, where the awareness of stakeholders, in general, and those in the private sector in particular, on the issue just started. In such a context where the identification of SMEs that have adopted SD in its three dimensions is difficult, we have chosen an exploratory methodological approach that combines the qualitative and quantitative approaches in order to understand the perception of the leaders of the concept and especially of identifying CSR practices in their businesses.

The interest of the qualitative approach for an emerging theme is to make it possible to apprehend the predispositions of the leaders and their motivations and fears as well as the obstacles to be lifted for their commitment in SD. The quantitative method, even if in this case is not intended to be extrapolated, would help to reinforce the explanation of the phenomenon as it was described during the discussions.

6.4.1 Collection Tools

Two tools were used for the collection of information: a questionnaire and an interview guide. In the absence of prior information on the theme, we drew on the theoretical and empirical review to identify the dimensions and practices of SD and the main themes that are common in the work in terms of the determinants of SME engagement in SD.

This information allowed us to build the questionnaire which includes five (5) components:

- the leader's profile, vision, and strategy;
- the environmental and social perception of the manager (perception of the environment, perception of the environmental impact of the action of the companies, and perception of the social dimension in the company);
- the level of knowledge and perceptions of the leader's SD (SD knowledge level, SD benefits, SD perception, motivation to adhere to SD, and brakes on SD commitment);
- SD practices within the organization (staff practices, customer practices, community practices, and environmental practices);
- General information about the company.

The questionnaire was designed using a five-level (5) Likert scale. In fact, for a given dimension, criteria have been identified and for each criterion, affirmations have been formulated in order to capture the attitude and/or the perception of the leader. The leader was asked to decide by ticking a number between 1 and 5: (1) strongly disagree; (2) disagree; (3) undecided; (4) agree; and (5) strongly agree.

With regard to the qualitative aspect of the research, an interview guide containing the same themes as the questionnaire was designed to serve as a basis for conducting semi-structured interviews with some of the leaders in the sample.

6.4.2 The Sample

As part of this research, we have adopted the definition of the SME contained in Decree No. 2012-05 of January 11, 2012 promulgated by the Ivorian government. This law identifies three categories of SMEs:

- Microenterprise is defined as a company that permanently employs less than ten people or has an annual turnover excluding taxes of CFAF 30 million or less;
- Small business is defined as a company that permanently employs less than 50 people or has an annual turnover excluding taxes of more than 30 million CFA francs and less than or equal to 150 million CFA francs;
- The medium-sized enterprise is defined as a company that permanently employs less than 200 people or has an annual turnover excluding taxes of more than 150 million CFA francs and less than or equal to 1 billion francs. CFA.

In the absence of any information allowing the "scientific" determination of the sample size for the quantitative component, we chose to send the questionnaire to one hundred (100) SMEs (a reasonable size to allow quantitative analyzes especially in an exploratory approach) of different sizes and different sectors of activity. We asked the support of the Ivory Coast Federation of SMEs (SMEs) and the SME Movement (MSME) to facilitate access to SMEs.

In the end, fifty-seven (57) SMEs have agreed to participate in the research, including 24 commercial enterprises, 9 in the industrial sector, 14 in the services sector, and 10 companies operating in several sectors at the same time. In addition, 37 of these enterprises are located in Abidjan, 7 in the suburbs of Abidjan, and the rest (13) in the interior of the country.

In terms of size, the SMEs that make up the sample of this research are spread among the three families of SMEs: 57.9% for microenterprises, 31.6% for small businesses, and 10.5% for medium-sized enterprises on the basis of staff numbers. The weight of microenterprises in the sample is proportional to the representativeness at national level of this category of enterprise within SMEs. Indeed, a report by the DGI on the census of Ivorian SMEs in 2000 estimated that the proportion of microenterprises in Ivorian SMEs was 98%. Regarding interviews, we selected 8 SMEs among those who filled in the questionnaire. In order to understand the motivations and the obstacles to SME engagement in SD, we have selected 4 SMEs registered in a certification process and 4 SMEs that are not in this process. We have adopted certification as a criterion, because not only does it take into account certain SD requirements, but above all because it is as constraining for SMEs as they are specified. This is why we postulate that a certified SME, regardless of the standard, is more predisposed to engage in SD than another that is not.

In both forms of collection, the person empowered to provide the answers to the questions is the leader of the SME.

6.4.3 Data Processing

The data collected using the questionnaire were processed using the method of scores and principal component analysis (PCA) which identifies latent constructs that could explain the attitude of Ivorian SME managers to SD. A total of 96 items were evaluated on a five-level Likert scale. The analysis of the responses to these items was done in three stages. The first was to verify the assumptions underlying the use of the PCA method. At this level, the analysis of the correlation matrix shows that many variables seem to correlate at least slightly. These results are confirmed by the KMO index and Bartlett's sphericity test and thus suggest the existence of association between the items. Then, the analysis of the table of the total variance explained and the graph of the eigenvalues led to the ten (10) first axes (with a part of 65.3% of the total variance explained).

In the second phase of data processing, the construction of the VARIMAX before and after component rotation matrix made it possible to identify not only the variables that saturate on each of the ten (10) axes, but also to identify those that saturate not negligible on several axes. In the category of items that are not properly positioned on a single factor, eight (8) items were recoded and forty-nine (49) removed. The analysis that was done on the basis of the remaining 47 items identified 36 variables that saturate significantly on the ten (10) axes.

The third and last stage of the treatment focused on the analysis of the internal coherence between the components of each factor. The purpose of this approach is to verify if the items that saturate each factor make it possible to measure the resulting construct and if the scale to measure this latent construct is stable. From this perspective, Cronbach's alpha analysis showed that out of the ten (10) factors, only six (6) have items that have an acceptable level of consistency with a value of the higher index or equal to 0.6. As for the qualitative data, they were exploited thanks to the content analysis method.

6.5 Results and Discussion

6.5.1 Profile of SME Leaders

The SMEs in our sample are headed mainly by men with a proportion of 81%. Overall, the leaders of these structures have a level of academic study. Indeed, 85.5% of them have at least a high school diploma. This level of study is sufficient to enable them to better understand SD issues in order to integrate them into the company's strategy. In addition, more than one-third (1/3) of business executives surveyed have proven experience in managing the SME.

The analysis of the strategic posture of the leader highlights the predominant vision of positioning on the local market with a proportion of 56% even if 73% of leaders believe that external pressures are sources of opportunity. This result reinforces the evidence from the interviews. Indeed, for most of the interviewed leaders, external pressures may be perceived as an opportunity in terms of tailoring offers, but the conquest of the external market remains a big challenge for SMEs because of the more or less modest means they have as evidenced by the comments of an uncertified SME executive: "It is more reasonable to seek a better positioning in the local market in order to improve its financial situation and later to conquer the international market."

In addition, it was noted that 9 of the 57 (or 16%) company managers who provided information on this issue adopt a passive attitude toward innovation, compared to 84% who make a lot of effort in the field of innovation matter by importing technology for 34% of them, or by constantly seeking to improve processes and offers internally for 50% of respondents.

For SMEs focused on innovation, the analysis of interviews indicates that this is in most cases an incremental innovation given that SMEs do not have substantial resources to invest in research and development. Indeed, the objective for SME managers interviewed, when they talk about innovation, is not to profoundly change the way the company operates, but to bring about an organizational or technical improvement that requires very little large-scale operations. However, to the question of what concrete actions have been taken in the company in terms of innovation, the leaders have remained largely elusive. All that leads us to the conclusion that while the leaders of SMEs are willing to make innovative decisions, they have had trouble identifying the right decisions in this area.

6.5.2 Ivorian SMEs and SD

At the end of the statistical treatments, six (6) factors or constructs are found which make it possible to apprehend the perception as well as the practices of SD as the brakes to the commitment at the level of the Ivorian SMEs: (i) perception of the environment; (ii) staff assessment and waste management; (iii) social equity; (iv) relationship with the community; (v) organizational climate; and (vi) barriers to DD.

These six factors can be grouped into three for the purposes of analysis: environmental perception, SD practices, and barriers to engagement.

6.5.3 The Perception of the Environment by SME Managers

Only one item was retained after the consistency analysis to characterize this construct, namely the inexhaustible or exhaustible nature of natural resources. At this level, the average score of 3.63 shows that the leaders of the SMEs surveyed are aware that the natural resources are exhaustible. This result is reinforced by the statements from the interviews, which highlight the urgency for men to adopt responsible attitudes in order to guarantee a peaceful life for future generations. This perception of the environment by these managers can be explained by the effects of climate change experienced more or less in different countries and particularly the Ivory Coast where studies show that the rate of deforestation could lead to a total loss of forest cover by 2034.

6.5.4 SD Practices in SMEs

Regarding the evaluation of SD practices, four (4) key factors emerge from data processing: "staff evaluation and waste management," "social equity," and "relationship with the community local 'and' the organizational climate." These factors measure the practice of SD/CSR in SMEs. The analysis of the average score of these various factors (respectively 3.68, 3.86, 3.56, and 3.92) highlights the existence of these practices among the Ivorian SMEs surveyed, which means that the SMEs in our study have a waste management policy that they produce. They claim to have a staff appraisal policy and a fair compensation practice. And all this in an organizational climate that seems favorable. In addition, the results indicate that managers support employees, take their advice into account in decision making, and make them accountable for their tasks.

In terms of community relations, leaders are aware that good community integration is necessary for their sustainability. As a result, they stated that they favor local/surrounding companies in awarding contracts.

These results allow us to observe that while not being part of a formal SD process, Ivorian SMEs have practices that conform to those of certain dimensions of CSR. The average overall score of the "practical SD" construct of 3.76 is a sign that many SMEs are in a favorable position for CSR practices with regard to the social component.

6.5.5 The Barriers to the Commitment in the SD of SMEs

Despite the existence of some CSR practices in SMEs, the data analysis highlights a number of elements identified by managers as barriers to their commitment to SD. In fact, the CPA construct, which measures the obstacles to SMEs joining the SD approach according to the leaders, includes three items, namely "lack of regulatory framework and support from the state," "lack of coaching structure," and "lack of information on the concept of DD/CSR" with respective scores score of 3.49; 3.51, and 3.80. From these data, it can be seen that for most SME managers in the sample, these three elements constitute the main causes of non-engagement of Ivorian SMEs in the CSR approach.

6.5.6 Discussion

The results of the study show that about 72% of SMEs surveyed say they know the notion of DD. The rate of knowledge of the concept DD, a little higher than that found in France which is 63%, by Berger-Douce (2008) is corroborated by their degree of perception with an average score of 3.63. These results mean that the concept of SD is not ignored by SME managers. However, comparing the level of knowledge of

SD with CSR reveals some confusion in the minds of leaders about these concepts. Indeed, some leaders think that the two concepts are different.

At the level of obstacles or barriers, SMEs in the sample mention the lack of information on the concept, the lack of a management structure, and the absence of a regulatory framework and support from the state. These obstacles, especially the last two, seem to corroborate the results found by Berger-Douce (2008) when she evokes the "classic" obstacles to SME engagement in SD.

Regarding SD/CSR practices in SMEs, the factors mentioned by managers are related to personnel management, social equity, the environment, the relationship with the local community, and the organizational climate. These companies consider taking into account in their daily management the evaluation of employees, a policy to stabilize jobs, and a waste recycling system. In addition, managers give equal importance to employees in terms of compensation and involvement in the company's decision making. They do not take gender into account when dealing with employees. These different sub-factors show that SMEs in the Ivorian context adhere to the practices of DD/CSR at least with regard to some of its dimensions. The main reason cited by SMEs is that related to the benefits of SD/CSR in terms of opportunities. Indeed, with the arrival of increasingly demanding multinationals on SD/CSR issues in the Ivorian economy, SME managers have stated that they are willing to adopt an alignment strategy to acquire and maintain markets.

Regarding the size effect, we note that the propensity to adopt SD/CSR increases with firm size (87%) for SMEs with less than 10 employees (100%) for SMEs with between 10 and 50 employees and (100%) with a high degree of commitment for SMEs with more than 50 employees. Our results are in line with the work of Udayasankar (2007); Pogutz et al. (2004). As for SMEs' ability to innovate, we observe that nearly 3/4 of the SMEs surveyed are making efforts to improve manufacturing processes and offers. We can say that they are inclined to innovate. This result, therefore, confirms those obtained by Spence et al. (2007). In the case of SMEs operating on the regional and even international market, we find that they are more inclined to be part of a process of sustainable development/social and environmental responsibility. In fact, selling internationally requires compliance with certain environmental, organizational, and societal standards.

6.6 Conclusion

This research aims to identify the determinants of the engagement of Ivorian SMEs in SD. Specifically, the research aims, first of all, to analyze the perception of the concept of DD by the Ivorian SMEs and then to identify the "responsible" practices that prevail in the Ivorian SMEs finally to analyze the motivating factors and the barriers on commitment to the sustainable development of Ivorian SMEs.

In terms of practices, the analysis of data from qualitative and quantitative approaches shows that Ivorian SMEs adopt a posture favorable to CSR. Indeed, the results of the research suggest the existence within SMEs of certain CSR practices, particularly in terms of personnel management, the environment, and the organizational climate. However, this commitment encounters obstacles precisely because of the lack of information on the DD/CSR concept, the absence of an accompanying regulatory framework and the management structure of managers.

The results of this research open up avenues for future research on the determinants of responsible practices within Ivorian companies in general and SMEs in particular. In this perspective, a greater number of SMEs could be selected and the methodology should integrate the perception that different stakeholders have of corporate CSR practices in order to reduce the biases related only to the declarations of company managers.

References

Adeoti J (2000) Small enterprise promotion and sustainable development: an attempt at integration. J Devel Entrepreneurship 5(1):57–71

Barney J (1991) Firm resources and sustained competitive advantage. J Manag 7(1):99–120

Berger-Douce S (2008) Rentabilité et pratiques de RSE en milieu PME premiers résultats d'une étude française. Manag Avenir 15: 9–29

Bonneveux E, Saulquin JY (2009) L'appropriation de la RSE par les dirigeants de PME. Le réseau comme vecteur de l'apprentissage managérial. Manag Avenir 23

Brundtland GH (1987) Notre avenir à tous, Commission mondiale de l'environnement et du développement, Editions du Fleuve, Montréal

Carroll AB (1979) A three-dimensional conceptual model of corporate performance. Acad Manag Rev 4(4):497–505

Dadene A, Atil A, Bouanini S (2012) The determinant of Csr in Algerian SMEs: an empirical and exploratory study. El-bahith Rev 13(13):677

Donaldson T, Preston LE (1995) The stakeholder theory of the corporation: concepts, evidence and implications. Acad Manag Rev 20(1):65–91

Friedman AL, Miles S (2002) SMEs and the environment: evaluating dissemination route and handholding levels. Bus Strat Environ 11: 324–341

Gendron C (2000) Le questionnement éthique et social de l'entreprise dans la littérature managériale, Cahier de recherche du CIRESS, UQAM, N°4

Gimenez LG, Casadesus FM, Valls PJ (2003) Using environmental management systems to increase firms' competitiveness. Corp Soc Responsib Environ Manag 10:101–110

Jenkins H (2009) A "Business opportunity" model of corporate social responsibility for small-and medium-sized enterprises business ethics. A Eur Rev 18(1):21–36

Julien P-A, Beaudoin R, Njambou R (1999) PME exportatrices et information en zones rurales ou zones urbaines. Revue internationale PME 12(1/2):107–126

Labelle F, Saint-Pierre J (2010) Les déterminants institutionnels, organisationnels et individuels de la sensibilité des PME au sujet du développement durable. 10èmeCIFEPME, Congrès International Francophone en Entrepreneuriat et PME, Bordeaux, France, 27–29 octobre

Lauriol J (2004) Le développement durable à la recherche d'un corps de doctrine. Revue française de gestion 2004 5(152): 137–150

Mercier S (1999) L'éthique dans les entreprises, Editions La Découverte, collection Repères, Paris

Mercier S (2001) L'apport de la théorie des parties prenantes au management stratégique: une synthèse de la littérature. Xième Conférence de l'Association Internationale de Management Stratégique, 13-14-15 juin 2001

Paradas A (2007) Le dirigeant comme levier de la RSE en TPE: approche exploratoire basée sur l'utilisation de récits et d'une cartographie cognitive, Revue internationale PME.: économie et gestion de la petite et moyenne entreprise 20(3–4): 43–67

Paradas A (2006) Perception du développement durable par des dirigeants de petites entreprises: résultats d'enquêtes. VIIIe Congrès international francophone en entrepreneuriat et PME, Fribourg, Suisse, 25–27 octobre

Pogutz S, Tencati A, Perrini F (2004) New tools to foster corporate socially responsible behavior. J Bus Ethics 173–190

Quairel F, Auberger MN (2005) Management responsable et PME: une relecture du concept de « responsabilité sociétale de l'entreprise ». La revue des Sciences de gestion. 40(211–212):111–126

Sassenou N (2006) Développement durable et responsabilité sociétale de l'entreprise: apport de la théorie économique. Revue d'économie financière 85(4): 49–62

Spence M, Ben Boubaker G, Biwolé V (2007) Développement durable et PME: une étude exploratoire des déterminants de leur engagement. Revue internationale PME: économie et gestion de la petite et moyenne entreprise 20(3–4): 17–42

Starik M, Rands G (1995) Weaving an integrated web: multilevel and multi system perspective of ecologically sustainable organizations. Acad Manag Rev 20(4):908–936

Temri L, Fort F (2009) Partage des bonnes pratiques de développement durable: le cas des PME agroalimentaires du Languedoc-Roussillon. Innovations 2009, 1(29): 103–125

Torres O (1999) Les PME, Edition flammarion

Torres O (2008) Comprendre le management des PME: le rôle clé de la procimité, conférence à l'EM LYO

http://localhistory.kingston.vic.gov.au/htu

Chapter 7
Environmental and Economic Burden of Sand Dredging on Artisanal Fishing in Lagos State, Nigeria

Fatai Abiola Sowunmi

7.1 Introduction

Fishery is an important agricultural subsector in Nigeria. There are more than 6 million coastal and riverine artisanal fisher folks fishing the 46,300 km^2 of maritime areas and 125, 470.82 km^2 of inland water bodies in Nigeria (Fish for All Submit 2005). Artisanal fishing is a subsistence fishing practice involving the use of simple tools such as throw nets and drag nets, rods and tackle as well as the traditional fishing boats. The catch is only sufficient for the family meal and occasionally for sale (Wikipedia 2014; Adesina and Ayanda 2005). According to Bada (2005), artisanal fishing accounted for more than 80% of total fish production in Nigeria. Aside from depending on fishing as their means of livelihood, 75% of the households' animal protein intake comes from fishing.

Lagos state is one of the Nigerian coastline states dotted with many fishing communities. The coastline is about 180 km long, and it is generally characterized by steep sandy beaches, offshore wave breakers and littoral drift. Some of important fishing communities in Lagos state are Badore, Agbowa, Ikosi and Oreta, Ibeshe, Ipakodo, Yovoyan, Moba, Majidun, Avijio and Itoikin. According to Sustainable Fisheries Livelihood Programme (SFLP) (2002), these fishing communities are characterized by high population densities which translate to higher pressure on the fisheries resources. Aside from high population density, other human activities such as sand dredging which has continued to spread in many fishing communities as a result of high demand for sand for construction purposes may also pose a difficult challenge to food security and employment opportunities in the fishing communities (SFLP 2002). In this study, sustainability refers to agriculture's capacity to maintain productivity while preserving the natural environment over the long run (Guo and Marchand 2012).

F. A. Sowunmi (✉)
Department of Agricultural Economics, University of Ibadan, Ibadan, Nigeria
e-mail: fa.sowunmi@ui.edu.ng

C. Atewamba and D. Yong Ngondjeb (eds.), *Inclusive Green Growth*, Advances in African Economic, Social and Political Development,
https://doi.org/10.1007/978-3-030-44180-7_7

Sand dredging is an activity of harvesting sand by excavation at least partly underwater (ask.com 2014). Sand and gravel are essential materials for construction, and high-quality material is often found in rivers and shallow seas (Whitehead 2007; Kondolf 1997). According to Kim et al. (2008a), sand is a critical input for construction in industrial as well as developing nations. Combined with aggregate and cement, the resulting concrete is used for buildings, roads and pipes, among many other uses. Dredge et al. (2009) posited that Lagos may be the place with the highest sand needs in Nigeria, if not in Africa, especially with the development of the World Bank-financed Lagos Mega City project, the Eko Atlantic City and innumerable residential and industrial estates, the proposed Eko Energy City, new roads, airports and seaports cropping up at the vast Lekki Peninsula, in Badagry and practically every conceivable part of the Lagos metropolis and suburban areas. Hence, the pressure on fishing sites for sand.

Among the various yardsticks for determining the quality of river's water for most aquatic habitat to thrive (dissolved oxygen, pH, water temperature, electrical conductivity, suspended solids among others), turbidity is a serious problem in sand dredging area. Turbidity refers to an optical property of liquids that measures the scattering and/or absorption of light due to material suspended in solution. Suspended material includes inorganic and organic solids as well as living organisms. Suspended solids have complex optical and physical properties that often make them hard to quantify (Moore 1977; Wilber 1983; Sigler 1990; Cone 1995). High turbidity is treated as an environmentally detrimental input. According to Dankwa et al. (2005), suspension of large quantities of solids in water column is one of the immediate physical effects resulting from sand dredging. Suspended solids may affect biological resources in various ways (Chansang 1988), including physical harm to fish, interference with self-purification of water by diminishing light penetration and, hence, photosynthesis reactions.

According to Balogun (2011), river sand though is vital for human use; the manner of mining has presented a multifaceted problem. Permit for dredging, preserving the environment, meeting the demand of sand for construction, as well as putting food on the table of the local fishermen are some of the nagging questions waiting to be addressed. Past efforts (Mafimisebi et al. 2013; Idowu 2010; Ekeke et al. 2008; Kim et al. 2008b; Anyanwu et al. 2009; Ogunniyi et al. 2012; Sesabo and Tol 2007) failed to account for the effect of water quality as environmental factors on artisanal fishing. Lack of data on environmental factors may be attributed for this. According to Organization of Economic Cooperation and Development report (1997), the supply of quantitative information about agro-environmental linkages is inadequate. Without such information, governments and other users cannot adequately identify, prioritize and measure the environmental impacts associated with agriculture, which makes it difficult to improve the targeting of agricultural and environmental programs and to monitor and assess policies. Currently, world order advocates for environmentally efficient economy that brings about sustainability of natural resources. The role of artisanal fishing as a means of livelihood and source of animal protein cannot be overemphasized. With the increasing consciousness about the environmental problems caused by sand dredging, the environmental performance of artisanal fishing has become increasingly important. With the increasing interest in environmental sustainability, another aspect of efficiency, environmental efficiency, has emerged. Environmental efficiency is defined as the ratio of minimum feasible to observed use of an environmentally detrimental input, conditional on observed levels of the desirable output and the conventional inputs (Reinhard 1999).

While there are substantial literatures (Mafimisebi et al. 2013; Idowu 2010; Ekeke et al. 2008; Kim et al. 2008b; Anyanwu et al. 2009; Ogunniyi et al. 2012; Sesabo and Tol 2007) on the economy of artisanal fishing (production, efficiency, cost and return), biological and chemical effects of sand dredging on aquatic habitat (Igben 2014; Ashraf et al. 2011; Fischer and Paukert 2009; Collins 1995; Burczynski 1991; Hay and McKinnell 2002; Anchor Environmental 2003); the same cannot be said on the effect of sand dredging on the economy of artisanal fishing as well as how this important agricultural subsector has been fairing in terms of input utilization bearing in mind the impact of environmentally detrimental input like turbidity. The study is set out to fill this gap in the literature and also to identify the variations in inputs associated to this environmental factor. Generally, this study contributes positively to the ongoing environmental sustainability of natural resources endowments in Africa.

7.1.1 Problem statement

Dredging activities is one of the ways through which aquatic habitat is disturbed. Rivers and coast (2010) stated that the economic consequences of sand dredging may include declines in fishery species populations and catch, impacts of increased turbidity or toxin release on aquaculture activities and increased shoreline erosion.

Increasing demand for sand for construction purpose and the supply gap created by dredging on land has made river/sea sand dredging a major threat to aquatic habitat. According to Ramilan et al. (2011), until recently, analysis of the performance of the agricultural subsector ignored negative externalities. The current emphasis on environmental issues makes it pertinent for farmers to target improvements in both environmental performance and productivity. They submitted that measuring the environmental performance of farms and integrating this information into farm productivity calculations should assist in making informed policy decisions which promote sustainable development.

While there are studies on effect of sand dredging on artisanal fishing in Nigeria, none of these studies incorporate environmental factors in their analyses. Reinhard et al. (1999) identified dairy farms which were both technically and environmentally efficient by treating nitrogen surplus as an environmentally detrimental input. This study utilized this approach in artisanal fishing by considering turbidity in sand dredging and non-dredging sites as an environmentally detrimental input. The study provided answers to the following research questions: Does environmental factor (water turbidity) significantly influence the quantity of fish caught per day in the study area? What are the environmental efficiencies of artisanal fishermen in sand dredging and non-dredging sites? Does sand dredging significantly affect environmental efficiency of fishermen in the study area? What are the factors influencing environmental efficiencies in the study area? Is there significant variation in the quantity of fish caught per day by fishermen in the sand dredging and non-dredging areas?

7.1.2 Objective of the Study

The broad objective of the study is to examine the environmental and economic burden of sand dredging on artisanal fishing in Lagos state, Nigeria. The specific objectives are to:

1. analyze whether there are variations or not in output and kilometers covered (fish caught) among the artisanal fishermen in sand dredging and non-dredging sites in the study area;
2. determine the environmental efficiencies of artisanal fishing in sand dredging and non-dredging sites;
3. determine the factors influencing environmental efficiency in the study area;
4. determine the costs and returns of fishermen in non-dredging and dredging areas.

7.1.3 Research Hypotheses

1. H_o: There is no significant difference in the quantity of fish caught per day between fishermen in sand dredging and non-dredging sites.
2. H_o: There is no significant difference in the environmental efficiencies of the fishermen in non-dredging and dredging areas.
3. H_o: Environmental efficiency is not influenced by sand dredging.

7.2 Theoretical Framework and Literature Review

The study was based on the economic theories of a common-property resource and production efficiency. Economic theory of common-property resource states that the ownership of resources is based on descent rights and age-long socio-cultural values which confer equal rights on the member. The owners demonstrate strict compliance with the inheritance rules and practices, maintain exclusive rights over the resources and uphold the principle of inalienability so as to ensure ease of transferability to their heirs (Olomola 1998).

Common-property natural resources are free goods for the individuals in the community and scarce goods for society. Under unregulated private exploitation, they can yield no rent; that can be accomplished only by methods which make them private property or public (government) property, in either case subject to a unified directing power (Gordon 1954; Olomola 1993). Regardless of who is governing a common-property resource, it is subject to basic concepts of production theory. Apart from being subject to law of diminishing returns, other human activities such as overexploitation, sand dredging may hasten the rate in which fish production reaches third stage of production (fish output decreases at decreasing rate).

The conventional definition of efficiency can be traced to the work of Farrell (1957) where the efficiency of a farm is measured directly from observed data. It refers to "how well" or "how effectively" a decision-making unit combines inputs to produce an output. Efficiency consisted of both technical and allocative efficiencies. Technical efficiency focuses on output produced from a given bundle of inputs and technology, while allocative efficiency focuses on the ability and willingness of an economic unit to minimize costs of production for a given set of input prices through substitutions or reallocation of inputs (Graham 2004). More recently, a third type of efficiency, environmental efficiency, is being defined and measured as a result of the impact agriculture has on the environment. Environmental efficiency is the ratio of minimum feasible to the observed use of an environmental detrimental input (Reinhard et al. 1999). The study followed Reinhard et al. (1999) concept. Other studies in the past that utilized this concept include Ramilan et al. (2011), Skevas et al. (2012), Tamini et al. (2011), Guo and Marchand (2012), Van Meensel et al. (2010), Hoang and Nguyen (2013), Wossink and Denaux (2006), Asmild and Hougaard (2006), Kamande (2010), Kumar and Gupta (2004) and Wu (2009). The considered turbidity is the environmentally detrimental variable. Among all the indicators of water quality, turbidity was used because it is a major problem in sand dredging area. Most aquatic animals are sensitive to change in water turbidity. Also, SFA was adopted because necessary assumptions on the environmental detrimental variable can be tested among other advantages.

7.3 Methodology

The study utilized primary and secondary data. The primary data were collected in July 2014 from two local government areas (LGA) in Lagos state known for artisanal fishing and sand dredging; namely Ikorodu and Epe. The two local government areas have rivers that empty into Lagos lagoon. The map of the two contiguous LGAs is shown in Fig. 7.1. The primary data were collected using two-stage sampling technique (purposive and simple random). The fishing communities sampled in the two LGAs were Oreta, Majidun, Itoikin, Ofin, Bayeku, Ijede, Ejinrin, Elubo and Iponmi via Agura Gberigbe. A total of 450 questionnaires that addressed the objective of the study were administered while 332 were returned (see http://www.ifad.org/gender/tools/hfs/anthropometry/ant_3.htm for detailed of how 450

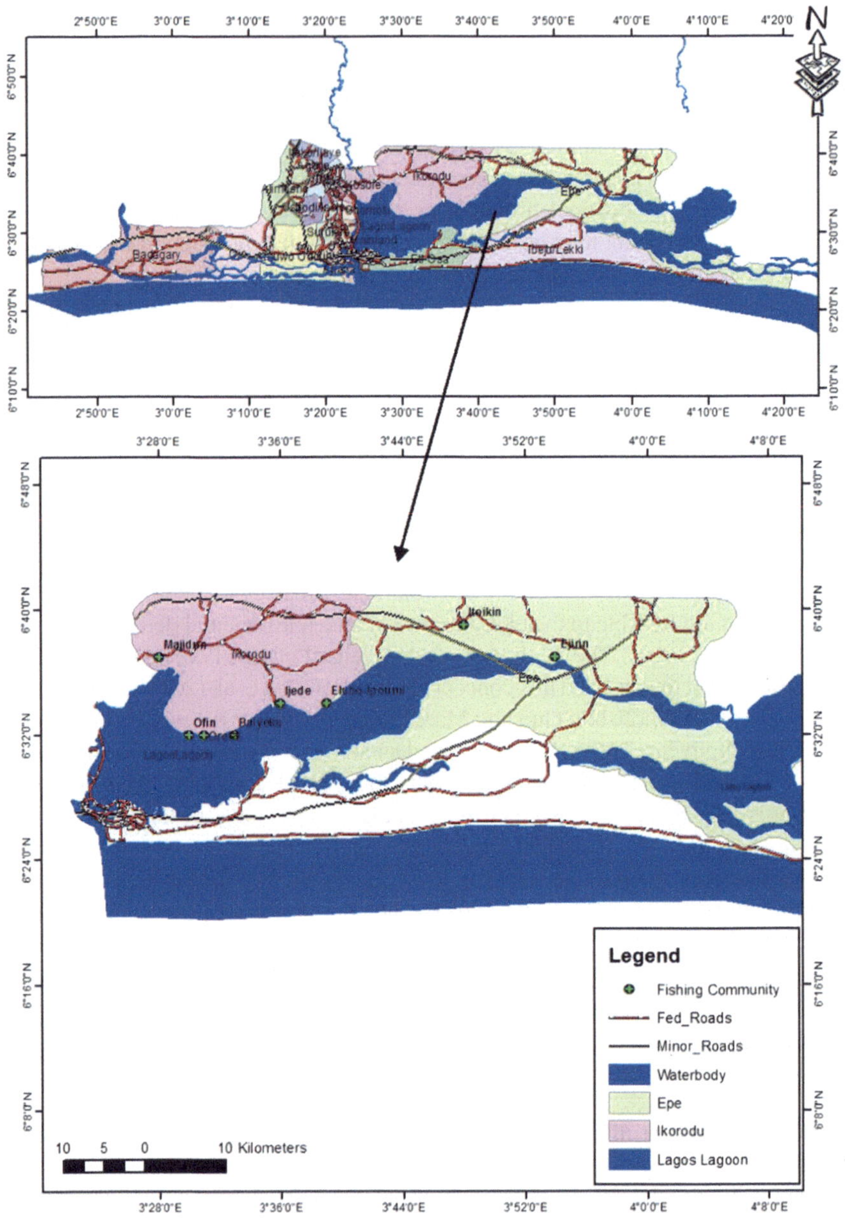

Fig. 7.1 Map of Ikorodu and Epe in Lagos state

respondents were arrived at). After processing, 314 of the questionnaires were appropriate for the analysis. Data collected from fishermen in sand dredging and non-dredging sites included socio-economic characteristics, the average quantity of fish caught per day (kg), price per kilogram of fish, average hours spent per fishing trip as well as the kilometers covered. Other data collected were the various cost items incurred in artisanal fishing (boat, net, rope, basket, paddle and other locally made traps used). Being cross-sectional data, it was possible to estimate the performance of each fisherman at a specific period in time, unlike panel data that estimate the time pattern of performance (Kumbhakar and Lovell 2000).

Secondary data on the water quality which included water turbidity of the fishing communities sampled were sourced from Odunaike et al. (2013), Nkwoji et al. (2010) and Idowu et al. (2014). The period between these studies and the current one was short; hence, it was expected that the differences in the water quality will be insignificant.

7.3.1 Analytical Framework

The environmental efficiency (EE) that this study estimated is different from the conventional technical efficiency (TE). Environmental efficiency is defined as the ratio of minimum feasible to observed use of an environmentally unfavorable input (water turbidity), based on observed levels of output and the traditional production inputs. The EE was calculated from TE with the classical approach of stochastic frontier approach (SFA). Determination of EE followed Reinhard (1999) two-step approach. EE was first calculated from TE using SFA. This was followed by regressing EE on variables that were not used in the estimation of TE. Following Reinhard et al. (2000), the non-radial environmental efficiency can formally be defined as:

$$\mathrm{EE}_i(x, y) = [\min\theta : F(X_i, \theta Z_i) \geq y_i] \tag{7.1}$$

where y_i is the quantity of fish caught per day, using X_i of the conventional inputs and Z_i the environmentally detrimental input. $F(.)$ is the best practice frontier with X and Z. Using Reinhard two-step approach, the environmental efficiency is given as:

$$\begin{aligned}\ln \mathrm{EE}_{it} = & \left[-\left(\beta_z + \sum_{j=1}^{m} \beta_{jz} \ln X_{ij,t} + \beta_{zz} \ln Z_{i,t} \right) \right. \\ & \left. \pm \left\{ \left(\beta_z + \sum_{j=1}^{m} \beta_{jz} \ln X_{ij,t} + \beta_{zz} \ln Z_{i,t} \right) - 2\beta_{zz} U_{i,t} \right\}^{0.5} \right] / \beta_{zz}\end{aligned} \tag{7.2}$$

As mentioned by Reinhard (1999), the output-oriented efficiency was estimated econometrically, whereas environmental efficiency was calculated from parameter estimates (β_z and β_{jz}) and the estimated error component ($U_{i,t}$). Since a technically efficient fisherman ($U_{i,t} = 0$) is necessarily environmentally efficient ($\ln \text{EE}_{it} = 0$). The sign $\sqrt[+]{.}$ is ideal.[1] The detailed derivation of EE is contained in Reinhard (1999).

7.3.2 *Empirical Model for Environmental Efficiency*

In the study, three traditional inputs (labor hour, depreciation value on fixed items used in production and bait used for setting traps for fish) and one environmental detrimental input (water turbidity) were identified for the production function. The tanslog production function is shown below:

$$\begin{aligned}
\text{Quantity}_{k,i,t} &= \beta_0 + \beta_1.\text{Labour hour}_{k,i,t} + \beta_2.\text{Capital}_{k,i,t} \\
&+ \beta_3.\text{Bait}_{k,i,t} + \beta_4.\text{Water Turbidity}_{k,i,t} \\
&+ \beta_5.\textit{Labour hour}^2_{k,i,t} + \beta_6.\text{Capital}^2_{k,i,t} + \beta_7.\text{Bait}^2_{k,i,t} \\
&+ \beta_8.\text{Water Turbidity}^2_{k,i,t} + \beta_9.\text{Labour hour}_{k,i,t}.\text{Capital}_{k,i,t} \\
&+ \beta_{10}.\text{Labour hour}_{k,i,t}.\text{Bait}_{k,i,t} \\
&+ \beta_{11}.\text{Labour hour}_{k,i,t}.\text{Water Turbidity}_{k,i,t} + \beta_{12}.\text{Capital}_{k,i,t}.\text{Bait}_{k,i,t} \\
&+ \beta_{13}.\text{Capital}_{k,i,t}.\text{Water Turbidity}_{k,i,t} \\
&+ \beta_{14}.\text{Bait}_{k,i,t}.\text{Water Turbidity}_{k,i,t} - U_{k,i,t} + V_{k,i,t} \ldots \qquad (7.3)
\end{aligned}$$

where the output is quantity of fish caught per day, three traditional inputs are the labor, capital and bait; and the environment detrimental input is water turbidity (from sand dredging and non-dredging areas). Note that all the variables are in natural log. The maximum likelihood estimator was used to estimate TE. The estimates of TE were substituted in Eq. (7.2) to obtain EE for each fisherman in the study area.

In the second stage of this study, factors influencing EE was determined as indicated in Eq. 7.4. Following Reinhard et al. (1999) approach, only variables that were not considered in stage one were incorporated into the model. A two-stage least squares (2SLS) method was chosen in order to determine the factors influencing EE in the study area. The choice of 2SLS is because many economic models have problems of endogeneity. As a general rule, when a variable is endogenous, it will be correlated with the disturbance term, hence violating the Gauss–Markov assumptions and making our ordinary least squares (OLS) estimates biased (Nagler 1999; Woodridge 2009). This problem often arises as a result of omitted variables or measurement error in variables associated with data collection. Instrumental variable (IV) estimator (two-stage least squares) method is designed to handle the consequences of omitted variables (or measurement error) unlike OLS (the test is based on

[1] The sign in front of the second term in (7.2) should be positive. Thus, if $U_{i,t} = 0$, then $\ln \text{EE}_{it} = 0$.

the hypothesis below). Also, 2SLS is preferred to the more conventional maximum likelihood (ML) method for structural equation model (SEM) because it does not require any distributional assumptions for RHS independent variables; they can be non-normal, binary among others (Oczkowski 2003).

$$H_0 = \mathrm{Cov}(y_2, u) = 0, \left(\hat{\beta}^{\mathrm{OLS}} \text{ and } \hat{\beta}^{\mathrm{2SLS}} \text{ are consistent but } \hat{\beta}^{\mathrm{OLS}} \text{ is more efficient}\right)$$

$$H_1 = \mathrm{Cov}(y_2, u) \neq 0, \left(\text{only } \hat{\beta}^{\mathrm{2SLS}} \text{ is consistent}\right)$$

Source Belloc (2009).

The instrumented variables (dredging status, distance covered by fishermen, experience in fishing and educational status of fishermen) are contained in Eq. (7.4).

However, it should be noted that other socio-economic variables are not considered among the variables influencing EE because problems of the detrimental input (water turbidity) go beyond what a fisherman's age, household size, marital status and gender can influence. It is the expertise of the fishermen and the condition of the environment that may influence EE.

$$\mathrm{EE}_c = \delta_0 + \delta_1 \ln \phi_1 + \delta_2 \ln \phi_2 + \delta_3 \ln \phi_3 + \delta_4 \ln \phi_4 + U \tag{7.4}$$

where

- EE_c Represents the environmental efficiency of each fisherman.
- ϕ_1 Represents status of fishing site (dredging site = 1, non-dredging site = 0)
- ϕ_2 Represents distance covered (km) while fishing
- ϕ_3 Represents experience in fishing (year)
- ϕ_4 Represents the educational status of respondent (educated = 1, no formal education = 0)
- δ Represents vectors of parameters to be estimated
- U Represents error component.

The instruments were age, gender, marital status, household size, other economic activities and price per kilogram of fish caught.

7.3.3 Cost and Return Analysis

Having determined the environmental efficiency of the fishermen in the study area, the cost and return analyses associated with fishermen in the non-dredging and sand dredging areas were estimated. Various cost items (variable and fixed) incurred by respondents in the two areas (dredging and non-dredging) were identified. The contribution of fixed items to fishing activity per day was determined using straight-line method of depreciation (see Eq. 7.5). Gross profit of fishermen in dredging and non-dredging sites was estimated using Eq. (7.6). The costs of the following fixed items, namely locally made canoe, paddle, net, baskets, trap and rope, were considered for each fisherman in the dredging and non-dredging areas. Bait was the only variable item used by the artisanal fishermen (Sowunmi et al. 2016).

$$\text{Depreciation Value}\,(N) = \frac{\text{Cost of the item}\,(N)}{\text{Economic life span (year)}} \tag{7.5}$$

$$\text{Average Gross Profit (AGP)} = \text{TR} - \text{TC} \tag{7.6}$$

where

AGP Average gross profit (N)
TR Total revenue (N)
TC Total cost (N)

$$\text{Total Revenue (TR)} = \text{PQ} \tag{7.7}$$

$$\text{Total Cost (TC)} = \text{Total Fixed Cost} + \text{Total Variable Cost} \tag{7.8}$$

where

P Price (N) per kg of fish
Q The quantity of fish sold.

7.4 Results and Discussion

The result reveals that 91.1% of the respondents are male. This confirms earlier studies that artisanal fishing is a male-dominated economic activity (Anyanwu et al. 2009; George et al. 2012; Solomon and Kerere 2013). Women are more involved in fish marketing and processing (Mafimisebi et al. 2013). Also majority (33.5%) of the respondents falls within the age bracket of 34–43 years (see Fig. 7.2) while the average age is 43.9 years. Most respondents in the non-dredging (31.4%) and dredging (42.2%) sites are within the age bracket of 44–53 and 34–43 years, respectively (see Fig. 7.2). These age brackets are the economically active ages of human population. However, the result shows that there is no significant difference between the average age of fishermen in non-dredging and dredging areas ($p > 0.05$).

Also, the result reveals that 70.6% of the total respondents are married while 64.2% and 77.3% are married among fishermen in non-dredging and dredging sites, respectively. From the result, there are more married fishermen in the dredging areas. Moreover, more than half of the respondents can read and write in the two sites. However, the level of literacy is marginally higher in the sand dredging areas (71%) compared with non-dredging areas (69%). The level of education among the literate fishermen and women ranges from Primary School Leaving Certificate to Ordinary

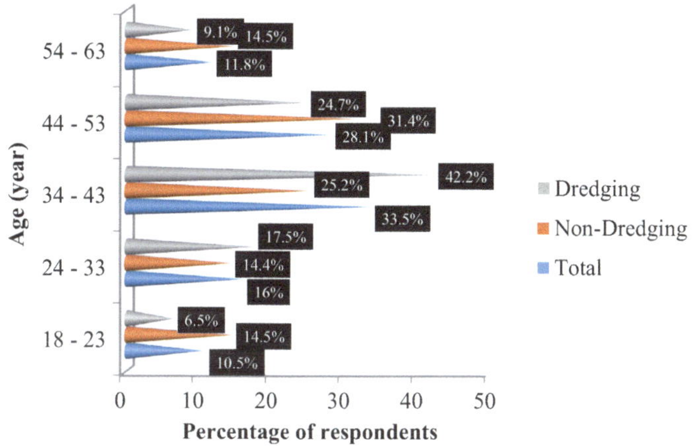

Fig. 7.2 Age distribution of respondents

National Diploma (OND). The average household size among the all respondents is 7.9 while the value is higher (8.6) among the respondents in non-dredging areas compared with sand dredging areas (7.2). There is also statistically significant difference in household size between respondents in non-dredging and dredging areas ($p < 0.01$) (see Table 7.1). The average household size in the two locations is more than the national average household size (5.2) and the average household size of Lagos state (3.8) (NBS 2012). This confirms high population among the fishing communities in the study area which encourages overexploitation of fish in order to meet the food needs of the household.

The high household size is typical of peasant farmers generally (Bester et al. 1998; Asogwa et al. 2014). They believe that large family assist in farming as shown in this result where 95.8% of the fishermen rely on family labor.

Table 7.1 shows that for all the respondents, the average quantity of fish caught per day is 8.05 kg with the standard deviation of approximately 2.9 kg. Moreover, the quantity of fish caught in the dredging area (8.9 kg) is significantly ($p < 0.01$) higher than that of the non-dredging area (7.3 kg). This may be attributed to the long distance traveled by fishermen whose communities are in the sand dredging vicinity in order to fish. Their survival as well as their family depends on the fish caught per day. However, the quantities of fish caught per day in relation to labor hours expended show that fishermen in non-dredging areas are more labor efficient. Specifically, fishermen in the non-dredging areas caught 3.45 kg of fish per labor hour compared with 1.4 kg per labor hour for fishermen in dredging areas. According to Tawari (2002), fishermen move in fulfillment of their occupation. They move in search of fish as dictated by the type of fish required and the movement of the tide which may be caused by sand dredging. The average distance covered by fishermen in the sand dredging area is significantly greater than that of fishermen in non-dredging areas (see Table 7.1). Traveling longer distance to places where the effect of dredging is minimal

Table 7.1 Descriptive statistics by fishing site

Output and input	Total (313)		Non-dredging site (161)		Dredging site (152)		Equalities
	Mean	SD	Mean	SD	Mean	SD	*p*-value
Quantity caught (g)	8054.3	2866.3	7250.2	2345.3	8959.4	3076.9	0.000
Labor hour (h)	4.2	2.4	2.1	1.1	6.4	2.0	0.000
Capital (₦)	255.2	152.4	281.9	174.4	227.6	120.3	0.001
Bait (N)	122.2	37.5	123.1	35.3	121.4	40.0	0.683
Water turbidity (NTU)	79.9	77.8	23.3	17.1	139.9	71.7	0.000
Household characteristics							
Experience in fishing (year)	16.5	6.7	16.5	7.1	16.6	6.2	0.888
Age (year)	43.9	11.5	44.3	12.6	43.6	10.1	0.597
Marital status	0.7	0.5	0.6	0.5	0.8	0.4	0.011
Household size	7.9	3.1	8.6	3.0	7.2	3.1	0.000
Educational status	0.7	0.5	0.69	0.46	0.71	0.46	0.759
Other economic activities	0.45	0.5	0.5	0.5	0.4	0.5	0.057
Area characteristics							
Distance covered (km)	6.8	6.7	2.4	2.3	11.4	6.7	0.000

Note For all tests of means, the null hypothesis is that there is no significant difference in the variables. The confidence level chosen is 5%

maybe their copy strategy. Also, the result shows that apart from traveling longer distance, fishermen whose community is located in the vicinity of sand dredging spent longer hours fishing ($p < 0.01$). Moreover, there was significant difference in the average turbidity in the sand dredging and non-dredging areas ($p < 0.01$). The high water turbidity areas have negative effect on fish activities.

The fixed items used in production by fishermen are unmotorized canoes, paddle, net, baskets, knife and plastic bowls. The contribution of each of these items to the fish caught per day was determined using straight-line method of depreciation and the number of days each fisherman work per week. The result shows that respondents (fishermen) at non-dredging areas spent ₦281.90 (average) on fixed capital per day while respondents having their community in the vicinity of sand dredging areas spent

₦227.62 per day. Fishermen in the non-dredging areas incurred higher amount on fixed capital because they spent more on local traps that are set in their surroundings ($p < 0.01$). The closeness of their bases to fishing arena afforded them the opportunity to check traps frequently unlike fishermen in non-dredging sites. Also, fishermen in the non-dredging areas invest more in new unmotorized boats.

Moreover, bait was the only variable item the fishermen used. This is common in environment where common-property theory is in place. Like forest product gatherers, artisanal fishermen continuously exploit the natural resources without adding anything in return. Increase in population increases the pressure on the exploitation of natural resources. Same reason may be adduced for fishermen engaging in other economic activities such as barbing, vulcanizing, tailoring, bricklaying, security among others to complement their little income from fishing. From the result, 50.3 and 39.6% of fishermen in non-dredging and dredging areas, respectively, are engaging in other economic activities.

7.4.1 *Determination of Environmental Efficiency (EE) from Stochastic Frontier Approach (SFA) Model*

In order to obtain the EE of each of the artisanal fishermen, the coefficients and the residuals obtained from stochastic production frontier model were substituted in Eq. (7.2). The stochastic production frontier model is shown in Table 7.2. The estimation of EE was preceded by ascertaining the theoretical consistency of the estimated efficiency model. The need for the marginal productivity of inputs to be positive as stipulated in microeconomic theory is germane.

Since translog functional form does not allow direct interpretation of the magnitude and the significance of the individual input elasticity as it is done in constant elasticity Cobb–Douglas case (Sharma and Leung 1999; Manchard and Guo 2014), the elasticity of each input (labor hour, capital, water turbidity and bait) was calculated at sample mean and median using formula[2] The result shows that all the elasticities are positive at sample mean and median except labor hours. From the table, quantity of fish caught depends more on labor hours and bait at sample mean. The negative coefficient for labor hours may be as a result of large populations of Nigerians taking up fishing as a means of livelihood which is reflected in the daily size of fish caught. Common property law gives every descendant of the fishing community the right to fish. Under common property law in artisanal fishing, marginal productivity of labor is not given due attention (Olomola 1993).

Furthermore, the returns to scale at sample mean (0.898) and sample median (0.803) are positive. This implies that artisanal fishing in the study area was in stage II of production where daily quantity of fish caught increases at decreasing rate. The

[2]Following Sharma and Leung (1999) and Manchard and Guo (2014), the elasticities of mean output with respect to the *j*th input variable are calculated at the mean/median of the log of the input variable and its second-order coefficients as follows.

Table 7.2 Stochastic production frontier model

Dependent variable: quantity of fish caught per day	Input elasticities			
	(1)	(2)	(3)	(4)
Variables	Coefficient estimate	Standard error	Sample mean	Sample median
Labor hour	−1.358**	0.674	0.371	0.156
Capital	0.044	0.513	−0.192	−0.161
Bait	0.083	0.665	0.370	0.386
Turbidity	−0.828*	0.427	0.349	0.422
Labor hour square	−1.137***	0.221		
Capital square	−0.032	0.033		
Bait square	0.035	0.041		
Turbidity square	0.094	0.069		
Labor hour * capital	0.230**	0.112		
Labor hour * bait	0.404*	0.230		
Labor hour * turbidity	0.385***	0.078		
Capital * bait	−0.068	0.163		
Capital * turbidity	0.036	0.036		
Bait * turbidity	−0.051	0.066		
Intercept	10.922***	3.256		
Sample size	313			
Log-likelihood	−52.642			
Sigma-squared (σ^2)	0.242*			
Gamma (γ)	0.848***			
Mu (μ)	−0.320*			
Sigma$_u$2	0.206			
Sigma$_v$2	0.037***			

Estimation method: maximum likelihood estimator. Note that *** means statistical significance at 1%, ** means statistical significance at 5% and * means statistical significance at 10%

significance of σ^2 ($p < 0.10$) shows the presence of inefficiency effects and random errors in artisanal fishing in the study area. The gamma value shows that 85% of the variability in the quantity of fish caught by fishermen is explained by their technical inefficiency.

Within the framework of translog stochastic production frontier, I predict TE and use the coefficients from the model to calculate EE. The average TE is 0.78 (see Table 7.3a). The score ranges from 0.37 to 0.96. Also, there is no significant difference in average score of TE of fishermen in the non-dredging and sand dredging areas ($p > 0.05$). The average TE score in this study is less than what Sharma and Leung (1999) and higher than what Sesabo and Tol (2007) obtained in similar studies

Table 7.3 (a) Descriptive statistics for technical and environmental efficiencies of respondents

Parameter	Technical efficiency			Environmental efficiency		
	Total	Non-dredging site	Dredging site	Total	Non-dredging site	Dredging site
Mean	0.776	0.780	0.772	0.305	0.497	0.102
Standard error	0.007	0.008	0.011	0.015	0.017	0.007
Median	0.811	0.811	0.813	0.239	0.461	0.061
Mode	0.788	0.788	0.803	0.723	0.723	0.051
Standard deviation	0.120	0.103	0.135	0.258	0.214	0.088
Skewness	−1.080	−1.344	−0.888	0.695	0.110	1.406
Minimum	0.371	0.413	0.371	0.001	0.060	0.001
Maximum	0.959	0.913	0.959	0.942	0.942	0.402

(b) Equality test for technical and environmental efficiencies of respondents

Parameter	Non-dredging site (161))		Dredging site (152)		Equalities	
	Mean	SD	Mean	SD	*z*-value	*p*-value
Technical efficiency	0.790	0.101	0.780	0.103	0.784	0.549
Environmental efficiency	0.497	0.214	0.102	0.088	3.564	0.000

Source Author's calculation

carried out in Hawaii and Tanzania, respectively. This may be attributed to differences in method of fishing, climate and the type of fishing input used.

However, the average EE in the study area is 0.31 while the average EEs of fishermen in non-dredging and dredging areas are 0.49 and 0.10, respectively. The result also affirms that there is significant difference in EE scores between the two areas ($p < 0.01$) (see Table 7.3a, b). The standard error of EE is greater than that of TE. The result suggests that most fishermen are not environmentally efficient. This implies that higher TE score does not guarantee high EE score.

The distribution of EE score in Fig. 7.3 shows that 65.1% of the fishermen in sand dredging area have EE score ranging from 0.001 to 0.099 while only 0.6% of the fishermen in non-dredging areas are in this category. Also, 29.8% and 5.3% of fishermen in the non-dredging and sand dredging areas, respectively, are within the environmental efficiency score of 0.31–0.50. Only 2.5% of the fishermen in non-dredging area have EE score of 0.91–0.99. Generally, the result shows a very low EE scores among fishermen in the dredging area. The solution for the control of the detrimental input (water turbidity) that will bring about increase in EE score is beyond the control of individual fisherman or group of fishermen.

This is unlike in crop farm where nitrogen or sulfur in the soil may be a detrimental input; all that is required is to control the amount of this nutrient in the soil through organic/inorganic fertilizer.

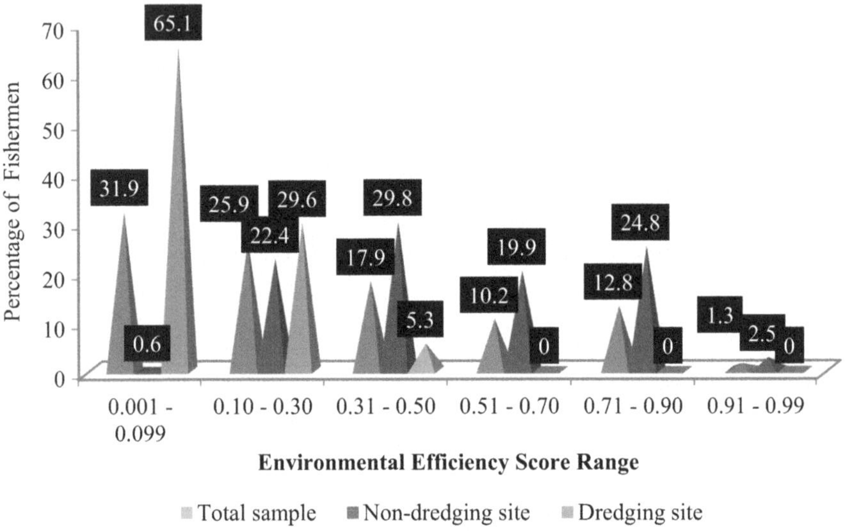

Fig. 7.3 Distribution of environmental efficiency among fishermen

7.4.2 Determinants of Environmental Efficiency

The Hausman's endogeneity test reveals that the parameter of 2SLS is consistent ($p < 0.01$), and hence, it is preferred to OLS. However, the results of the two analyses (OLS and 2SLS) are given in Table 7.4. The result (2SLS) shows that dredging status (dredging area = 1, non-dredging = 0), experience in fishing (year) and educational status of fishermen significantly influenced environmental efficiency. Specifically, the

Table 7.4 Determinants of environmental efficiency

Parameter	OLS estimator		2SLS estimator	
	Coefficient	SD error	Coefficient	SD error
Const	0.4159***	0.0326	0.1978*	0.1136
dreStat	−0.3672***	0.0248	−0.4109***	0.1337
Discov	−0.0035*	0.0019	−0.0097	0.0115
Expyr	0.0055***	0.0014	0.0141***	0.0034
Edstat	0.0042	0.0203	0.2044***	0.0786
Sample size	313		313	
Adjusted R^{-2}	0.6182		0.5419	
F statistics	127.2711		23.9126	
Hausman statistics			28.6691	
Hausman p-value			0.0000	

Source Author's computation

result reveals that sand dredging has a negative causal relationship with environmental efficiency. That is, sand dredging reduces environmental efficiency of fishermen. This result supports the earlier result that shows that environmental efficiency is generally low in sand dredging (0.10) areas and by extension affects the quantity of the fish caught by the fishermen in the dredging area.

However, as a coping strategy, the fishermen usually move far away from their base where dredging is taking place. The result also shows that experience in fishing ($p < 0.01$) and educational status ($p < 0.01$) are significant and positively related to environmental efficiency. That is, education and experience of fishermen increase environmental efficiency.

7.4.3 Breakdown of Costs and Returns of Fishermen in the Study Area

Table 7.4 shows the breakdown of the costs incurred per day as well as the average daily revenue of fishermen in the study area. The table reveals that fishermen in the dredging area incurred higher cost per day. This may be attributed to cost incurred on long distance traveled to catch fish in order to avoid dredging area. Fishermen in the non-dredging area incurred more costs on canoes, trap and net. Specifically, the higher cost on locally made trap may be due to their closeness to the fishing water which gives them opportunity to inspect the trap easily. This is unlike fishermen residing in the sand dredging areas that they have to travel a long distance to fish. Hence, few traps were set. However, the fishermen in the sand dredging area spent more on miscellaneous items (knife, plastic bowl among others).

Generally, average total cost per day for fishermen in the sand dredging areas is higher than that of non-dredging area. However, the average daily revenue from fish is higher among fishermen in the sand dredging areas (see Table 7.4). The distance traveled away from the dredging site for fishing and the smaller household size may be attributed to higher quantity of fish caught and the revenue accruable. The averages of the cost of distance covered are N110.58 and N23.28 for dredging and non-dredging sites, respectively. However, fishermen in non-dredging areas have higher returns from other economic activities they engaged in. From Table 7.4, the average gross profit per day is higher among the fishermen in the non-dredging areas. Moreover, the average per capita gross profit is higher among fishermen in the dredging areas. This may be attributed to smaller average household size among fishermen in the dredging areas. While overexploitation (large household size) alone may be the reason for small per capita gross profit among fishing households in the non-dredging area, it is the combination of overexploitation and dredging activities in the sand dredging area (Sowunmi et al. 2016b) (Table 7.5).

Overexploitation is the result of increasing population encouraged by common property theory and the unemployment situation in the country. The average per capita gross profit is far below the national per capita income of ₦1339.72 per day

Table 7.5 Breakdown of costs and return for fishermen

Fixed cost per day	Average amount (₦)		
	Total sample site (313)	Non-dredging site (161)	Dredging site (152)
Canoe	152.34	156.46	148.22
Paddle	5.21	5.52	4.84
Net	81.48	90.32	72.02
Basket	4.57	4.21	4.85
Trap	5.68	7.26	4.1
Rope	6.68	6.93	6.58
Miscellaneous (knife, plastic bowl/bucket, etc.)	19.11	11.2	16.99
Cost of distance covered (kilometers)	65.96	23.28	110.58
Average total fixed cost	341.03	305.18	368.18
Variable cost			
Bait	122.24	123.24	122.81
Average total cost per day	377.44	405.14	350.41
Average revenue per day from fish	4309.94	3952.88	4695.08
Revenue accruable for extra hours (other economic activities)	660	1290.00	–
Average gross profit per day	4628.91	4937.70	4326.90
Average household size	7.9	8.6	7.2
Average per capita gross profit	585.94	574.15	600.96

Source Author's computation

(The Guardian Global Development Professional Network 2014)[3]. Hence, the need for fishermen to engage in other economic activities to augement what is obtained from fishing.

[3]Per capita income of $3000 (₦489,000 at $1 = N163). This is equivalent to ₦1339.72.

7.5 Conclusion and Recommendation

The study revealed that environmental factor (water turbidity) has a negative effect on the environmental efficiency of fishermen in the study area. The negative effect of sand dredging on environmental efficiency captured by water turbidity was more pronounced in the sand dredging fishing communities. Moreover, the large population in the fishing communities encourages overexploitation of fish as confirmed by the sizes of fish caught. As a coping strategy, fishermen residing in the dredging vicinity usually move far away from dredging site in order to fish. The negative effect of sand dredging did not manifest in environmental efficiency alone but also in the average gross profit of the fishermen. The fishermen in the non-dredging area incurred less cost and higher gross profit. The finding attributed the low per capita gross profit (lower than national per capita income) among the fishermen in non-dredging areas to mainly overexploitation (large household size) while dredging and overexploitation are the reasons for small per capita gross profit among fishermen in sand dredging areas.

However, while this study has been able to include environmental factor as production input; environmental inefficiency among the fishermen may not be attributed to just one environmental factor (water turbidity). One environmental factor was used due to data limitations.

In order to bring about harmonization between sustainability of natural resources and human survival, the following are recommended based on the findings of the study:

In order to bring about harmonization between sustainability of natural resources and human survival, the following are recommended:

- The activities of the dredging firms should be properly monitored by government by ensuring that dredging license is restricted to non-fishing communities. This will not only reduce the environmental degradation but also help to sustain the aquatic habitat.
- The need for the artisanal fishermen to be properly integrated into government's Agricultural Transformation Agenda (ATA) is imperative not only by assisting them with fishing inputs but also by encouraging fishermen to rear fish or engage in other agricultural ventures suitable to their environment (example is cultivation of swampy rice). This will go a long way in addressing overexploitation and also help to improve their well-being.
- The family planning unit of the ministry of health Lagos state should ensure that their activity in the fishing communities is intensified. This will help not only to reduce the household size but also reduces overexploitation of aquatic animals.
- The fishery department of the ministry of agriculture should ensure that fishing nets used by fishermen are of sizable mesh that will allow small fish to grow to the table size before being caught. This can be achieved through periodic inspection; and any fisherman culpable should pay fine.

References

Adesina EM, Ayanda JO (2005) Fishery management plans and the development of artisanal fisheries in Nigeria: a review. In: 19th annual conference of the Fisheries Society of Nigeria (FISON), 29 Nov–03 Dec 2004, Ilorin, Nigeria, pp. 219–224. Availabe on http://aquaticcommons.org/3988/

Adesoji SA, Kerere FO (2013) Assessment of the knowledge level of fishers and fish farmers in Lagos State, Nigeria. Int J Knowl Innov Entrepr 1(1–2):41–56

Anchor Environmental LLC (2003) Literature review of effects of resuspended sediments due to dredging operations. iii + 87 pp + App. A–C. June 2003. Prepared for Los Angeles Contaminated Sediments Task Force

Anyanwu DC, Mkpado M, Ohaka CC (2009) Economic analysis of artisanal fishing at River Niger. J Trop Agric Food Environ Ext 8(3):175–179

Ashraf MA, Maah MJ, Yusoff I, Wajid A, Mahmood K (2011) Sand mining effects, causes and concerns: a case study from Bestari Jaya, Selangor, Peninsular Malaysia. Sci Res Essays 6(6):1216–1231

Ask.com (2014) What is sand dredging? Available on www.ask.com

Asmild M, Hougaard JL (2006) Economic versus environmental improvement potentials of Danish pig farms. Agric Econ 35(2):171–181

Asogwa BC, Abu O, Ochoche GE (2014) Analysis of peasant farmers' access to agricultural credit in Benue State, Nigeria. Br J Econ Manag Trade 4(10):1525–1543

Bada AS (2005) Strategies for bridging the supply-demand gap in fish production in Nigeria. In: Proceedings of the 19th annual conference of the farm management association of Nigeria, pp 329–337

Balogun I (2011): Sand miners eroding fishermen's livelihood. Available on http://www.vanguardngr.com/2011/04/sand-miners-eroding-fishermen%E2%80%99s-livelihood/#sthash.jEomIY2b.dpuf

Belloc M (2009) Econometrics: specification tests. Available on www.w3.uniroma1.it/belloc

Bester BJ, Belete A, Kelembe SG (1998) Assessing peasant farmers' risk attitudes and factors that influence decisions on risky alternatives: a case study in the Border Region of South Africa. Agrekon 37(3):268–282

Brian C (nd) Riverine gravel mining in Washington State, physical effects with implications for Salmonid Habitat, and summary of Government regulations. Report prepared for: US Environmental Protection Agency 1200 Sixth Avenue Seattle, WA 98101 Completion report for Grant X-000694-01-0

Burczynski J (1991) Hydroacoustic survey of fish distribution and reaction to dredge disposal activities in San Francisco Bay. Final report. [vi] + 19 pp. + app. A–B. BioSonics, Inc., Seattle, WA. Feb 21 1991

Chansang H (1988) Coastal Tin mining and marine pollution in Thailand. Ambio 17:223–228

Collins B (1995) Riverine gravel mining in Washington state, overview of effects on salmonid habitat, and a summary of government regulations. Unpublished report to the US EPA, Seattle

Committee on Ecosystem Effects of Fishing (2002) Effects of trawling and dredging on seafloor Habitat. Retrieved from http://www.nap.edu/catalog.php?record_id=10323

Cone LD (1995) A review of the potential impacts of mechanical harvesting on subtidal and intertidal shellfish resources. Marine Resources Research Institute, SC Department of Natural Resources, Charleston, SC

Dankwa HR, Biney CA, deGraft-Johnson KAA (2005) Impact of mining operations on the ecology of river Offin in Ghana. West Afr J Appl Ecol 7:17–30

Dredge, Drill and Haul (2009) The Changing dredging scene. Available on http://www.ddhmag.com/2ndqtr09editorial.htm

Dredging (2014) Wikipedia, the free encyclopedia

Economic Analysis of Environmental Issues (2005) Common-property resources and public goods. Available on www.ase.tufts.edu/gdae/publications/textbooks/jhtext_ch04.pdf

Ekeke BA, Davies OA, Alfred-Oekiya JF (2008) Sand dredging on the fish catch in Bonny River Estuary, Nigeria. Env Res J 2(6):299–305

Farrell MJ (1957) The measurement of productive efficiency. J R Stat Soc A CXX:253–290

Fischer JR, Paukert CP (2009) Effects of sampling effort, assemblage similarity, and habitat heterogeneity on estimates of species richness and relative abundance of stream fishes. Can J Fish Aquat Sci 66(2):277–290

Fish for All Summit (2005) Presentation of Nigeria's Fisheries resources, Abuja, Nigeria 22–25 Aug 2005

George F, Odulate D, Idowu A (2012) Climate change effects and adaptation strategies in a Nigerian coastal Agro-ecological zone. A paper presented at 2nd international symposium on effects of climate change on the world's oceans. Yeosu, South Korea, 13–20 May, 2012

Gordon HS (1954) The economic theory of a common-property resource: the fishery. J Polit Econ 62:124–142

Graham M (2004) Environmental efficiency: meaning and measurement and application to Australian. Available on www.Ageconsearch.Umn.Edu/Bitstream/58450/2/2004_Graham.Pdf

Guo H, Marchand S (2012) The environmental efficiency of organic farming in developing countries: a case study from China. Serie Etudes et Documents Du Cerdi (CERDI). Études et Documents No. 38. Nov 2012

Hay DE, McKinnell SM (2002) Tagging along: association among individual Pacific herring (*Clupea pallasi*) revealed by tagging. Can J Fish Aquac Sci 59:1960–1968

Hoang VH, Nguyen PM (2013) Environmental awareness and attitude of Vietnamese consumers towards green purchasing. VNU J Econ Bus 29(2):129–141

Idowu AA (2010) Effect of some socio-economic activities on fish diversity of lagoon systems in Ogun waterside Local Government of Ogun State, Nigeria. Int J Biol Chem Sci 4(1):88–99

Idowu ET, Amaeze NH, Adie PI, Otubanjo OA (2014) Heavy metal bioaccumulation and biomarkers of oxidative stress in the wild African tiger frog, *Hoplobatrachus occipitalis*. Afr J Env Sci Technol 8(1):6–15

Igben JL (2014) Inland sand dredging and economic activities in the Niger Delta, Nigeria. Sacha J Env Stud 4(1):1–7

International Fund for Agricultural Development (IFAD). Calculating the sample size. Available on http://www.ifad.org/gender/tools/hfs/anthropometry/ant_3.htm

Kamande M (2010) Technical and environmental efficiency of Kenya's manufacturing sector: a stochastic frontier analysis. Available on https://www.csae.ox.ac.uk/conferences/2010-edia/papers/197-kamande.pdf

Kim YT, Kim HJ, Lee GH (2008a) Mechanical behaviour of lightweight soil reinforce with waste fishing net. Geotext Geomembr 26:512–518

Kim TG, Grigalunasa TA, Hanb K-N (2008b) The economic costs to fisheries because of marine sand mining in Ongjin Korea: Concepts, methods, and illustrative results. Ecol Econ 6(5):498–507

Kondolf GM (1997) Hungry water: effects of dams and gravel mining on river channels. Environ Manage 21:533–551

Kumar S, Gupta S (2004) Resource use efficiency of US electricity generating plants during the SO_2 trading regime: a distance function approach. National Institute of Public Finance and Policy Working Paper No. 17/04. Available at SSRN: http://ssrn.com/abstract=552582

Kumbhakar S, Lovell C (2000) Stochastic frontier analysis. Cambridge

Mafimisebi TE, Ikuemonisan ES, Mafimisebi OE (2013) Comparative profitability of women dominated Fish-based livelihood activities in Southwest, Nigeria. Available on https://genderaquafish.files.wordpress.com/2013/04/ppt_01.pdf

Marchand S, Guo H (2014) The environmental efficiency of non-certified organic farming in China: a case study of paddy rice production. China Econ Rev 31:201–216

Moore PG (1977) Inorganic particulate suspensions in the sea and their effects on marine animals. Oceanogr Mar Biol Ann Rev 15:225–363

Nagler J (1999) Notes on simultaneous equations and two stage least squares estimates. Available on http://www.nyu.edu/classes/nagler/quant2/notes/2slsnotes_oh.pdf

National Bureau of Statistics (2012) Social statistics in Nigeria. Available on www.nigerianstat.gov.ng/pages/download/168
Nkwoji JA, Onyema IC, Igbo JK (2010) Wet season spatial occurrence of phytoplankton and Zooplankton in Lagos Lagoon, Nigeria. Sci World J 5(2):1–8
Nnodim O (2014) Nigeria spends N125bn annually on fish importation. In: The punch. Retrieved from http://www.punchng.com/news/nigeria-spends-n125bn-on-fish-import-annually-fg/
Oczkowski E (2003) Two-stage least squares (2SLS) and structural equation models (SEM). Available on http://csusap.csu.edu.au/~eoczkows/home.htm
Odunaike K, Laoye JA, Adeniyi TLP, Akinyemi LP, Adeyeloja A (2013) Physico-chemical characterization of water samples at Egbin thermal station in Ijede and its environs in Lagos state of Nigeria. J Env Earth Sci 3(14)
Ogunniyi LT, Ajao AO, Sanusi WA (2012) Technical efficiency in Artisanal Fisheries (AF) in Badagry local government area of Lagos State, Nigeria. Cont J Fish Aquat Sci 6(1):1–8
Olomola AS (1993) The traditional approach towards sustainable management of common property fishery resources in Nigeria. Marine Anthropology Studies, vol 6 (l/2):92–109
Olomola SA (1998) Sources and resolution of conflicts in Nigerian artisanal fisheries. Soc Nat Res 11:121–135
Organization for Economic CoOperation and Development (1997) Environmental indicators for agriculture: concepts and framework. Available on http://www.oecd.org/greengrowth/sustainable-agriculture/40680795.pdf
Pavelescu F (2011) Some aspects of the Translog production function estimation. Institute of National Economy, Romanian Academy. Available on www.revecon.ro/articles/2011-1/2011-1-8.pdf
Ramilan T, Scrimgeour F, Marsh D (2011) Analysis of environmental and economic efficiency using a farm population micro-simulation model. Math Comput Simul 81(7):1344–1352
Reinhard S (1999) Econometric analysis of economic and environmental efficiency of Dutch dairy farms. Ph.D. thesis, Wageningen Agricultural University
Reinhard S, Lovell CAK, Thijssen G (1999) Econometric estimation of technical and environmental efficiency: an application to Dutch dairy farms. Am J Agr Econ 81(1):44–60
Reinhard S, Knox Lovell CA, Thijssen GJ (2000) Environmental efficiency with multiple environmentally detrimental variables; estimated with SFA and DEA. Eur J Oper Res 121:287–303
Rivers and Coast (2010) Shallow water dredging. Cent Coast Resour Manage 5(1):1–8
Sesabo JK, Tol RSJ (2007) Technical efficiency of small-scale fishing households in Tanzanian coastal villages: an empirical analysis. Afr J Aquat Sci 32(1):51–61
Sharma KR, Leung PS (1999) Technical efficiency of the longline fishery in Hawaii: an application of stochastic production frontier. Mar Resour Econ 13:259–274
Sigler JW (1990) Effects of chronic turbidity on anadromous salmonids: recent studies and assessment techniques perspectives. In: Simenstad CA (ed) Effects of dredging on anadromous Pacific Coast fishes: workshop proceedings. University of Washington Sea Grant Program, Seattle, pp 26–37
Skevas T, Oude Lansink A, Stefanou SE (2012) Measuring technical efficiency in the presence of pesticide spillovers and production uncertainty: the case of Dutch arable farms. Eur J Oper Res 223:550–559
Sowunmi FA, Haogarh JN, Omigie CO, Idiaye CO (2016) Economic Burden of Sand dredging on artisanal fishing in Lagos State, Nigeria. Poult Fish Wildl Sci 4(2):1–8. https://doi.org/10.4172/2375-446x.1000171
Sustainable Fisheries Livelihoods Programme (SFLP) (2002) Committee on fisheries research joint working party on poverty in small-scale fisheries: promoting the contribution of the sustainable livelihoods approach and the code of conduct for responsible fisheries in poverty alleviation, vol 678. FAO fisheries report, Food and Agriculture Organization, Rome, 10–12 Apr 2002
Tamini LD, Larue B, West G (2011) Technical and environmental efficiencies and best management practices in agriculture. Appl Econ 44(13):1659–1672

Tawari F (2002) Dissemination of research findings on Nomadic Education in Nigeria (The Migrant Fishermen Education Experience): Issues and Directions; at the International Conference organised by International Extension College (IEC) Cambridge and Sponsored by the Department for International Development (DFID) at Rock View Hotel Abuja-Nigeria, 16th–19th Jan 2002
The Guardian Global Development Professional Network (2014) Nigeria becomes Africa's largest economy. Available on http://www.theguardian.com/globaldevelopment/datablog/2014/apr/07/nigeria-becomes-africa-largest-economy-get-data
Van Meensel J, Van Lauwers L, Huylenbroeck G, Van Passel S (2010) Comparing frontier methods for economic–environmental trade-off analysis. Eur J Oper Res 207:1027–1040
Whitehead GJ (2007) Land and environment: sand mining. Available on http://localhistory.kingston.vic.gov.au/htu
Wikipedia, Free encyclopedia (2014) Artisan fishing. Retrieved from en.wikipedia.org/wiki/Artisan_fishing
Wilber CG (1983) Turbidity in the aquatic environment: an environmental factor in fresh and oceanic waters. Charles Thomas Publications, Springfield
Wooldridge JM (2009) Introductory econometrics, 4th edn. South Western-Cengage
Wossink A, Denaux ZS (2006) Environmental and cost efficiency of pesticide use in transgenic and conventional cotton production. Agric Sys 90(1–3):312–328
Wu Y (2009) Environmental efficiency and its determinants in China's regional economies. Available on https://www.web.uwa.edu.au/__data/assets/pdf_file/0003/98661/07_21_Wu.pdf

Chapter 8
An Assessment of the Impact of Grazing Livestock on Cereal and Tuber Crops Production in Abuja, Nigeria

Julius Ajah

8.1 Introduction

Nigeria is one of the countries in sub-Saharan Africa (SSA) that depends on agriculture (Ijirshar 2015) to feed her population of about 163 million (British Council Nigeria 2012). Agriculture is an important source of income and a livelihood option for many households. The major actors in the agricultural value chain are small-scale farmers who constitute substantial proportion of the population and are scattered all over the habitable regions of the country. They constitute about 70% of those who are in agricultural production (Idowu et al. 2011), and they produce about 90% of the total agricultural output in the country (IFAD 2007). The small-scale farmers are diverse and comprise of both male and female farmers in different age categories. They have different abilities depending on circumstances that lead them into agriculture. Their significant contribution to agriculture and sustainable development is one of the reasons they are placed at the center stage of research and development programmes by the Nigerian agricultural policy-makers and planners. Because of their role in the economy, the small-scale farmers are key players in agricultural development as they dominate and determine the total output and productivity of the country. They have the freedom and privileges to produce whatever crop or livestock they desire except those banned by law. They can, as a matter of occupation or interest, embark on crop and/or livestock production. Some engage in agriculture as hobby, while others regard it as their main occupation or business. These farmers are given recognition to the extent that almost all international and national programmes aimed at promoting agriculture and rural development are designed to capture their interest and motivate them to improve their scale of operations and overall productivity. The significant role that small-scale farmers play in Nigeria's agriculture and economy shows that

J. Ajah (✉)
Department of Agricultural Economics and Extension, Faculty of Agriculture, University of Abuja, Abuja, Nigeria
e-mail: juliusajah2@yahoo.com

C. Atewamba and D. Yong Ngondjeb (eds.), *Inclusive Green Growth*, Advances in African Economic, Social and Political Development,
https://doi.org/10.1007/978-3-030-44180-7_8

they are indispensable (particularly in the short run) hence, they deserve continued and unconditional support.

The importance of small-scale farmers in agricultural development cannot be overemphasized. While every effort is being made to encourage and improve their productivity in both livestock and crops in the country, one of the greatest challenges being experienced is the destruction of crops by grazing livestock (Gefu and Gills 1990; Adebayo and Olaniyi 2008; Ofuoku and Isife 2009; Adogi 2013; Okoli and Atelhe 2014). The grazing livestock (cattle, goat, sheep) can encroach into crop fields and destroy crops at any developmental stage—they can eat the stems, seeds, flowers, leaves, tubers or the roots of the crops. However, not all livestock production systems result in the destruction of crops. The destruction depends on whether it is intensive, semi-intensive or extensive system that is adopted by the farmers.

Extensive system of livestock production refers to a system whereby the animals are allowed to roam and look for food unrestricted (Ezeibe 2010). On the other hand, semi-intensive system allows for good control of feeding, management and the animals are more protected. Intensive system is a total confinement of the livestock (Ezeibe 2010; Devandra and Fuller 1989). In some instances, cattle, goats and sheep can be reared by nomadic herdsmen (Blench 2010) who guide the animals as they graze on pasture, while in some, the livestock move freely with little or no form of restriction. Of all the systems, the predominant among the small-scale farmers in Nigeria is the extensive or free-range system (Ezeibe 2010; Nweze et al. 2003; Ovwigho et al. 2009). This poses a serious challenge, as the livestock graze without proper guidance, with crop fields being an easy target. Unfortunately, some of the livestock can destroy crops even when they are guided by nomadic herdsmen. The problem is compounded because most crop fields are not fenced and this results in extensive destruction of crops. The inability of crop farmers to fence their farms coupled with the cultivation of small portions of land that are characterized by low yields per hectare (Kolawale and Ojo 2007) makes them highly vulnerable.

Ironically, while the livestock farmers have nothing to lose in some cases, the crop farmers are at the receiving end as they suffer crop failure and/or low productivity (Ofuoku and Isife 2009). Contrastingly, in advanced economies with organized farms, the symbiotic relationship between crop farmers and livestock producers is very encouraging and maintained to the extent that both enjoy mutual benefits. For those who practice extensive system of livestock production, there will be proper grazing routes that make it very difficult for grazing livestock to damage crops on farmlands. But in Nigeria, the adoption of extensive system by livestock farmers coupled with the absence and/or non-adherence to grazing routes has created problems between crop and livestock farmers. The resultant effect of this scenario is crop damages (and vice versa) with multiple socioeconomic consequences. In other words, the predominance of extensive system of livestock production coupled with the absence and/or non-adherence to grazing routes has become major threats to crop production, and this is contrary to the concept of sustainable agricultural development and green business.

In this context, another worrisome issue is the difficulty in computing the extent and quantity of crops damaged annually in Nigeria. This is partly due to lack of farm records reflecting such crop damages. The extent of damage can only be imagined by looking at the stock populations for cattle, goats, sheep and poultry birds which were estimated by the National Bureau of Statistics (NBS 2012) as 16,722,170, 57,937,176, 36,372,233 and 201,928,991, respectively. Though these stock figures do not give specific populations of livestock under different grazing systems, they help to give an idea of the number of livestock that is reared in Nigeria.

Apart from crop damages on farmland, several other studies have shown that grazing livestock have other negative consequences on the environment. First, they destroy young trees, herbaceous plants and grasses thereby causing losses in biodiversity (FAO 2002; Marty 2005; Enete and Amusa 2010; Ayanda 2013). The empirical evidence shows that grazing livestock affects biomass (Clary and Kinney 2002), decreases species richness and composition (Champion et al. 2001; Touzard and Clement 2001; Greviliot and Muller 2002) and kills birds by trampling on their nests (Popotnik and Giuliano 2000). The impact on biodiversity is enormous, for instance, Khanal (2009) estimated that 25–42% of species habitat have been lost, with effects on both food and non-food crops. This is in line with the argument by FAO (2002) which indicated that the extent and method of agriculture, forestry and fishing are leading causes of loss of the world's biodiversity. FAO further noted that over-stocking land with grazing animals causes desertification and loss of plant biodiversity. Unfortunately, the livestock farmers may not be too concerned with the implications of loss of biodiversity. They are more likely to be conscious of the pastures, water availability, disease-free and a conducive environment for their herds.

Second, studies also showed that grazing livestock affect soil, output and productivity (Ofuoku and Isife 2009; Gefu and Gills 1990). A study by McDowell et al. (2005) showed that cattle grazing pastures on sloping land contribute to run off, soil erosion and loss of nutrients. Similar studies by Drewry et al. (2008), Greenwood and McKenzie (2001) showed that treading on soil by livestock may lead to increased soil strength and bulk density, reduced soil porosity and hydraulic conductivity. Bell and Gilkes (2010) further demonstrated that rainfall infiltration was increasingly affected by grazing livestock with up to 10% reduction in grain yield. They attributed the effects to the reduction in soil surface conductivity due to compaction of soil surface by grazing livestock. A similar report by Radford et al. (2008) also indicated that there was a significant reduction in subsequent crop growth and production after treading by livestock. The study added that grazing livestock increased soil compaction resulting in restricted movement of water into and through soil profile. In one way or the other, all these studies concur with the fact that inappropriate farming methods, deforestation and overgrazing land result in soil depletion and erosion (FAO 2002; Aruleba and Ajayi 2011). This calls for an elaborate approach to properly manage livestock production so as to minimize its negative impact on the environment.

Third, the most critical issue identified by numerous scholars (Nyong and Fiki 2005; Tonah 2006; Adebayo and Olaniyi 2008; Blench 2010; Abbass 2012; Adogi 2013; Ayanda 2013; Audu 2013, 2014; Okoli and Atelhe 2014) is the conflict arising from the destruction of crops by grazing livestock which threatens the unity and peaceful coexistence in various communities across the country. Other studies also indicated that grazing livestock contaminate water sources in rural communities in Nigeria (FAO 2002; Ofuoku and Isife 2009). It is noteworthy to point out that the impact of grazing livestock on water points is not only in Nigeria but even in countries such as the USA. For instance, USDI (1994) showed that about 80% of the damage incurred by streams and riparian systems in the arid environments of the USA arose from grazing livestock. According to Belsky et al. (1999), stream and riparian damage resulting from livestock grazing occurred as a result of alterations in watershed hydrology, soil compaction and erosion, riparian vegetation destruction and water quality impairments. It also leads to indiscriminate bush burning that negatively affects the environment (Ofuoku and Isife 2009).

From the foregoing, there is no doubt that scholars have tried to reveal the impact of grazing livestock on the environment and its socioeconomic consequences, but none specifically addressed how small-scale farmers perceived the impact of grazing livestock on crop production. Most of the studies were focused mainly on the impact of cattle, goats and sheep that are reared by pastoralists with no emphasis on domestic fowls which also have the potential to destroy crops and induce conflict in the society. Hence, the need for this study to assess the impact of grazing livestock on cereal and tuber crops production in Abuja, Nigeria.

8.1.1 Objectives of the Study

The broad objective of the study is to assess the impact of grazing livestock on cereal and tuber crops in Abuja, Nigeria. Specific objectives are to:

1. assess farmers' perception of impact of grazing livestock on some cereal and tuber crops,
2. determine the cereal and tuber crops that are mostly affected by grazing livestock,
3. determine the livestock (cattle, goat/sheep, domestic fowls) that farmers perceived as most destructive during grazing,
4. determine if there are locational differences in impact of grazing livestock on cereal and tuber crops production,
5. estimate the costs of damage by grazing livestock on cereal and tuber crops, and

8.1.2 Research Questions

The following questions were addressed in the study:

1. How do small-scale farmers perceive the impact of grazing livestock on the production of cereal and tuber crops?
2. Which livestock impacts more on the cereal and tuber crops?
3. Which of the cereal and tuber crops is most affected by grazing livestock?

4. Are there locational differences on the impact of livestock on cereal and tuber crops?
5. Does the impact of grazing livestock depend on the type of crop cultivated as well as the type of livestock reared in a given location?
6. What are the estimated costs of crops destroyed by grazing livestock?

8.1.3 Significance of the Study

Agriculture, be it in the form of crop or livestock production, is one of the major components of green growth business in rural communities in Nigeria and other countries. For example, in Ethiopia, agriculture is a major component of green business and they are targeting increased productivity, enhanced food security, job creation and stability of export income (Federal Democratic Republic of Ethiopia 2011). This implies that in Ethiopia, any factor that affects agriculture has affected a major component of green business. In many African countries, agricultural production is one of the major sources of livelihood in rural communities, and it can be tailored to be an important vehicle towards achieving a green economy.

UNU-INRA is interested in identifying factors that affect the conservation of biodiversity in Africa and other parts of the world. In this regard, there is a need to determine if grazing livestock significantly affect crop production. This calls for serious attention because crops are major components of biodiversity in any environment. Any activity that affects crop production will directly or indirectly affect biodiversity conservation.

Every country is working toward preventing, controlling or eliminating factors that negatively affect their ecosystems and also factors that hinder them from achieving sustainable development. In order to design appropriate measures to achieve this, there is a need to have a proper understanding of such factors. This requires proper research that integrates both qualitative and quantitative data so as to have an in-depth understanding of some of the factors that affect the ecosystem and sustainable development, hence the importance of this study. The findings of this study will help in the formulation of policies, strategies and programmes that will help in tackling the associated challenges and contribute to the overall green economy objective in Nigeria and other countries facing similar challenges. The outcome will have many benefits including poverty alleviation effects. As noted by Smulders and Withagen (2012), measures that improve natural resource management, prevent pollution and reduce vulnerability to environmental risks will tend to benefit the poor.

Finally, the study will help (especially) agricultural policy-makers and planners in Nigeria and other countries to appreciate the challenges confronting crop production in an environment where there is absence and/or non-adherence to grazing routes. That is, rather than realizing mutual benefits between crop and livestock production, they become significant constraints for each other.

8.2 Methodology

The study area is Abuja, Nigeria. It is located between latitudes 8° 25′ and 9° 25′ North of the Equator and longitudes 6° 45′ and 7° 45′ East of Greenwich. The territory covers an area of 760,700 ha of land (NBS 2012) lying in the center of the country. Abuja is bordered on all sides by four states, namely: Niger, Nasarawa, Kogi and Kadunna (Dawan 2000). The territory is currently made up of six area councils—Gwagwalada, Abuja Municipal, Abaji, Kuje, Bwari and Kwali. The choice of Abuja is very important because it is situated within the Savannah region with moderate climatic conditions which have tremendous potential for supporting both crop and livestock production (Balogun 2001). The target population includes all rural small-scale crop farmers. For effective coverage of the study area, a multistage sampling technique was adopted while questionnaire was used for data collection. In the first stage, 5 out of the 6 Local Government Areas (LGA) also called Area Councils were purposively selected because they were rural communities dominated by both livestock and crop farmers. In the second stage, five rural communities were randomly selected from each of the five LGAs giving a total of 25 communities. From each of the 25 rural communities (third stage), 20 crop farmers were randomly selected and interviewed giving a total of 500 respondents. To quantify the impact of grazing livestock on cereal and tuber crops, a rating scale of 0–4 was used. In the questionnaires, farmers were asked to rate the impact of grazing livestock on cereal and tuber crops production using: very highly destructive (4), highly destructive (3), moderately destructive (2), least destructive (1) and not destructive at all (0). Shah and Madden (2004) highlighted the importance and validity of using the rating scale in impact assessment. The responses were used to run analysis in line with the method adopted by Ajah (2014, 2015), Robert (2011), Gray and Kinnear (2011) and Field (2005). Also, the farmers were asked to estimate the cost of damages caused on their crops by grazing livestock. To reach the farmers, Agricultural Extension Officers (AEO) were contacted and they served as enumerators.

A three-way mixed analysis of variance (ANOVA) (Field 2005) was used for the analysis and it is expressed mathematically as:

$$Y_{ijtk} = \mu + L_i + C_i + S_t + LC_{ij} + LS_{it} + CS_{it} + LCS_{ijt} + e_{ijtk} \tag{8.1}$$

where

Y_{ijtk} Individual farmer's response on the impact of grazing livestock on cereal crops production;
i denotes the level of factor *L*,
j denotes the level of factor *C*,
t denotes the level of factor *S*,
and *k* denotes the kth observation in cell or treatment (*i, j, t*)

μ Population mean

L_i Impact as a result of differences in location (Kwali, Kuje, Gwagwalada (Gwa), Abaji, Bwari). This measures the main effect of location

C_i Crop type—this measures the main effect of crop type, i.e., impact due to the type of cereal crops cultivated (e.g. maize, rice, sorghum, millet)
S_t Livestock type—this measures the main effect of livestock type, i.e., impact due to the type of grazing livestock (cattle, goat/sheep, domestic fowls)
LC_{ij} Interaction between location and crop type
LS_{it} Interaction between location and livestock type
CS_{it} Interaction between crop type and livestock type
LCS_{ijt} Interaction of location, crop type and livestock type
e_{ijtk} Error term.

The model hypothesizes that the impact of grazing livestock on cereal (or tuber) crops production depends on three factors—the location of the farmer in Abuja, type of livestock reared and types of cereal (or tuber) crop cultivated. Crop type and livestock type were measured repeatedly; hence, they are called "within factor variables." Crop type for cereals has four levels (maize, rice, sorghum and millet), while the tuber has three levels (yam, cassava and cassava). Livestock type has three levels (cattle, goat/sheep and domestic fowls). Location (a between factor variable) has five levels (Kwali, Kuje, Gwagwalada (Gwa), Abaji and Bwari). By implication, the model states that the impact of grazing livestock on cereal (or tuber) crops production (Y_{ijtk}) depends on location of the farmer in Abuja (L_i), type of crop cultivated (C_i), type of livestock reared (S_t), both location and type of crop cultivated (LC_{ij}), both location and type of animal reared (LS_{it}), both livestock type and crop type (CS_{it}), and the joint effect of location, livestock type and crop type (LCS_{ijt}). The μ is the population mean which has no effect on the scores obtained and does not contribute to any variation in the observed differences (Aggarwal 2002). The error term is given by e_{ijtk}. SPSS 21.0 was used to run the analysis, and mean separation was done using the Bonfenori model (Field 2005). It was tested at 5% probability level.

8.3 Results and Discussion

8.3.1 Analysis of Impact of Grazing Livestock on Cereal Crops Production

Table 8.1 shows results of the three-way mixed ANOVA carried out to assess the impact of grazing livestock on the cereal crops. The model provides the opportunity to look at the data from two perspectives, that is, assessing the main effects of the factors and their interaction effects. It should be noted that each row in Table 8.1 provides answers to each of the hypotheses stated and the objectives of the study. Each of the factors and their interactions were interpreted separately, in order to enhance the comprehension of the results.

Table 8.1 ANOVA results of the impact of grazing livestock on cereal crops production

Sources of variation	df	SS	MS	F-cal	*p* value
Livestock type	2	269.10	134.55	142.00	0.00
Livestock type * location	8	143.52	17.94	19.07	0.00
Error (Livestock type)	858	807.23	0.94		
Crop type	3	1004.98	334.99	512.89	0.00
Crop type * location	12	1113.16	92.76	142.03	0.00
Error (crop type)	1287	840.60	0.65		
Livestock type * crop type	6	122.30	20.34	60.43	0.00
Location * livestock type * crop type	24	131.31	5.47	16.22	0.00
Error (livestock type * crop type)	2574	868.31	0.34		
Location	4	171.29	42.82	147.71	0.00
Error (location)	429	124.37	0.29		

Source Survey data analysis, 2015

8.3.1.1 Impact of Grazing Livestock on Cereals Production in Different Locations

The results of the impact of grazing livestock on cereal crops production in different locations (main effect of location) are shown in Table 8.1. It shows how the farmers rated the impact of livestock on cereal crops production in different locations. The question is: Are there significant locational differences in the impact of grazing livestock on cereal crops production? That is, without reference to any particular crop or livestock, how did farmers in each area council rated or perceived the impact of grazing livestock on cereals production? It tests the hypothesis which states that there are no significant locational differences in the impact of grazing livestock on cereal crops production ($\mu_{\text{Kwali}} = \mu_{\text{Kuje}} = \mu_{\text{Gwagwalada}} = \mu_{\text{Abaji}} = \mu_{\text{Bwari}}$). The result, $F(4, 429) = 147.71$, $p = 0.00$, showed that there were significant locational differences ($p < 0.01$) in impact of grazing livestock on cereal crop production hence the rejection of the null hypothesis. The mean responses (Fig. 8.1) showed that, of all the locations, Kwali Area Council was the most affected by grazing livestock as the mean response is significantly ($p < 0.05$) higher than the perceived impact in other locations. The farmers' ratings indicated that the least affected Area Council was Gwagwalada followed by Kuje. The implication of the magnitude of the mean response in Gwagwalada Area Council is that impact of grazing livestock was not a major constraint limiting the production of cereal crops. But, no matter the magnitude of impact, it should be seen as a threat to sustainable agricultural development in the area. The locational differences may be attributed to the availability and use of grazing routes. Abbass (2012) stated that there were 4125 grazing routes in Nigeria including Abuja. In terms of policy formation, this result is very important if the government wants to take decisions without reference to any particular crop or livestock in each

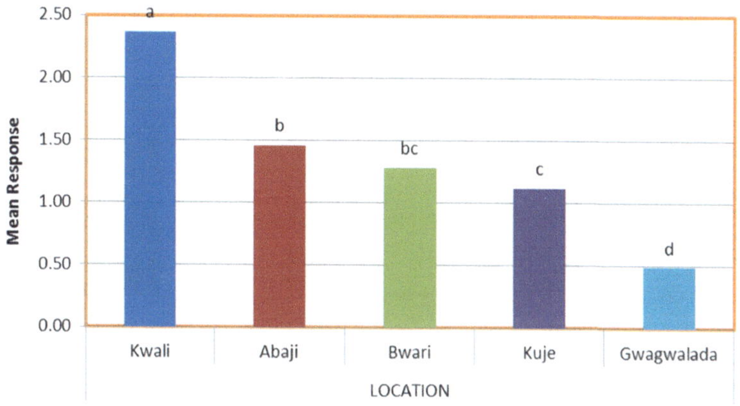

Fig. 8.1 Farmers' rating of impact of livestock on cereal crops production in different locations (main effect of location). *Source* Survey data analysis, 2015

location. However, subsequent results will reveal peculiarities in some locations with respect to livestock and crops.

8.3.1.2 Impact on Cereal Crops Production Due to Differences in Livestock Type

The effect of different livestock types on cereal crops production was also assessed. The question is: Regardless of location (Area Councils) and crop type, does the impact of grazing livestock on cereal crops depend on livestock type? It tests the hypothesis which states that impact of grazing livestock on cereal crops production does not depend on the type of livestock reared ($\mu_{\text{cattle}} = \mu_{\text{goat/sheep}} = \mu_{\text{domestic fowls}}$). The result, F(2, 858) = 142.00, $p = 0.00$, indicated that the impact of grazing livestock on cereal crops production significantly ($p < 0.01$) depends on the type of livestock reared. In other words, irrespective of location and crop type, the impact of some animals on cereal crops production significantly differed, hence the rejection of the null hypothesis. The mean separation (Fig. 8.2) showed that the impact of cattle and goat/sheep on cereal crops production did not significantly differ ($p > 0.05$) from each other, but they are significantly higher than that of domestic fowls. This shows that farmers rated the impact of cattle and goat/sheep on cereal crops production almost the same. Again, cattle and goat/sheep were the most destructive animals, while domestic fowls were the least. This is in line with the findings of Ofuoku and Isife (2009) as well as Adebayo and Olaniyi (2008) who indicated that the impact of cattle, goat and sheep on crops was a major source of conflict between the pastoralists and the crop farmers. Although other sources of conflict have also been identified by other scholars (Abbass 2012; Okoli and Atelhe 2014), but crop destruction is a major factor.

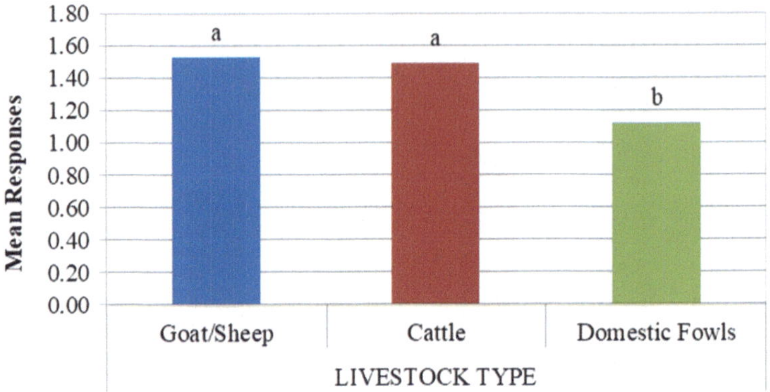

Fig. 8.2 Farmers' rating of impact of each livestock on cereal crops production (main effect of livestock type). *Source* Survey data analysis, 2015

8.3.1.3 Impact of Interaction Between Location and Livestock Type on Cereals Production

Table 8.1 also shows the result of interaction between location and livestock type (livestock type * location). It provides information on how the farmers rated the impact of each livestock type on cereal crop production in each area council. The question is: Without reference to crop type, does the impact of each livestock type on all cereal crops differ in each location. It tests the hypothesis which states that there is no significant interaction effect between location and livestock on cereal crops production ($\mu_{\text{Kwali.Cattle}} = \mu_{\text{Kwali.Goat/Sheep}} = \mu_{\text{Kwali.DomesticFowls}} = \cdots = \mu_{\text{Bwari.Cattle}} = \mu_{\text{Bwari.Goat/Sheep}} = \mu_{\text{Bwari.Domestic Fowls}}$). The result, $F(8, 858) = 19.07, p = 0.00$, shows there was significant ($p < 0.01$) interaction between location and the type of livestock reared, hence the rejection of null hypothesis. This shows that the impact of some animals on cereal crops production significantly ($p < 0.01$) differed in some Area Councils (locations). Mean separation was done separately to identify the livestock that was rated as most destructive in each location. The result (Fig. 8.3) indicates that, in Kwali Area Council, the impact of cattle on cereal crops production was significantly ($p < 0.05$) higher than that of goat/sheep, while the impact of goat/sheep was also significantly ($p < 0.05$) higher than that of domestic fowls. This shows that cattle were rated as the most destructive livestock in Kwali Area Council while domestic fowls were the least. Results from Gwagwalada Area Council were similar to that in Kwali Area Council, but on the contrary, it differed with those of Kuje, Abaji and Bwari Area Councils. In Kuje and Bwari Area Councils, goat/sheep were rated as the most destructive animals, while in Abaji, both cattle and goat/sheep were rated the same. In all the Area Councils, farmers rated the impact of domestic fowls on cereal crops production as the least. This result is similar to the report by Ajah (2012) though the study differed because of non-inclusion of crop type. The result suggests that livestock rearing especially cattle and goat/sheep may be more predominant in some area councils compared to others. Ofuoku and Isife (2009) stated that livestock

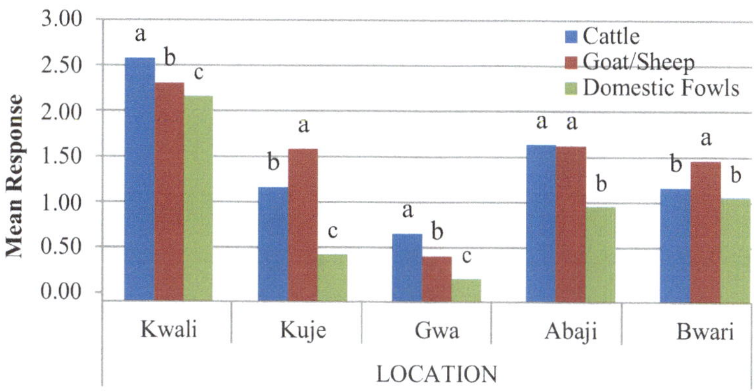

Fig. 8.3 Farmers' rating of impact of each livestock on cereal crops production in each location (interaction of location and livestock type). *Source* Survey data analysis, 2015

farmers are transhumance in their quest for pasture and other necessities of life, but the more the number of livestock in a location, the more likelihood of interaction with crop farms.

8.3.1.4 Impact of Grazing Livestock on Cereal Crops Production by Crop Type

The impact of grazing livestock was also assessed on the type of cereal crop cultivated, and it is denoted by "Crop type" in Table 8.1. The question here is: Does the impact of grazing livestock depend on the type of cereal crop cultivated? It tests the hypothesis which states that the impact of grazing livestock on cereal crops production does not depend on the type of cereal crops cultivated ($\mu_{\text{Maize}} = \mu_{\text{Rice}} = \mu_{\text{Sorghum}} = \mu_{\text{Millet}}$). This is more like a test of preference, which relates to the observation by Kie and Boroski (1996) that cattle selectively forage on grasses. Cereal crops are different in many aspects, and there is a likelihood that animals may prefer one or some of the cereal crops to others. The result, $F(3, 1287) = 512.89$, $p = 0.00$, indicates that the main effect of crop type was significant ($p < 0.01$), leading to the rejection of the null hypothesis. The implication of this is that the impact of grazing livestock significantly depends on the type of cereal crop cultivated. On the other hand, it also implies that some of the animals preferred some cereal crop to others or that some are by nature, more vulnerable to destruction by grazing livestock. Mean separation (Fig. 8.4) indicates that maize was the most affected cereal crop. The impact on maize was significantly ($p < 0.05$) higher than any other cereal crop. The second most affected crop was sorghum, and the impact on sorghum was significantly ($p < 0.05$) higher than that of millet. The least affected was rice, and this may be attributed to the fact that rice is mostly planted in *fadama* areas. These areas are relatively swampy or water-logged (Dalil and Nsini 2014), of which livestock like goats and sheep tend to avoid water during grazing (Forsyth et al. 2002). Looking at the results, it can be seen

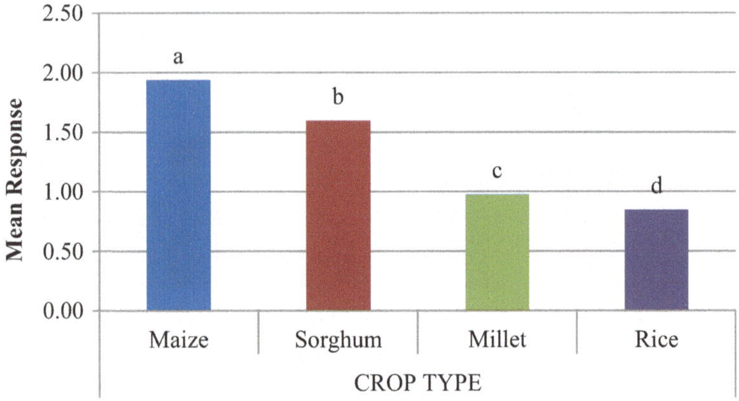

Fig. 8.4 Farmers' rating of impact of livestock on each cereal crop (main effect of crop type). *Source* Survey data analysis, 2015

that the farmers rated maize as the most affected crop followed by sorghum, millet and rice. If we look at the result on the basis of preference, it can be inferred that grazing livestock generally preferred the maize crop to other cereals or that maize was more vulnerable to destruction by grazing livestock.

8.3.1.5 Impact of Interaction Between Location and Crop Type on Cereals Production

The effect of interaction between location and crop type on cereals production (location * crop type) is shown in Table 8.1. This interaction shows farmers' rating of impact of livestock on each cereal crop in each area council. The question is: Irrespective of livestock type, in which of the Area Councils (locations) was each of the cereal crops mostly affected (destroyed) by grazing livestock? It tests the hypothesis which states that there is no significant interaction between location and crop type. In other words, it tests the hypothesis that the impact of grazing livestock on each cereal crop does not significantly differ in each Area Council. The result, $F(12, 1287) = 142.03$. $p = 0.00$, shows that there was a significant ($p < 0.01$) interaction effect between location and crop type; hence, the null hypothesis was rejected. Mean separation (Fig. 8.5) carried out for each location indicates that in Kwali Area Council, the most affected crop was sorghum and the impact on sorghum was significantly ($p < 0.05$) higher than the impact on maize. The impact on maize was significantly higher than that of rice. Similarly, the impact on rice was significantly ($p < 0.05$) higher than that of millet. On the contrary, maize was the most affected crop in Kuje Area Council, and the impact on maize was significantly ($p < 0.05$) greater than that of millet (the second most affected crop). The impact on millet was significantly ($p < 0.05$) higher than that of sorghum, while that of sorghum was higher than that of rice. Looking at Gwagwalada, Abaji and Bwari Area Councils, it can be seen that

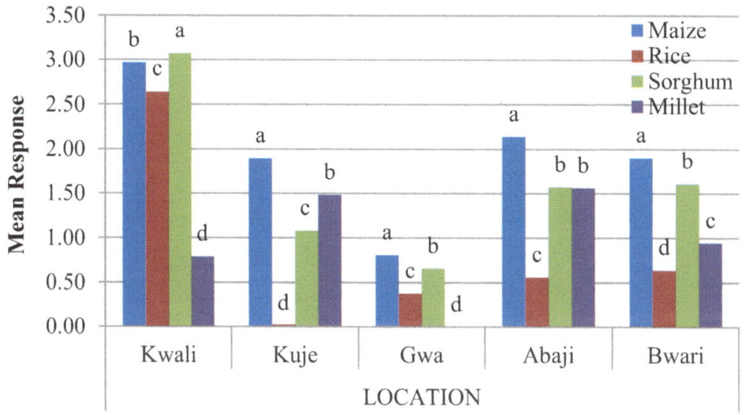

Fig. 8.5 Farmers' rating showing the most affected cereal crop by grazing livestock in each location (interaction of location and crop type). *Source* Survey data analysis, 2015

there were variations in the impact of grazing livestock on the different cereal crops. It is important to note that apart from Kwali Area Council, the most affected crop was maize while the least affected crops varied from one location to another. For instance, in Kwali and Gwagwalada Area Councils, the least affected cereal crops were millet, while in Kuje, Abaji and Bwari Area Councils it was rice. The results suggest that there were differential challenges in the production of cereal crops in all the locations. This is discouraging because Nigeria is one of the countries that depends on the importation of cereal crops to meet her local demand (Vaughan et al. 2014). It is noteworthy to highlight that small-scale farmers in Nigeria are the major producers of these crops (Oyeyinka and Bolannwa 2009).

8.3.1.6 Impact of Interaction Between Livestock Type and Crop Type on Cereals Production

The effect of interaction between livestock type and crop type (livestock type * crop type) on cereals production is shown in Table 8.1. This interaction provides information on how the small-scale crop farmers rated the impact of each animal on each cereal crop. The question here is: In the entire study area (irrespective of location), which livestock affected or was associated with the destruction of each cereal crop more than the other? In other words, was a particular cereal crop more vulnerable to destruction by a particular grazing livestock? It tests the hypothesis which states that there is no significant interaction effect between livestock type and crop type. The result, $F(6, 2574) = 60.43$, $p = 0.00$, shows that there was a significant ($p < 0.01$) interaction effect between livestock type and crop type. This implies that some of the animals were known to destroy (or eat) some of the crops more than others in the study area. Mean separation (Fig. 8.6) revealed that maize was more vulnerable to

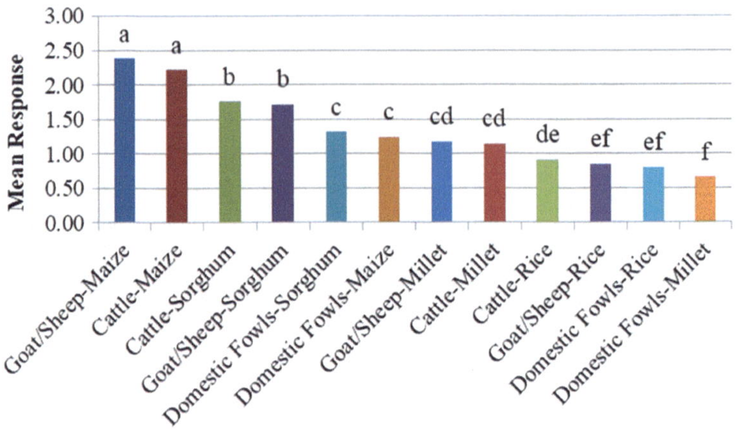

Fig. 8.6 Farmers' rating of impact of each livestock on each cereal crop (interaction of livestock types and crop type). *Source* Survey data analysis, 2015

destruction by cattle and goat/sheep than domestic fowls. The second most affected crop was sorghum, and the mean rating indicated that it was damaged mostly by cattle and goat/sheep. The impact of domestic fowls on sorghum and maize was also relatively high, but in the whole interaction, its impact on millet was the least. The relative vulnerability of maize and sorghum to destruction by the animals presents a big challenge which needs immediate action. FAO (2002) predicted that developing countries like Nigeria would produce only 85% of her cereal needs with net import of about 265 million tonnes annually. This import bill can be reduced if the negative impact of grazing livestock on crop productivity and other challenges is addressed.

8.3.1.7 Impact of Interaction of Location, Livestock Type and Crop Type on Cereals Production

Table 8.1 contains the result of interactions of *location, livestock type and crop type* (location * livestock type * crop type). Here, the impact of the three factors is jointly assessed; hence, it shows how the farmers rated the impact of each livestock on each cereal crop in each location sampled. The question is: In each location sampled, how did each livestock affect each crop? It tests the hypothesis which states that there is no locational difference in the impact of each grazing livestock on each crop. The result, $F(24, 2574) = 16.22$, $p = 0.00$, showed that there were significant ($P < 0.01$) interaction effects of location, livestock type and crop type; hence, the null hypothesis was rejected. This implies that in some locations, there were significant ($p < 0.01$) differences in the impact of some livestock on some crops. The interaction of location, crop type and livestock type is represented in Figs. 8.7, 8.8, 8.9, 8.10 and 8.11.

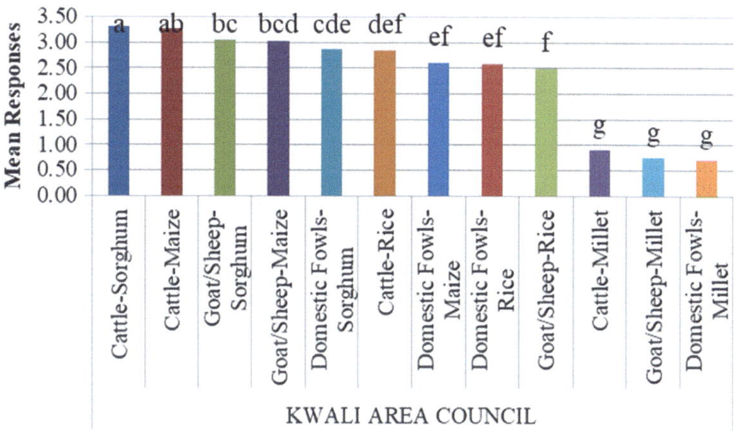

Fig. 8.7 Farmers' rating of impact of each livestock on each cereal crop in Kwali Area Council. *Source* Survey data analysis, 2015

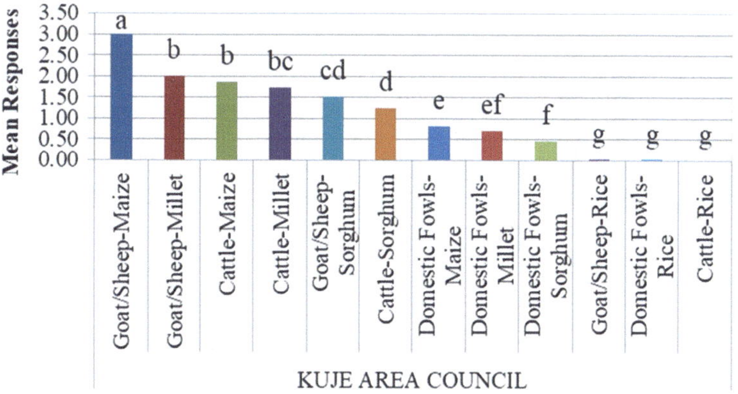

Fig. 8.8 Farmers' rating of impact of each livestock on each cereal crop in Kuje Area Council. *Source* Survey data analysis, 2015

8.3.1.8 Impact of Grazing Livestock on Cereal Crops Production in Kwali Area Council

The result of mean separation of the interaction of location, livestock type and crop type for Kwali Area Council is presented in Fig. 8.7. It is worthy to note that the farmers rated the impact of cattle on sorghum, maize and rice very high. The mean responses showed that the impact of cattle on sorghum and maize was not significantly ($p > 0.05$) different from each other implying that sorghum and maize were highly vulnerable to destruction by cattle compared to other crops. This is discouraging and may have contributed to the low productivity of these crops in Nigeria compared to India according to Ahmad et al. (2015). On the other hand, sorghum and maize

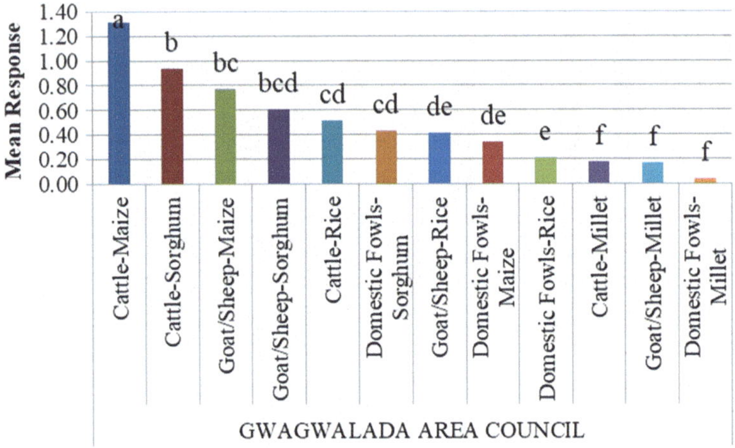

Fig. 8.9 Farmers' rating of impact of each livestock on each cereal crop in Gwagwalada Area Council. *Source* Survey data analysis, 2015

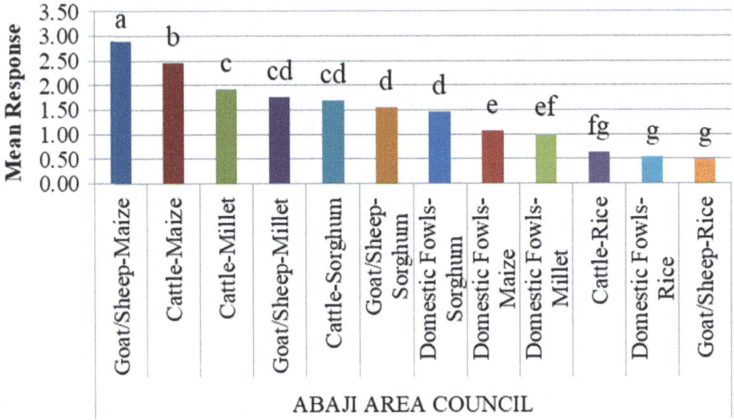

Fig. 8.10 Farmers' rating of impact of each livestock on each cereal crop in Abaji Area Council. *Source* Survey data analysis, 2015

were also rated as being vulnerable to destruction by domestic fowls. Based on the mean response, rice was also affected by cattle, goat/sheep and domestic fowls, but the impact was not as high as that of sorghum and maize. The least affected crop in the area was millet. This means that among the four cereal crops studied, millet was one of the crops that could be produced in this location with minimal destruction by grazing livestock. The destruction of crops by grazing livestock is contrary to the concept of sustainable agricultural development and has to be addressed so that it does not add to the numerous challenges hindering farmers' productivity as outlined by Ohen and Ajah (2015) and Ismaila et al. (2010). Comparing this result to the one

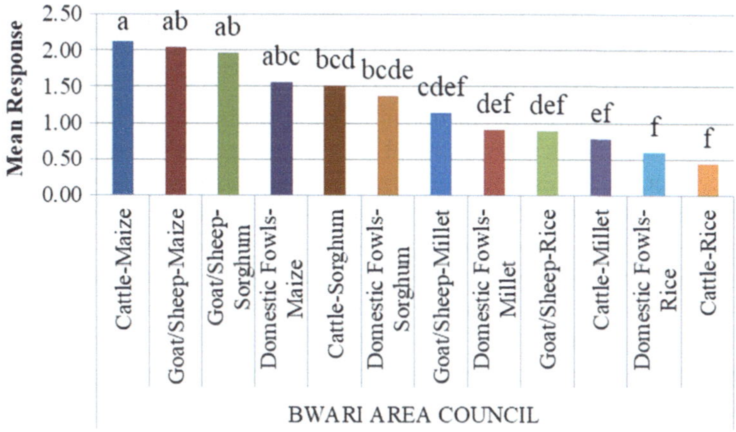

Fig. 8.11 Farmers' rating of impact of each livestock on each cereal crop in Bwari Area Council. *Source* Survey data analysis, 2015

in Fig. 8.4 which showed that maize was the most affected crop while rice was the least affected, it can be seen that sorghum was the most affected crop while millet was the least affected crop in Kwali Area Council. If policies were to be made based on the result in Fig. 8.4, emphasis would have been placed generally on maize and rice as the most and least affected crop respectively hence, the importance of interaction of location, crop type and livestock type in the analysis.

8.3.1.9 Impact of Grazing Livestock on Cereal Crop Production in Kuje Area Council

The mean separation of interaction of location, livestock type and crop type for Kuje Area Council is presented in Fig. 8.8. The mean responses showed that the farmers rated maize as most vulnerable to destruction by goat/sheep. The impact of goat/sheep on maize was significantly ($p < 0.05$) higher than the impact of cattle and domestic fowls on maize and other cereal crops studied. The second most affected crop was millet, and this is contrary to the result in Fig. 8.4 where sorghum was second to maize with rice being the least. This further highlights the importance of using three-way mixed analysis of variance in the study. It provided the true picture of the impact of livestock on cereal crops production in different locations, thereby opening the window for different policy options. Generally, while maize was most affected crop, rice was the least (Fig. 8.4) but this is not the same in all the locations. The impact of domestic fowls, cattle and goat/sheep was very minimal on rice thus, it can be inferred that rice is one of the crops that could be produced in Kuje Area Council with minimal destruction by grazing livestock.

8.3.1.10 Impact of Grazing Livestock on Cereals Production in Gwagwalada Area Council

The mean separation of the interaction of location, livestock type and crop type for Gwagwalada Area Council is shown in Fig. 8.9. The result showed that maize was the most vulnerable to destruction by grazing cattle. The mean responses indicated that the impact of grazing cattle on maize was significantly ($p < 0.05$) higher than the impact of goat/sheep and domestic fowls on other crops. Similarly, the farmers' rating indicated that the impact of grazing cattle on sorghum was also high although it did not significantly differ from the impact of goat/sheep on maize. Judging from the mean responses, millet was the least affected crop implying that it is one of the crops that can be produced in this location with minimal destruction by grazing livestock. It is worthy to note here that even though there were significant differences in the mean responses, the mean ratings for the impact of grazing livestock on cereal crops production were generally low compared to other locations. This calls for further investigation to find out if, they have grazing routes and are adhering to them. Further inquiry should be made to know if they have by-laws or customary laws that regulate the behavior of crop and livestock farmers resulting in minimal interaction of livestock and crops in the farms.

8.3.1.11 Impact of Grazing Livestock on Cereal Crops Production in Abaji Area Council

Like other Area Councils, the mean separation of the interaction of location, livestock type and crop type for Abaji Area Council is presented in Fig. 8.10. The result showed that maize was the most vulnerable to destruction by grazing goat/sheep. The mean responses indicated that the impact of goat/sheep on maize was significantly ($p < 0.05$) higher than the impact of other animals on maize and other crops. Similarly, the impact of cattle on maize was significantly ($p < 0.05$) higher than the impact of domestic fowls. The farmers rated the impact of cattle and goat/sheep on millet the same. Looking at the mean responses, it is evident that rice was the least affected crop. This result is similar to the one obtained in Kuje Area Council (Fig. 8.8), and it could be that these Area Councils cultivated rice in *fadama* (water-logged) areas that are very difficult for the grazing livestock to access. It could also mean that rice is not a major crop cultivated in these area councils. Generally, it is clear that cattle and goat/sheep were the most destructive animals and their impact on crops have been noted as a major source of conflict between pastoralists and crop farmers in Nigeria (Blench 2010).

8.3.1.12 Impact of Grazing Livestock on Cereal Crops Production in Bwari Area Council

The result in Fig. 8.11 is the mean separation of interaction of location, livestock type and crop type for Bwari Area Council. The result indicated that the most affected crop was maize followed by sorghum. The impact of cattle on maize was not significantly ($p > 0.05$) different from that of goat/sheep. Although the mean responses were not the same, it shows that cattle and goat/sheep were the most destructive livestock. The rating of impact of domestic fowls on maize was also relatively high compared to other crops. Like other area councils, farmers in this location rated the impact of grazing livestock on rice and millet very low. In other words, among the cereal crops, millet and rice were some of the crops that could be produced in Bwari Area Council with minimal destruction by grazing livestock. However, no matter the situation, if the impact of grazing livestock on cereals production is allowed to continue, the total output and productivity of farmers may likely fall below an economically acceptable level and crop production in Nigeria may not cope with population growth rate of 3.2% per annum (OPEC 2013). This implies more importation to meet the demand of the growing population.

8.3.2 Analysis of the Impact of Grazing Livestock on Tuber Crops Production

Table 8.2 shows the three-way mixed analysis of variance (ANOVA) results of the impact of grazing livestock on tuber crops (yam, cassava, potato) production. Like cereal crops, the model gives the opportunity to look at the data in different ways and

Table 8.2 ANOVA results of the impact of grazing livestock on tuber crops production

Sources of variation	df	SS	MS	F-cal	*p* value
Livestock type	2	605.35	302.67	521.06	0.00
Location * livestock type	8	241.15	30.14	51.96	0.00
Error (livestock type)	830	483.06	0.58		
Crop type	2	84.39	42.19	131.31	0.00
Location * crop type	8	16.25	2.03	6.34	0.00
Error (crop type)	830	268.19	0.32		
Crop type * livestock type	4	46.73	11.68	40.24	0.00
Location * livestock type * crop type	16	56.67	3.54	12.20	0.00
Error (Crop type * livestock type)	1660	495.85	0.29		
Location	4	86.54	21.63	270.37	0.00
Error (location)	415	34.71	0.08		

Source Survey data analysis, 2015

take appropriate decisions. For a better understanding of the results, the interpretation is based on the three factors and their interactions.

8.3.2.1 Impact of Grazing Livestock on Tuber Crops Production in Different Locations

Table 8.2 shows the impact of grazing livestock on tuber crops production in different locations (main effect of location). It shows the differences in impact of grazing livestock on tuber crops (yam, cassava, potato) in different area councils. The question is: Without reference to a particular livestock or tuber crop, how did farmers in different locations rated the impact of grazing livestock on the tuber crops? It tests the hypothesis which states that there is no significant locational difference in the impact of grazing livestock on tuber crops production ($\mu_{\text{Kwali}} = \mu_{\text{Kuje}} = \mu_{\text{Gwagwalada}} = \mu_{\text{Abaji}} = \mu_{\text{Bwari}}$). The result, $F(4, 415) = 270.37$, $p = 0.00$, shows that there were locational differences in impact of grazing livestock on tuber crops production, hence the rejection of the null hypothesis (Fig. 8.12). Furthermore, mean separation was carried out, and the result shows that the impact of grazing livestock was felt more by farmers in Kuje and Abaji Area Councils. The impact of grazing livestock in Kuje and Abaji Area Councils was statistically the same ($p > 0.05$) but significantly higher than that of Kwali Area Council, which in turn was significantly higher ($p < 0.05$) than that of Bwari Area Council. Similarly, the impact of grazing livestock in Bwari Area Council was significantly ($p < 0.05$) greater than that of Gwagwalada Area Council. Although there were significant locational differences, the magnitude of the mean responses conveyed an important message. It shows that the farmers rated the impact of grazing livestock on tubers very low, especially in Kwali, Bwari and Gwagwalada Area Councils. The policy implication of the magnitude of the mean responses is that grazing livestock did not constitute a major threat to tuber crops production in

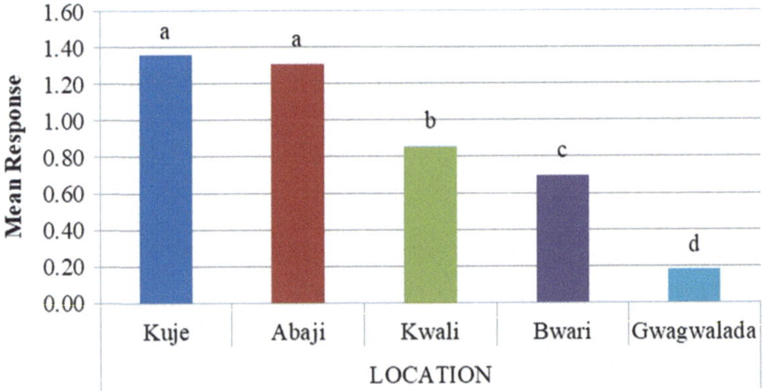

Fig. 8.12 Farmers' rating of impact of livestock on tuber crops production in each location. *Source* Survey data analysis, 2015

these locations. But, using this result to take decision may be misleading hence the need to look at the interaction results of factors which are presented in subsequent analyses.

8.3.2.2 Impact of Grazing Livestock on Tuber Crops Production Due to Livestock Type

The impact of grazing livestock on tuber crops production due to livestock type (main effect of livestock type) is shown in Table 8.2. Here, emphasis is laid on the most destructive livestock regardless of whether it is yam, cassava or potato. The question is: Without reference to any particular location or crop type, does the impact of cattle, goat/sheep and domestic fowls on tuber crops differ from each other? It tests the hypothesis which states that the impact of grazing livestock on tuber crops production does not depend on the type of livestock reared ($\mu_{\text{Cattle}} = \mu_{\text{Goat/Sheep}} = \mu_{\text{Domestic Fowls}}$). The result, F(2, 830) = 521.06, $p = 0.00$, indicated that there was significant ($p < 0.01$) difference in the impact of cattle, goat/sheep and domestic fowls on tuber crops production, hence the rejection of the null hypothesis. This means that the impact of livestock on tuber crops production depends on whether it is cattle, goat/sheep or domestic fowls (livestock type). Mean separation (Fig. 8.13) showed that, among the three livestock types, cattle were perceived or rated as the most destructive compared to goat/sheep and domestic fowls. The mean responses showed that the impact of cattle on tuber production was significantly ($p < 0.05$) higher than the impact of goat/sheep, while that of goat/sheep was significantly higher ($p < 0.05$) than that of domestic fowls. Based on magnitude of the mean response, cattle and goat/sheep were the most destructive while domestic fowls were the least. The magnitude of the mean response connotes that the impact of domestic fowls was not a major factor limiting the production of tuber crops in the study area.

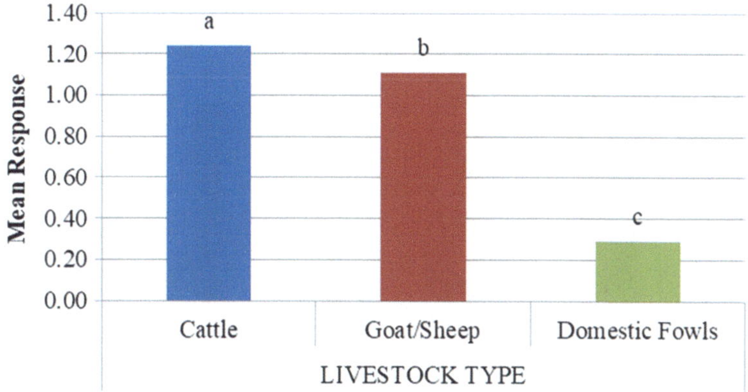

Fig. 8.13 Farmers' rating of impact of each livestock on tuber crops production (main effect of livestock type). *Source* Survey data analysis, 2015.

Although the impact of domestic fowls on tuber crops was rated low, it is important to highlight that domestic fowls also feed on harvested crops during processing (sun drying), which further reduces total output.

8.3.2.3 Impact of the Interaction Between Location and Livestock Type on Tuber Production

The result of interaction between location and livestock type (location * livestock type) is shown in Table 8.2. The question is: In each area council sampled, how did the farmers rate the impact of each livestock on tuber crops production? Answering this question led to the test of hypothesis which states that there is no significant interaction between location and livestock type. The result, $F(8, 830) = 51.96$, $p = 0.00$, shows that there was significant ($p < 0.01$) interaction between location and livestock type, hence the rejection of the null hypothesis. This implies that the impact of cattle, goat/sheep and domestic fowls on tuber crops production was not the same in some of the area councils. The mean separation (Fig. 8.14) indicated that in Kwali and Bwari Area Councils, cattle were rated as the most destructive and its impact on tuber crops was significantly ($p < 0.05$) higher than that of goat/sheep, while that of goat/sheep was significantly ($p < 0.05$) higher than that of domestic fowls. On the contrary, in Kuje Area Council, goat/sheep was rated as the most destructive followed by cattle. The mean response indicated that the impact of goat/sheep was significantly ($p < 0.05$) higher than that of cattle while that of cattle was significantly higher than that of domestic fowls. In Abaji Area Council, the farmers rated the impact of cattle and goat/sheep the same, but their impact was significantly ($p < 0.05$) higher than that of domestic fowls. The scenario in Gwagwalada Area Council was different. Although the rating of impact of livestock on tuber crops was not the same in Gwagwalada, the magnitude of mean response was generally very low. This is

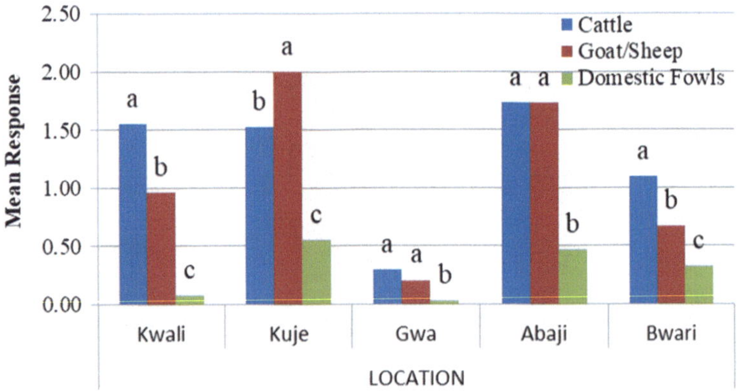

Fig. 8.14 Farmers' rating of impact of each livestock on tuber crops production in each location (interaction of location and livestock type). *Source* Survey data analysis, 2015

why the analysis of interaction between factors is very important. If we had used the result in Fig. 8.13 to take decision, we would have assumed that all the area councils faced similar challenges, but it can be seen that the impact on the tuber crops varied from one location to another.

8.3.2.4 Impact of Grazing Livestock on Tuber Production Due to Crop Type

The result of the main effect of crop type is shown in Table 8.2. Here, emphasis is placed on the most affected tuber crop regardless of whether it is done by cattle, goat/sheep or domestic fowls. The question is: In the entire study area, does the impact of grazing livestock depend on the type of tuber crop cultivated? It tests the hypothesis which states that the impact of grazing livestock on tuber crops production does not depend on the type of tuber crop cultivated ($\mu_{\text{Yam}} = \mu_{\text{Cassava}} = \mu_{\text{Potato}}$). The result, $F(2, 830) = 131.31$, $p = 0.00$, shows that the impact of grazing livestock on tuber crops production depends on crop type, hence the rejection of null hypothesis. In other words, some tuber crops were more vulnerable to destruction by grazing livestock than others. Mean separation (Fig. 8.15) was done, and it showed that cassava was the most affected tuber crop. This implies that the livestock preferred cassava to other tuber crops. The impact of grazing livestock on cassava was significantly ($p < 0.05$) higher than that on yam, while that on yam was significantly higher ($p < 0.05$) than that on potato. This shows that the nature of a crop is an important factor in studying the impact of grazing livestock on crop production. The fact that cassava was the most affected tuber crop calls for attention because cassava is important both as a food crop as well as a source of raw material for many industries (Balami et al. 2011). It also serves as feed for farm animals. With the exception of few farmers, Liverpool-Tasie et al. (2011) stated that Nigerian farmers across all regions were below their

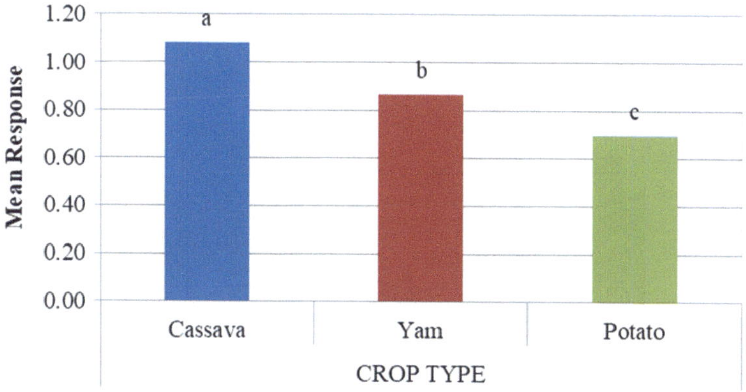

Fig. 8.15 Farmers' rating of impact of livestock on each tube crop (main effect of crop type). *Source* Survey data analysis, 2015

production frontiers and the impact of grazing livestock should be considered as one of the causes.

8.3.2.5 Impact of Interaction Between Location and Crop Type on Tuber Crops Production

The result of interaction between location and crop type (location * crop type) is shown in Table 8.2. In this interaction, emphasis is on the most affected tuber crop in each location regardless of whether the impact is from cattle, goat/sheep or domestic fowls; hence, the question is: Does the impact of grazing livestock on each tuber crop differ in each location and which tuber crop was mostly affected (destroyed)? It tests the hypothesis which states that there is no significant interaction between location and crop type. The result, $F(8, 830) = 6.34$, $p = 0.00$, showed that there was significant interaction between location and crop type. This implies that in some area councils, some tuber crops were more vulnerable to destruction than others, hence the rejection of the null hypothesis. Mean separation (Fig. 8.16) shows that, apart from Bwari Area Council, cassava was the most affected tuber crop in four locations. The impact of grazing livestock on cassava in the four locations was significantly ($p < 0.05$) higher than the impact on yam and potato. In Bwari Area Council, the impact on cassava and yam was rated the same. In Kwali, Kuje, Gwagwalada and Bwari Area Councils, the impact of livestock on yam was significantly ($p < 0.05$) higher than that of potato, but in Abaji, there was no difference ($p > 0.05$). Generally, the mean responses indicated that the farmers rated cassava as the most affected tuber crop while the least was potato. No matter the level of impact, solving the problem of grazing livestock will help a lot because the fear of cultivating crop only to be destroyed by animals is a disincentive to investment in agriculture (Audu 2014). Some crop farmers have resorted to planting small hectares of land as a coping strategy, and this is reported in a study conducted by Adisa (2012) which showed

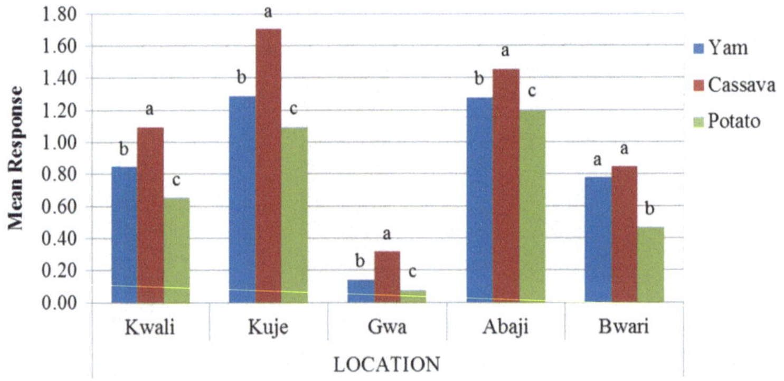

Fig. 8.16 Farmers' rating showing the most affected tuber crop by grazing livestock in each location (interaction of location and crop type). *Source* Survey data analysis, 2015

that 24.2% of crop farmers in Kwara State, Nigeria, adopted sowing less in order to cope with the effects of the mutual conflict arising from crop damages.

8.3.2.6 Impact of Interaction Between Livestock and Crop Type on Tuber Production

Table 8.2 contains the result of the interaction between crop type and livestock type (crop type * livestock type). It shows how the farmers rated the impact of cattle, goat/sheep and domestic fowls on each tuber crop in the whole study area. In other words, in the entire study area, how did the farmers perceive the impact of each livestock type on each tuber crop? The hypothesis states that there is no significant interaction between crop type and livestock type. The result, $F(4, 1660) = 40.24$, $p = 0.00$, showed that there was significant interaction between crop type and livestock type, thus the rejection of the null hypothesis. This implies that a particular livestock was associated with greater destruction of some tuber crops. Mean separation (Fig. 8.17) indicated that cassava and yam were more vulnerable to destruction by grazing cattle compared to other crops. In other words, cassava and yam were mostly affected by grazing cattle. The impact of cattle on cassava and yam was not significantly ($p > 0.05$) different from each other but significantly greater than the impact of goat/sheep and domestic fowls on other tuber crops. The mean responses also showed that there was no significant difference ($p > 0.05$) in the impact of domestic fowls on the three tuber crops. As a matter of emphasis, the impact of grazing livestock on crop production may have contributed to low technical efficiency recorded by farmers in northern Nigeria (Ibrahim et al. 2014) because the attendance of technical efficiency in production is only feasible if all the factors that

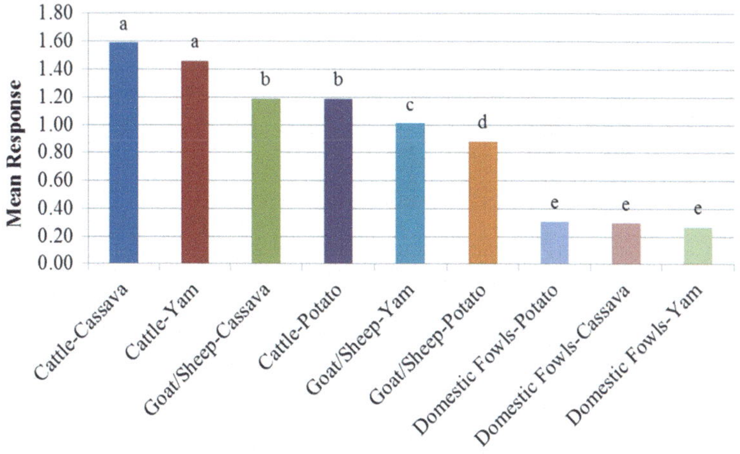

Fig. 8.17 Farmers' rating of impact of each livestock on each tuber crop (interaction of livestock type and crop type). *Source* Survey data analysis, 2015

influence output and productivity are identified and addressed within the limits of human error and knowledge.

8.3.2.7 Impact of Interaction of Location, Livestock Type and Crop Type on Tuber Crops

The result of interaction of location, livestock type and crop type is shown in Table 8.2. This is one of the most important results because it shows how the farmers rated the impact of each livestock on each tuber crop in each location. The question is: Does the impact of each livestock type on each tuber crop differ in each location? It tests the hypothesis which states that there is no significant interaction effect of location, livestock type and crop type. The result, $F(16, 1660) = 12.20$, $p = 0.00$, showed that there was significant ($p < 0.01$) interaction of location, livestock type and crop type hence, the null hypothesis was rejected. Analyzing the data down to each area council is vital because Boyle (2004) argued that data and research at the rural level provide local administrators with relevant information needed for making decisions. The result of the interaction is shown in Figs. 8.18, 8.19, 8.20, 8.21 and 8.22.

In Kwali Area Council, mean separation (Fig. 8.18) indicated that the impact of cattle on cassava was significantly ($p < 0.05$) higher than the impact of goat/sheep and domestic fowls. Yam was next to cassava, and it is mostly destroyed by cattle. Similarly, the impact of goat/sheep was significantly ($p < 0.05$) higher than that of domestic fowls. The impact of domestic fowls on the tuber crops approximates to no impact.

Similarly, in Kuje Area Council, mean separation (Fig. 8.19) showed that cassava was the most vulnerable to destruction by grazing goat/sheep. The impact of goat/sheep on cassava was significantly ($p < 0.05$) greater than that of cattle and

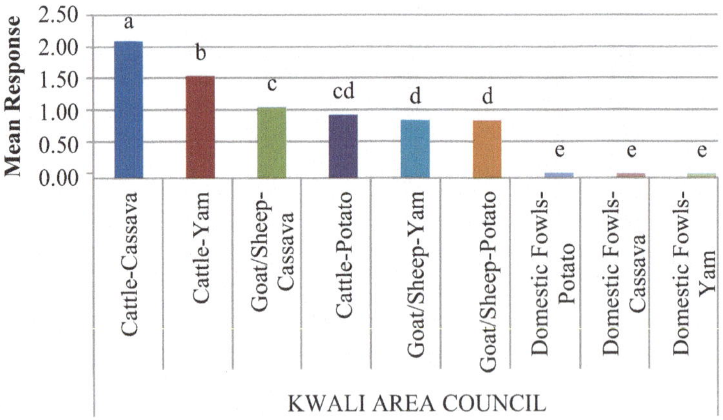

Fig. 8.18 Farmers' rating of impact of each livestock on each tuber crop in Kwali Area Council. *Source* Survey data analysis, 2015

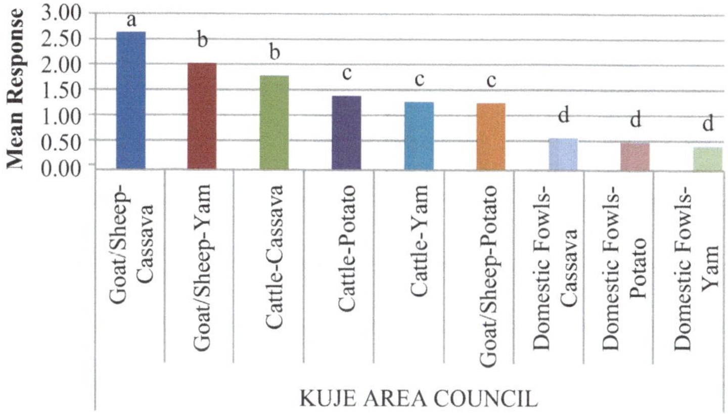

Fig. 8.19 Farmers' rating of impact of each livestock on each cereal crop in Kuje Area Council. *Source* Survey data analysis, 2015

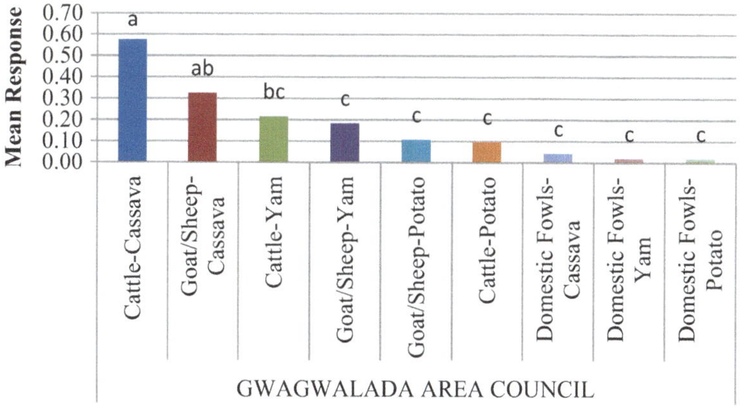

Fig. 8.20 Farmers' rating of impact of each livestock on each cereal crop in Kuje Area Council. *Source* Survey data analysis, 2015

domestic fowls. The second most affected tuber crop was yam, and this was also associated with goat/sheep. The impact of cattle on cassava comes third, and it is not significantly ($p > 0.05$) higher than that of goat/sheep on yam. There was no significant difference ($p > 0.05$) in the impact of domestic fowls on cassava, yam and potato.

Looking at Gwagwalada Area Council, the mean separation (Fig. 8.20) indicated that cassava was also rated as the most affected tuber crop with cattle and goat/sheep causing the greatest damage. The impact of cattle on cassava was significantly ($p < 0.05$) higher than that of goat/sheep and domestic fowls on yam and potato. The magnitude of the mean responses showed that the farmers rated the impact of livestock on tuber crops production very low, especially domestic fowls. This is similar

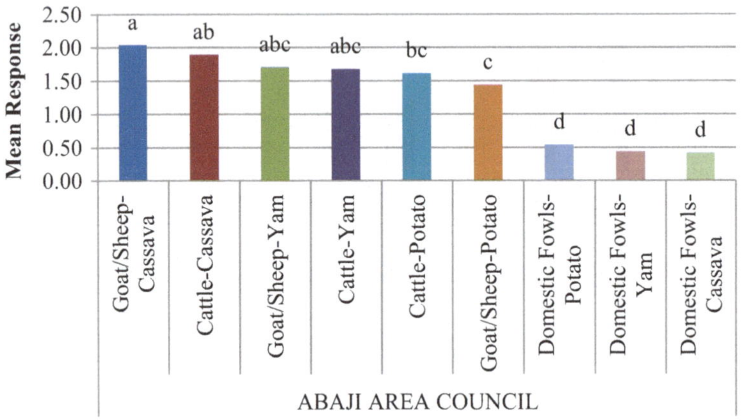

Fig. 8.21 Farmers' rating of impact of each livestock on each cereal crop in Abaji Area Council. *Source* Survey data analysis, 2015

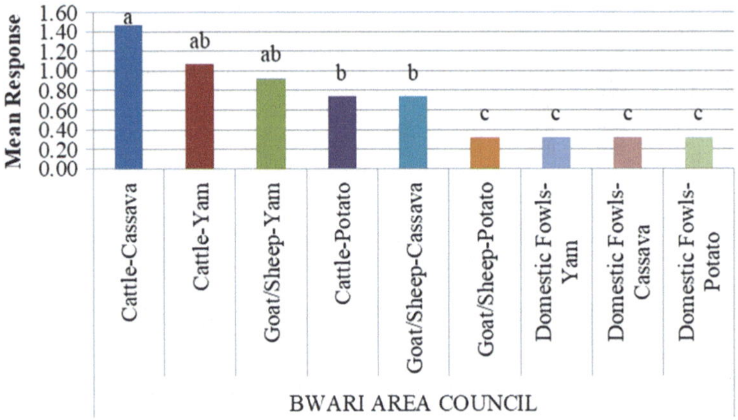

Fig. 8.22 Farmers' rating of impact of each livestock on each cereal crop in Bwari Area Council. *Source* Survey data analysis, 2015

to the result obtained for cereal crop production in the same area (see Fig. 8.9). This result calls for further investigation to determine why the impact of grazing livestock on tuber and cereal crops production was very low in Gwagwalada Area Council. Is it because the farmers are not into tuber crops production? Is it because they have grazing routes and are adhering to them or do they have by-laws guiding the behavior of both livestock and crop farmers? Unlike other locations, it can be inferred that grazing livestock are not major factors limiting the production of cassava, yam and potato in Gwagwalada Area Council. In other words, farmers in this location did not see grazing livestock as a major threat to the production of the tuber crops.

In Abaji (Fig. 8.21), cassava was also the most affected tuber crop with the damage attributed to goat/sheep and cattle. The impact of goat/sheep and cattle on cassava was not significantly ($p > 0.05$) different from each other. The second most vulnerable to destruction was yam, and the mean response showed that the impact of goat/sheep and cattle on yam was not significantly ($p > 0.05$) different from each other but significantly higher than the impact of domestic fowls on the three tubers. The impact of domestic fowls on the tuber crops was statistically the same ($p > 0.05$) and the rating approximates to no impact.

Furthermore, cassava is also the most affected tuber crop in Bwari Area Council (Fig. 8.22) and it was mostly damaged by grazing cattle. The impact of cattle on cassava and yam was not significantly different ($p > 0.05$) from each other. This is followed by the impact of goat/sheep on yam, though the impact was not significantly different ($p > 0.05$) from that of cattle on cassava and yam. The mean responses on the impact of cattle and goat/sheep on cassava, yam and potato were significantly ($p < 0.05$) higher than that of domestic fowls suggesting that they were the most destructive livestock. The impact of domestic fowls on the tuber crops was statistically the same and approximates to no impact.

Generally, it can be inferred that the rearing of domestic fowls under extensive system does not pose much threat to the production of cassava, yam and potato in the study area. However, cattle and goat/sheep negatively impact the production of the tubers to a reasonable extent.

8.3.2.8 Cost Estimates of Damaged Crops

During data collection, farmers were asked to estimate the cost of crops damaged by grazing livestock on their farms. Only the farmers whose crops were damaged provided the cost estimate, and the result is shown in Fig. 8.23. First, looking at the cereal crops alone, it can be seen that the most affected crop was rice, followed by

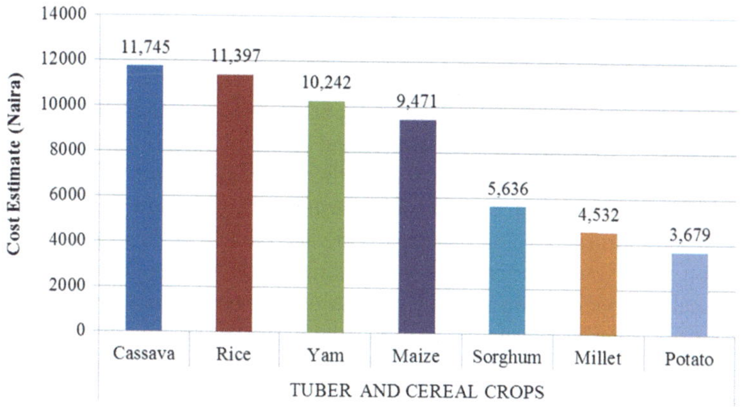

Fig. 8.23 Cost estimates of cereal and tuber crops destroyed by grazing livestock

maize, sorghum and millet. This is contrary to the result in Fig. 8.4 where the mean responses indicated that the most affected crop was maize followed by sorghum and millet while the least affected was rice. The difference can be attributed to the relatively high cost of rice per kilogram compared to other cereals in Nigeria. Apart from rice, the rating of impact of livestock on maize, sorghum and millet tallied with the cost estimate of damaged cereal crops. This shows the validity of application of rating scale on impact assessment and supports the report by Shah and Madden (2004) who applied ordinal data in designed factorial experiments. On average, affected rice farmers lost ₦11,397 (US$71). This is discouraging as Nigeria is a major importer of rice (FAO 2013) which can actually be produced in the country. On the other hand, maize, sorghum and millet farmers lost ₦9471 (US$59), ₦5636 (US$35) and ₦4532 (US$29) respectively to grazing livestock in 2014. Any small-scale farmer that suffers such losses due to grazing livestock is bound to be aggressive. It is evident that the most affected tuber crop was cassava (₦11,745 ~ US$73) followed by yam (₦10,242 ~ US$64) and then potato (₦3679 ~ US$23). This again, tallied with the mean response in Fig. 8.15 which showed that the farmers rated cassava as the most affected tuber crop followed by yam and potato. It should be noted that most of the damages on both cereal and tuber crops were caused by grazing cattle and goat/sheep although domestic fowls were not completely exempted from the damages. Rahji and Omotesho (2006) stated that the main issue in Nigerian agriculture is that of low productivity, and this result clearly suggests that grazing cattle and goat/sheep may have contributed to low productivity of the tuber crops in the study area. The farmers cultivate both cereal and tuber crops that are very important in every society (Mairiga 2014; Balami et al. 2011) and they depend on them for survival. No farmer will be comfortable to lose a stand of his/her crop and watch the pastoralists move freely with their herds.

8.4 Conclusion and Recommendations

In Nigeria, crops and livestock are produced mainly by small-scale farmers that are distributed all over the nation. The crop farmers cultivate small portions of land without fencing while their livestock counterparts rear few animals under extensive system, which lets the animals move around and feed with little or no form of restriction. This system of livestock production results in the destruction of crops, and this causes conflict between crop and livestock farmers. This scenario is detrimental to sustainable agricultural development hence the need to assess the impact of grazing livestock on cereal and tuber crops production in Abuja, Nigeria. The study focused on small-scale crop farmers in agrarian communities from five (out of six) area councils in Abuja. The farmers were asked to rate the impact of grazing livestock on crop production using a rating scale of 0–4. The data were analyzed using both inferential and descriptive statistics with impressive results. First, there were significant locational differences in the impact of grazing livestock on cereal and tuber crops production. For cereal crops, the mean responses showed that kwali Area Council was the most affected location while Kuje and Abaji Area Councils had the highest impact on tuber crops. This implies that the impact of grazing livestock on cereal and tuber crops depends on the dynamics of livestock and crop production in each location and therefore can be addressed based on prevailing conditions in such locations. The result also indicated that the level of destruction significantly depended on the type of livestock and crop produced. For cereal crops, cattle and goat/sheep were rated as the most destructive livestock, while for tuber crops, it was cattle. Generally, the least destructive livestock was domestic fowls. The policy implication is that, while stricter measures to control the extensive system of livestock production are required, this might not be the case with domestic fowls.

Again, the mean response showed that maize and cassava were the most affected cereal and tuber crops while the least affected were rice (cereal) and potato (tuber). This also connotes that rice and millet could be produced in some of the locations with minimal destruction from grazing livestock. In terms of cost estimate of damaged crops, the findings show that cassava farmers who were affected by grazing livestock lost an average of ₦11,745 (U$73), while potato farmers were the least affected, they lost an average of ₦3679 (US$23). Based on the findings, it can be concluded that the impact of grazing livestock is one of the key factors limiting cereal and tuber crops production, though the magnitude of the impact depends on location, livestock type and crop cultivated. It is clear that the estimated cost of damaged crops was relatively high and capable of eliciting aggressive response from the crop farmers.

Based on the findings, the following recommendations are made. The Federal Government of Nigeria should conduct a study to determine if the livestock farmers prefer grazing routes (Option A) or are willing to adopt intensive system of livestock production (Option B). Furthermore, the study should also find out how the various stakeholders perceive and prefer any of the options, and how they can be planned and implemented. This will help in addressing the impact of grazing livestock on crop production. The two options are spelled out below.

Option A If grazing routes are preferred by the farmers (stakeholders):

Government should establish a Rangeland and Grazing Routes (RGR) Commission that will be solely responsible for the demarcation, development, coordination and management of rangelands and grazing routes in Nigeria. The commission shall establish offices at the local government, state and federal levels and shall partner with the private sector. Bottom-up approach should be adopted in the demarcation of the routes so that it will emanate from communities. In addition, there is a need to establish Livestock Grazing Route Committees at community, local government and state levels.

Government should set up a task-force involving government and non-governmental organizations (private sector) to create awareness and inform stakeholders of the locations, size and importance of the use of new and existing grazing routes. The information should be disseminated through mosques, churches, radio, TV, schools, chats, maps and meetings with crop and livestock farmers' unions in their various communities.

Option B If intensive system of livestock production is preferred:

Government should teach farmers to adopt intensive system of livestock management through Agricultural Extension Agents (Livestock Subject Matter Specialist). Various information and communication technologies (ICTs) can be used to document and disseminate simple and locally adaptable ways of intensive system of livestock production so that they can watch and learn at their convenient time, e.g., using home videos and also mobile phone applications. This will make them to develop interest, try and practice intensive livestock production systems at their homes.

The documentation should be done in various languages including English, Hausa, Yoruba, Igbo and any other language that the farmers can understand. It shall cover all livestock that have the potential to damage crops and cause problem in the society. Agricultural extension agents who are livestock subject matter specialist can be trained to handle the documentation.

Acknowledgements This chapter is the outcome of a four-month engagement as a visiting scholar at the United Nations University–Institute for Natural Resources in Africa (UNU-INRA), Mineral Resources Unit, University of Zambia, Lusaka, Zambia. My unreserved thanks go to the entire Management and Staff of UNU-INRA wherever they may be for providing me this wonderful opportunity.

References

Abbass IM (2012) No retreat no surrender conflict for survival between Fulani pastoralists and farmers in northern Nigeria. Eur Sci J 8(1):331–346

Adebayo OO, Olaniyi AO (2008) Factors associated with pastoral and crop farmers conflict in derived Savannah Zone of Oyo State, Nigeria. J Hum Ecol 23(1):71–74

Adisa RS (2012) Land use conflict between farmers and Herdsmen–Implications for agriculture and rural development in Nigeria. Rural development–contemporary issues and practice. In: Adisa RS (ed) Intech Europe, University Campus, STeP Ri, Croatia

Adogi M (2013) Fulani-farmers conflicts in Nasarawa State: the ecology, population and politics. Murry Greens Consult, Abuja

Aggarwal YP (2002) Statistical methods, concepts, applications and computations, 2nd edn. Sterling Publishers Ltd., New Delhi

Ahmad IM, Samuel E, Makama SA, Kiresur VR (2015) Trends of area, production and productivity of major cereals: India and Nigeria scenario. Res J Agric For Sci 3(2):10–15

Ajah J (2012) Small-scale farmer's perception of the impact of grazing livestock animals on crop production in Abuja, Nigeria. Trends Agric Econ 5(4):115–123

Ajah J (2014) An evaluation of the impacts of cooperative membership and farmers' level of education on access to land in Abuja, Nigeria. J Sustain Dev Afr 16(2):147–156

Ajah J (2015) Comparative analysis of cooperative and non-cooperative farmers' access to farm inputs in Abuja, Nigeria. Eur J Sustain Dev 4(1):39–50

Aruleba JO, Ajayi AS (2011) Use of models in assessing the impact of system, land type and suitability on land degradation in South western Nigeria. Afr J Agric Res 6(5):1090–1096

Audu SD (2013) Conflicts among farmers and pastoralists in northern Nigeria induced by freshwater Scarcity. Dev Country Stud 3(12):25–32

Audu SD (2014) Freshwater scarcity: a threat to peaceful coexistence between farmers and pastoralists in northern Nigeria. Int J Dev Sustain 3(1):242–251

Ayanda IF (2013) Assessment of effect of climate change on the livelihood of pastoralists in Kwara State, Nigeria. J Dev Agric Econ 5(10):403–410
Balami DH, Ogboru I, Talba DM (2011) The cereal economy in Nigeria and the Sub-regional dimension. In: SSSG series, vol 1(29), Benue State University, Destiny Ventures, Makurdi, Nigeria
Balogun O (2001) The Federal Capital Territory of Nigeria: geography of its development. University Press, University of Ibadan, Ibadan
Bell LW, Gilkes RJ (2010) Impacts of soil compaction by livestock on crop production: a modeling analysis. In: Proceedings of the 19th world congress of soil science: soil solutions for changing world, symposium 3.1.2 farm system and environment impacts. International Union of Soil Sciences, Brisbane, Australia, 1–6 Aug 2010, pp 117–120
Belsky AJ, Matzke A, Uselman S (1999) Survey of livestock influences on stream and riparian ecosystems in the Western United States. J Soil Water Conserv 54(1):419–431
Blench R (2010) Conflicts between pastoralists and cultivators in Nigeria. Review paper prepared for DFID, Nigeria, Kay Williamson Educational Foundation, Cambridge, UK, pp 1–16
Boyle TB (2004) Urbanization: an environmental force to be reckoned with. In: Population reference Bureau, viewed 2 June 2015, from www.prb.org/articles/2004
British Council Nigeria (2012) Gender in Nigeria Report (2012) improving the lives of girls and women in Nigeria: issues, policies, action, 2nd edn. British Council Nigeria
Champion PD, Beadel SM, Dugdale TM (2001) Turf communities of Lake Whangape and experimental assessment of some potential management techniques. In: Science for conservation, vol 186. Department of Conservation, Wellington, p 54
Clary WP, Kinney JW (2002) Streambank and vegetation response to simulated cattle grazing. Wetlands 22(1):139–148
Dalil M, Nsini EE (2014) An analysis of landuse/landcover change distribution in Kuje Area Council, Federal Capital Territory, Abuja, Nigeria. Ethiop J Env Stud Manag 7(2):210–217
Dawan PD (2000) Brief history of the creation of Federal capital territory (FCT). In: Dawan PD (ed) Geography of Abuja federal capital territory. Famous/Asanlu Publishers, Minna, pp 1–8
Devandra G, Fuller F (1989) Pig production in the tropics. Oxford University Press, London
Drewry JJ, Cameron KC, Buchan GD (2008) Pasture yield and soil physical property response to soil compaction from treading and grazing—a review. Aust J Soil Res 46:237–256
Enete AA, Amusa TA (2010) Challenges of agricultural adaptation to climate change in Nigeria: a synthesis from the literature. Field Action Sci Rep 4:1–12
Ezeibe ABC (2010) Profitability analysis of pig production under intensive management system in Nsukka local government area of Enugu State, Nigeria. Int J Econ Dev Res Invest 1(2):48–54
FAO (2002) World agriculture: towards 2015/2030: summary reports. Food and Agriculture of the United Nations, Rome
FAO (2013) Rice market monitor, vol XVI, Issue no 2. Trade and Market Division, Food and Agriculture Organization of the United Nations, Rome, Apr 2013, pp 1–36
Federal Democratic Republic of Ethiopia (2011) The path to sustainable development: Ethiopia's climate resilient green economy strategy. Federal Democratic Republic of Ethiopia
Field A (2005) Discovering statistics using SPSS, 2nd edn. Sage Publications, London
Forsyth DM, Coomes DA, Nugent G, Hall GMJ (2002) Diet and diet preferences of introduced ungulates (order: Artiodactyla) in New Zealand. N Z J Zool 29(4):323–344
Gefu JO, Gills JL (1990) Pastoralist, ranchers and the states in Nigeria and North America: a comparative analysis. Nomadic Peoples 25–27:34–50
Gray CD, Kinnear RP (2011) IBM SPSS statistics 19 made simple. Psychology Press, East Sussex
Greenwood KL, McKenzie BM (2001) Grazing effects on soil physical properties and the consequences for pastures: a review. Aust J Exp Agric 41:1231–1250
Greviliot F, Muller S (2002) Grassland ecotopes of the upper Meuse as references for habitants and biodiversity restoration: a synthesis. Landscape Ecol 17(1):19–33

Ibrahim K, Shamsudin MN, Yacob R, Radam AB (2014) Technical efficiency in maize production and its determinants: a survey of farms across agro-ecological zones in Northern Nigeria. Trends Agric Econ 7(2):57–68

Idowu AA, Ayoola SO, Opele AI, Ikenweiwe NB (2011) Impact of climate change in Nigeria. Iran J Energy Environ 2(2):145–152

IFAD (2007) Rural poverty in Nigeria: agriculture in the Federal Republic of Nigeria. In: International fund for agricultural development, viewed 11 Oct 2016, from http://www.ruralpovertyportal.org/web/guest

Ijirshar VU (2015) The empirical analysis of agricultural exports and economic growth in Nigeria. J Dev Agric Econ 7(3):113–122

Ismaila U, Ghana AS, Tswanya NM, Dogara D (2010) Cereals production in Nigeria: problems, constraints and opportunity for betterment. Afr J Agric Res 5(2):1341–1350

Khanal RC (2009) Climate change and organic agriculture. J Agric Env 10:100–110

Kie JG, Boroski BB (1996) Cattle distribution, habitats and diets in the Sierra Nevada of California. J Range Manag 49:482–488

Kolawale O, Ojo SO (2007) Economic efficiency of small-scale food crop production in Nigeria. J Soc Sci 14(2):123–130

Liverpool-Tasie L, Kuku O, Ajibola A (2011) A review of literature on agricultural productivity, social capital and food security in Nigeria. Nigeria strategy support program working paper 21. International Food Policy Research Institute, Washington, DC

Mairiga TH (2014) Trend analysis of productivity of some selected cereal crops in Nigeria: 1983–2008. Res Humanit Soc Sci 4(8):110–116

Marty JT (2005) Effect of cattle grazing on diversity in Ephemeral Wetlands. Soc Conserv Biol 19(5):1626–1632

McDowell RW, Drewry JJ, Muirhead RW, Paton RJ (2005) Restricting the grazing time of cattle to decrease phosphorus, sediment and *E. coli* losses in overland flow from cropland. Soil Res 43(1):61–66

NBS (2012) Annual abstract of statistics, 2012. National Bureau of Statistics, Federal Republic of Nigeria

Nweze BO, Otuma MO, Ekwu LG, Oga GUEO (2003) Agro pastoral system in humid tropics. A case study of Rice Muturu integrated farming practice in Ebonyi State. In: Proceedings of the 8th annual conference of animal science association of Nigeria, Maiduguri, Nigeria, 17–19 Sept 2003, pp 139–149

Nyong A, Fiki C (2005) Draught related conflicts, management and resolutions in the West African Sahel: human security and climate change. An international workshop, GECHS, CICERO and PR 20, 21–23 June 2005. Holmen Fjord Hotel, Oslo

Ofuoku AU, Isife BI (2009) Causes, effects and resolution of farmers-nomadic cattle herders conflict in Delta State, Nigeria. Int J Soc Anthropol 1(2):049–054

Ohen SB, Ajah EA (2015) Cost and return analysis in small scale rice production in Cross River State. Int Res J Agric Sci Soil Sci 5(1):22–27

Okoli AC, Atelhe GA (2014) Nomads against natives: a political ecology of Herder/Farmer conflicts in Nasarawa State, Nigeria. Am Int J Contemp Res 4(2):76–88

OPEC (2013) Annual statistical bulletin: organization of the petroleum exporting countries. Helferstorferstrasse, Vienna

Ovwigho BO, Mmereole FUC, Udeh I, Akporhuarho PO (2009) Comparison of constraints to poultry producers in Delta State Nigeria. Int J Poult Sci 8(5):480–484

Oyeyinka RA, Bolannwa KK (2009) Using Nigerian agricultural cooperate and rural development bank small holder direct loan scheme to increase agricultural production in rural Oyo State, Nigeria. Int J Agric Econ Rural Dev 2(1)

Popotnik GJ, Giuliano WM (2000) Response of birds to grazing of riparian zones. J Wildl Manag 64(4):976–982

Radford BJ, Yule DF, Braumack M, Playford C (2008) Effects of grazing sorghum stubble on soil physical properties and subsequent crop performance. Am J Agric Biol Sci 3(4):734–742

Rahji MAY, Omotesho OA (2006) Technical inefficiency and competitiveness in production: The case of rice farmers in Niger State, Nigeria. Agrosearch 8(1&2):67–79

Robert AY (2011) Mixed analysis of variance models with SPSS. Statistics, Social Sciences and Mapping Group, Information Technology Services/Academic Computing Services, viewed 11 Oct 2016, from www.myu.edu/its/soc.sci//Docs/spssmixed.pptforRobert

Shah DA, Madden LV (2004) Nonparametric analysis of ordinal data in designed factorial experiments. Phytopathology 94(1):33–43

Smulders S, Withagen C (2012) Green growth: lessons from growth theory. World bank policy research working paper 6230. The World Bank

Tonah S (2006) Managing farmer-herder conflict in Ghana Volta Basin, Ibadana. J Soc Sci 4(1):33–45

Touzard B, Clement B (2001) Plant diversity dynamics in an eutrophic alluvial reed bed after experimental small-scale disturbances. Bot Helv 111(1):45–58

USDI (1994) Rangeland Reform '94: draft environmental impact statement. Bureau of Land Management, United States Department of the Interior, Washington, DC

Vaughan OI, Afolami AC, Oyekale TO, Ayegbokiki AO (2014) An analysis of Nigeria food imports and bills. Int J Econ Comm Manag 11(9):1–14

Chapter 9
Greening Farm–Family Business in Nigeria: Does Climate Change-Induced Migration Matter?

Oluwafunmiso Adeola Olajide

9.1 Introduction

The Green Economy Initiative looks promising for Africa because it focuses on green investments in priority sectors like natural resources. In the case of Sub-Saharan Africa, natural resources particularly land is the "gold mine" from which over 60% of the population directly or indirectly eke out a living. The use of land in production activities which lead to national food supplies and foreign exchange earnings is the center piece of rural livelihoods. Several forms of small- and medium-scale processing and marketing activities developed around it which form alternative income-generating sources for rural households, particularly smallholder farmers. Smallholder farmers form the backbone of the agricultural economy in most SSA countries, but in most countries, the sector is a struggling one even though it has a lot of potentials. For example, in Nigeria, it was neglected for several decades while the economy depended on crude oil; yet, during these years, it supplied 40% of Nigeria's gross domestic product.

Apart from direct macroeconomic planning neglect, a major problem confronting the sector across Africa is climate change. Climate change was in the past thought of as a future occurrence, but the future is here. The evidence of climate change is reflected in extreme temperature, droughts, heavy rainfall, floods, and severe weather storms. These shocks lead to loss of farm lands, livestock resources, homes, and livelihoods. These translate to reduced food production, income, and asset base. In other words, the poverty trap is tightened and the cycle reinforced. This appears as a bi-directional cyclical relationship because most rural people live below national poverty lines and in environments lacking proper amenities that support a high standard of living, as such the level of vulnerability is high. The occurrence of climate variability and shocks threatens their very existence.

O. A. Olajide (✉)
Department of Agricultural Economics, University of Ibadan, Ibadan, Nigeria
e-mail: funso.olajide@ui.edu.ng; adeola.olajide@daad-alumni.de

C. Atewamba and D. Yong Ngondjeb (eds.), *Inclusive Green Growth*, Advances in African Economic, Social and Political Development,
https://doi.org/10.1007/978-3-030-44180-7_9

The impact of climate shocks has severe effects on rural communities because production systems which include tree felling, bush burning, and other ecosystems unfriendly approaches are often practiced. According to Edame et al. (2011), climate change will negatively affect crop yield, stability of food supplies, and the ability of people to access and utilize food in many parts of the developing world, where much of the population, especially the poor, rely on local supply systems that are sensitive to climate variation. The disruption of existing food and water systems will have devastating implications for development and livelihoods (Adger et al. 2009). In the event of severe shocks, households in rural communities tend to seek off-farm/off-household opportunities to tie them over difficult seasons. Traditionally, they diversify risks by planting several crops, developing a side-trade, taking on jobs with regular payment and migrating. It is, however, becoming more common to have more households opt for migration as a result of climate shock impact on the farm and other means of livelihoods (Machiori et al. 2012). Migration, defined as the movement of people from one location to the other for economic, social, or political reasons (Sinha 2005; Jolly and Reeves 2005) could be the end result of a process or induced by an event such as extreme climate shocks. Ravenstein (1889) was one of the first to suggest that various factors, including "an unattractive climate" tend to push persons from one area to another area. Migration is considered a viable option because the migrant, if successful, can support the family at difficult times while helping at least one other member have an opportunity for a better life. It leads to structural changes in rural communities which may not support community development or growth.

While the challenges associated with climate change are obvious, one must not be oblivious of its many opportunities which include developing Green SMEs. Taking advantage of these opportunities requires the knowledge of how climate change affects the farm–family–household system and what opportunities and challenges exist for transitioning into a rural green economy. This calls for a detailed understanding of the current development in order to offer future strategies that will lead to a successful transition to a Green Economy in rural Africa.

The Green Economy Initiative includes activities that will lead to poverty reduction and sustainable agriculture, two-key issues which are relevant to the country. Also several studies linking migration and climate change have been qualitative in nature (Janssen and Ostrom 2006); this study adds to the body of knowledge by taking a quantitative approach. Also, the farming and rural systems approach is used (Doppler 2000, 2002). The approach puts the people at the center. Finally, this research is necessary for rural transformation and agricultural development not only in Nigeria but also in Africa (Nwajiuba and Ejiogu 2008).

The rest of this chapter presents the theoretical background for the study, the conceptual framework of how climate change could induce migration within the farm–family–household system (latter end of Sect. 9.1) as well as the methodological approach and the tools of analyses applied in the study (Sect. 9.2). These findings, discussions, concluding remarks, and policy recommendations are presented in the latter end of the chapter.

9.2 Theoretical Framework

The approaches to rural migration studies have revolved around some key models: the classical two-sector model, the neoclassical and expected income (Todaro) two-sector models, human capital models and the new economics of labor migration (NELM); the NELM is reviewed below for the purpose of this study.

New Economics of Labor Migration (NELM)

The fundamental view of the new economics of labor migration is presented in Stark (1991) and Stark and Bloom (1985). Under NELM, the decision to migrate is not taken by an individual but is agreed on by the household or family members. The NELM posits that people act collectively to minimize risks, loosen constraints created by market failures such as missing or incomplete capital, insurance, and labor markets; to maximize income. Migration often generates remittances as such it is seen as an intermediate investment that facilitates the transition from familial to commercial production by providing the rural households with capital and a means to reduce their risks (Mensah-Bonsu and Burger 2008). This perspective distinguishes the NELM from other theories of migration because it changes the unit of analysis from the individual to the household and brings about new motives for migration other than maximizing income.

Among other things, households want to minimize risk by diversifying their source of income (De Haas 2007); they achieve this by sending some member(s) of the household to another place where the income from the place of origin and destination are uncorrelated or are negatively correlated. Hence, the migrant member will support the family during bad times (like crop failure) while the family covers her migration cost and also supports her during bad economic conditions at the destination (Stark and Bloom 1985; Taylor 1999). Stark (1982, 1978) cited in Stark (1991) argues that an implicit contractual arrangement exists between migrant and household. Mutual altruism reinforces this implicit contract, as do inheritance motives (i.e., non-remitting migrants stand to lose their rural inheritance) and migrants' own aversion to risk, which encourages them to uphold their end of the contract in order to be supported by the rural household should they experience an income shock (e.g., unemployment) or other misfortune in the future. Fletcher and Clark (1997) and Rouse (1991) also point to the importance of rural households-of-origin as refuges for migrants who fall ill or suffer other sorts of misfortune (e.g., trouble with the law, substance dependence, etc.) that prevent them from working or residing at the migrant destination for extended periods of time.

Though the NELM model has many noble contributions for the understanding of migration, it is also criticized for totally ignoring the role of individuals in the migration process. While the situation of the household might affect the likelihood of individuals' migration, it is also worth noting that the individual can have an important role in the decision to migrate (Hoddinott 1994).

NELM Theoretical Model[1]

Migration involves human beings who not only own "labor" and move along with it but also possess feelings and independent wills, as such migration cannot be treated as traded goods. This is because the decision to migrate is borne out of several factors which include interpersonal "livelihood" measured by income comparisons within own reference groups. The feeling of relative deprivation or satisfaction arrived at as a result of the cost–benefit comparison can trigger migration for the purpose of changing the relative position in the same reference group, changing the reference group, and or showing preference for membership in a low relative deprivation reference group. The final decision to migrate is expected to be heightened by the extent of deprivation experienced by the individual as well as the degree of income inequality in a reference group. As migration occurs, the relative deprivation perceived by non-migrants may change, thereby inducing second waves of migration. If relative deprivation is gauged through a comparison with a reference group statistic such as average income, migration by low income relatively deprived individuals will cause this statistic to increase and thereby induce migration by other individuals who become increasingly relatively deprived.

Migration behavior of individuals also differs based on skills level; it may be assumed then that skills differ and employers have imperfect skills information. For a given profession, workers with skill S receive wages $W_P(S)$ and $W_R(S)$ from employers at P and R (representing the place of origin and destination, respectively) assuming that skill follows a uniform distribution along a unit interval that the functions $W_P(S)$ and $W_R(S)$ are non-decreasing and linear, and S is known by employers at P and R. Based on these, for levels of S below a threshold, S^*, the wage at P will be higher than the wage at R. That is, if $S < S^*$, then $W_P(S) > W_R(S)$; for $S \geq S^*$, then $W_P(S) < W_R(S)$. There will be no incentive for the lowest skilled person to migrate. But, if the assumption holds that R employers cannot observe the true skill level of individual P workers but know the distribution of S and will pay migrants from P a wage that is equal to average productivity of migrant's groups, then the threshold S^* is not required, and it vanishes and one of two corner solutions occurs: either no migration at all or migration by all.

This is because highly skilled workers who migrate under perfect information may not do so if the pooled wage is too low. If they do not, the pooled wage is lowered so that the next highly skilled group does not find it advantageous to migrate. Climate change affects labor use on farm, output, and income generated. Migration confers benefits upon those who stay behind in addition to those associated with a left-ward shift in labor supply curve. Employers, opportunities in other places, and images of success attract migrants. For as long as the alternative option is perceived to be better, migration response will be generated.

[1] Based largely on Stark and Bloom (1985).

9.3 Conceptual Framework

The conceptual framework below indicates that the farm–family–households are in a system where the family makes the decision about the allocation of resources which they own in order to meet their goals. The system is governed by their norms and culture. It is within this system that the families experience climate shocks which are as a result of climate variability in the locality or environment. The shocks include drought, flooding and heavy rains, and extreme temperature. The shocks affect the farm–family–household systems in different ways because they depend on nature and may find it difficult to insure their own food security and living standard (see Fig. 9.1). The situation presents both opportunities and challenges to the farm family. In the case of adverse situations, families must re-evaluate their circumstance, make decisions that will ensure their continued survival. The process results in either a reallocation of resources (labor quite often) to off-farm/non-farm activities or migration of some household members—the household head or someone who can be economically productive. It could also be a case of hedging for the future where young adults are sent to the city to be educated, learn a trade, or get an employment to support him and perhaps, hopefully the family he left (Olajide 2013). These are the two traditionally documented options.

However, Green SMEs (G-SMEs) development within the Green Growth Initiative presents a third possibility which could be offered to farm families as an economically viable option while adapting to climate variability shocks and mitigating the effects of climate change. Since the family makes decisions over its own resources, introducing several G-SMEs alternatives through several policy varieties offer them more choices for resource allocation and use efficiency for a sustainable livelihood. But migration seems to be the choice option because of the continued effect of high temperature, drought and flood on agricultural activities, the family, and the household. In extreme temperature cases, farm activities are delayed leading to loss of crop output, loss of birds, reduced income, reduced household food supply, and reduced nutrition, and ultimately a poor standard of living. This cycle of events could be mitigated by the

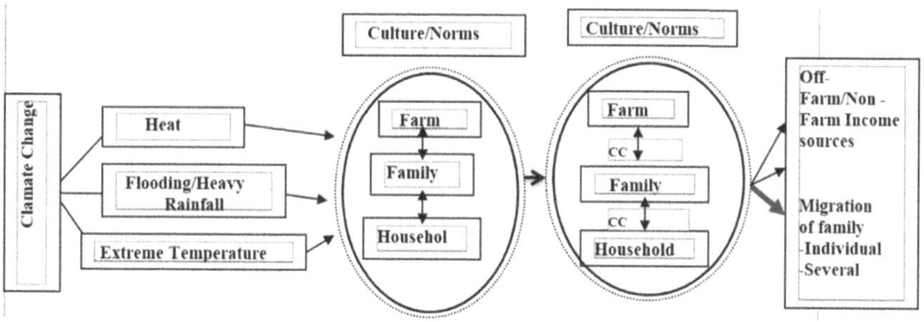

Fig. 9.1 Flow of climate shock effect within the farm–household–family system. *Source* Author's Concept

introduction of G-SMEs. The appropriate setting, however, can be created with a proper understanding of how climate change drives migration through its direct and indirect effects on the family's goals and decision making.

9.4 Research Methodology

The Farming and Rural Systems Approach (FRSA) was used in the study. The FRSA provides the philosophy, the concept, and the strategy for developing and introducing solutions offered to families, communal, and regional decision-making bodies to solve problems at farm, household, family, village, and regional levels (Doppler 2000, 2002; Akinsanmi 2005). It takes into consideration farmer's condition and development over time. It considers methodologies in the systems linkage and integrative potential; the strategies for farming and rural systems development; how local resources and local knowledge can contribute to this; and what roles institutions play in defining and implementing the strategies. The approach considers the farm–family–household as a system and ensures that the interventions offered are relevant to them.

Farming and rural systems are characterized by different problems, objectives, and decision-making bodies and levels; hence, require different systems with relations to each other. At the different level of systems, there is also a complex situation with respect to horizontal dimensions. Decision making takes place wherever individuals or decision-making bodies of a society will decide following their objectives and needs. At the family level (the lowest social unit in a society), the family decides on the allocation of the family resources for the best mix of activities in those units of action for which families are responsible. Since the objectives of a family are not restricted to the farm only, but related to the needs of the family, decisions in a family with respect to resource allocation to improve living standard cannot be dealt with considering the farm only. Whether family labor or family capital goes into the farm, the household, or into outside jobs depends on the expectation in each of the alternative and competing sectors of a family and the respective contribution to family objectives (Doppler 2000, 2002; Akinsanmi 2005).

It is for this reason that a system approach has to include the farm (production and agricultural resource use); the household (consumption and general resource use such as capital, energy, water, services); storage, processing and transportation sector (adding value to the farm products for better market prices and preparing for consumption continuously over time); and off-farm/off-household alternatives (family resource in external use and cash income to the family). At the village level, it could be the physical village infrastructure, social services, and allocation of resources. The same applies for water shed or regional levels (Fig. 9.2).

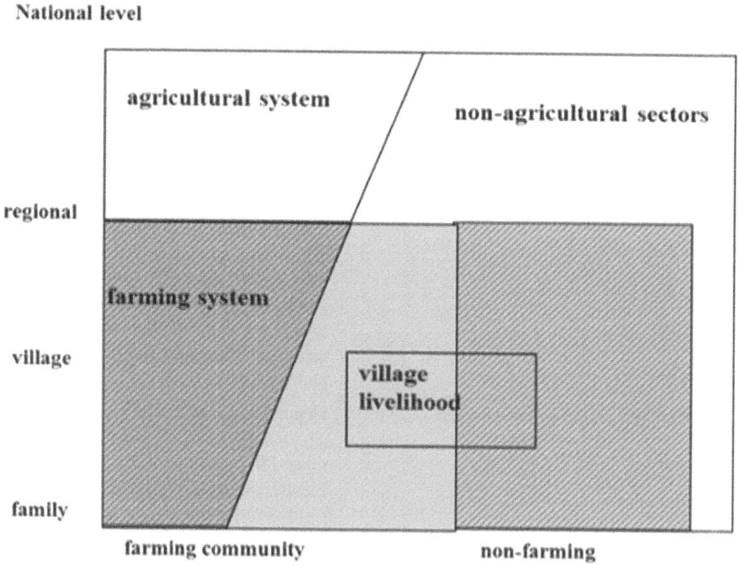

Fig. 9.2 Agricultural, farming, and rural systems. *Source* Doppler (1998, p. 5) cited in Doppler (2002) and Akinsanmi (2005)

9.4.1 Criteria for Living Standard

Living standard is a complex parameter and is composed of several aspects (Doppler 1997, cited in Akinsanmi 2005). This is determined by a set of criteria which consists of family income, cash and liquidity, degree of independence from resource owners, food supply and food security, supply of housing, sanitary equipment, energy and clothing as well as health condition, education and qualification, and social security. These criteria involve both social and economic aspects and both quantitative and qualitative values (Akinsanmi 2005). The living standard criteria will be used to measure the economic success and security of the farming systems. A summary of these will give the final socio-economic context of the system (Table 9.1).

9.4.2 Survey Design

The survey was designed to obtain information which would inform strategy or policy decisions and provide the kind of knowledge that will make interventions suitable to the conditions of respondents. The survey was carried out according to a statistical plan embodying random selection of households. The questionnaires were designed to give information on different aspects of respondent's lives and were administered using the participatory approach.

Table 9.1 Living standard criteria

S/N	Components	Description
1	Family income	This is the total of farm, household and off-farm/off-household income. Cash family income is the cash part of family income (excluding subsistence and imputed values)
2	Cash and liquidity	Ensuring that cash is available at the point in time when essential duties require cash, such as ensuring an existence minimum, ensuring payment for external resources (land, water, credit) if otherwise, those needed resources would not be available in the future or own resources (e.g., land, livestock) would be lost; ability to make payment for health services
3	Independency from resource owners	Dependencies of families on other individual persons, families, or organizations are often related to provision of resources (land, water, credit), means of production (seeds, fertilizers, pesticides), or to selling or processing of products
4	Food supply and food security	This includes the amount and quality of supply of food from the farm as well as from the market over time
5	Supply of water, housing, sanitary equipment, energy, and clothes	Amount and quality of water as well as the resource required to ensure availability is central to the level of living standard. While housing and clothes are often of less importance, the sanitary conditions require attention
6	Health conditions of the family	In insuring against minimal health problems, the amount and quality of food and water supply as well as the sanitary behavior are the main prophylactic sectors which will have to be related to curative measures using own or local knowledge and natural potential as well as modern medicine
7	Education and qualification	Education and qualifications add to new dimensions and possibilities to experience, and knowledge from the family and own society. It is a long-term issue and is relevant for decision making for all remaining in the farming business as well as for those seeking employment outside their farms
8	Social security and safety	To ensure the survival or well-being of older people, widows, orphans and handicapped, many family decisions are made to provide the economic base for them. This is done by accumulating capital in different forms, higher education for children, and social norms

9.4.3 Study Area

The study was carried out in Abia state. The state lies approximately within latitudes 4° 40′ and 6° 14′ north, and longitudes 7° 10′ and 8° east (Fig. 9.3). It covers a land mass of 6320 km^2 and has a population of 2,833,999 according to the 2006 population censors; the population density is 450/km^2. There are three agricultural development zones, namely Aba, Umuahia, and Bende. Over the years, the State has witnessed increased manifestations of climate events including flooding and soil erosion.

The worst affected areas include Umuahia, Ikwuano, Ohafia, Abariba, Nkporo, Igbere, Isikwato, and parts of Arochukwu in Bende zone; others include Umuezeukwu and other surrounding villages like Umuodeche, Nbawsi, Umuogu, Ikputu, and Agburuike in Isiala Ngwa North L.G.A.[2] Since the people of the area are predominantly farmers and depend on their farmlands for subsistence, these events which may affect their crop productivity could ginger up the people to adjust, migrate, or find alternative sources of livelihood. This study sought to empirically determine the event of households migrating in response to this climate change-related events and how they interact with the people's living standard.

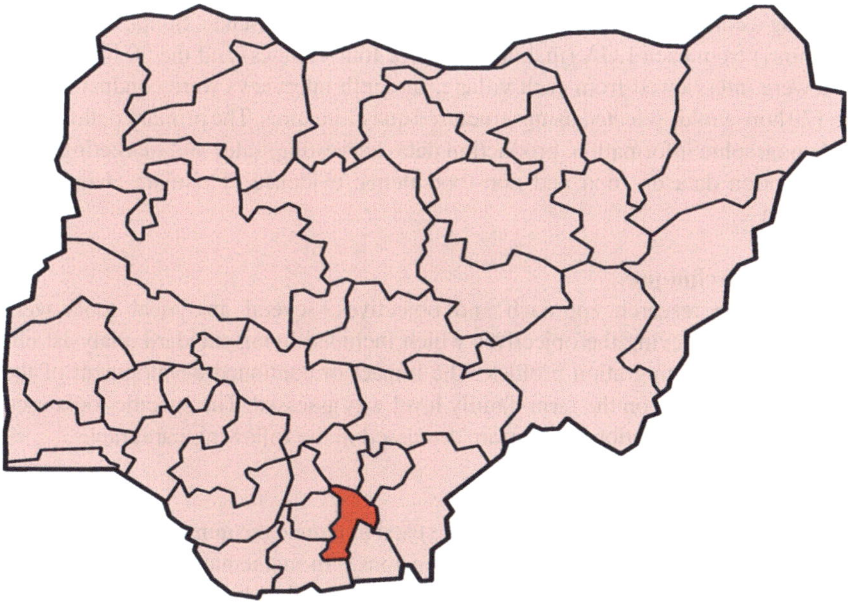

Fig. 9.3 Map of Nigeria with Abia state insetted. *Source* https://commons.wikimedia.org/wiki/File:NigeriaAbia.png

[2]A map of Abia state showing local government areas can be accessed through the weblink: http://www.abiastate.gov.ng/wp-content/uploads/2011/08/abiamap_Labels-e1314525045300.png.

The study was carried out in Abia state because of the knowledge of the prevailing environmental situation in the area. Also, although the per capita income is about ₦30,000, a large proportion of the population depends on the natural environment for their sustenance.

9.4.4 Data Source, Collection, and Analysis

Primary data were used for the study. Primary data were obtained from households in the study area. Collection of data was through a structured questionnaire designed to collect information on the socio-economic characteristics of the households, household migration status, asset ownership, etc.

Sampling Technique

Multistage sampling technique was used in the survey. Abia state and the ADP zone (Bende ADP zone) where the study was carried out were purposively chosen in the first stage based on the severity of the environmental issues in the areas; then in the second stage, two local government areas (LGAs Bende and Ikwuano) were selected from the zone using simple random sampling techniques. Simple random sampling technique was also applied to select two villages (Bende, Itunta, Ohafia, and Abommri) from each LGA (making a total of four villages) and the 30 households who were interviewed from each village. In-depth interviews were conducted with the 120 households selected using structured questionnaires. The primary data consist of demographic information, production data, harvesting, sales and marketing data; consumption data on food and non-food items; evidence of climate change and migration.

Analytical Techniques

Based on the research approach and objectives, several analytical tools were employed in achieving the objectives which included living standard analysis; climate shocks, and migration profiles. The impact of continued development of the situation or changes on the farm–family level was assessed. The specific tools used and their implicit functional forms are discussed in the following paragraphs.

Principal Component Analysis

PCA is a multivariate statistical technique used to reduce the number of variables in a data set into a smaller number of "dimensions." In mathematical terms, from an initial set of n correlated variables, PCA creates uncorrelated indices or components, where each component is a linear-weighted combination of the initial variables. Mathematically, the transformation is defined by a set of p-dimensional vectors of weights or loadings (1) that map each row vector $X(i)$ of X to a new vector of principal component scores (2) given by (3)

$$W(k) = \left(\omega_1, \ldots \omega_p\right)(k) \tag{1}$$

$$t(i) = \left(t_1, \ldots t_p\right)(i) \tag{2}$$

$$t_k(i) = X(i) \cdot W(k) \tag{3}$$

in such a way that the individual variables of **t** considered over the dataset successively inherit the maximum possible variance from **x**, with each loading vector **w** constrained to be a unit vector (Bro and Smilde 2014).

The tool was applied in selecting the variables or components which were used to cluster the households. The initial variables were 9 and they were selected to cover the farm–family household system. The initial variables were standardized before carrying out the principal component analysis. After the PCA was applied 4 components were identified and the factor scores for 3 out of the 4 components were used to cluster the data using a hierarchical clustering procedure. The PCA approach was also used to define the components of climate change-induced migration and living standard. The 4 components identified for CC-IM are the migrant component, the remittance component, the flood shock component, and the Erosion shock component. Those identified for living standard, derived from the 8 criteria setup are the household head economic viability component, the asset/income component, the food security component, and the vulnerability component. The CC-IM regression scores generated from the PCA were regressed against productivity measures, production outcome variables, and each of the living standard components to estimate the effects on these.

Hierarchical Clustering

Cluster analysis or clustering is the task of grouping a set of objects in such a way that objects in the same group are more similar to each other compared with those in other groups (clusters). The first stage of the analysis involved a hierarchical clustering technique to group the sampled households into clusters of farming systems which are homogenous within but heterogeneous between one another.

The selection of variables to be included in clustering analysis was done with regard to both theoretical/conceptual and practical considerations; only those variables that characterize the objects being clustered, and relate specifically to the objectives of the cluster analysis were included. The final set of variables used in the clustering procedure was selected through a principal component analytical technique.

The *k*-clustering approach was used in building the final clusters in order to have clear systems that are uncorrelated. The resulting farming systems (clusters) were described and labeled based on the variables and components used. Three clusters were obtained; the description and differences are presented in Table 9.2. All the farming systems identified derive their income mainly from farming but allocate their resources differently. The RRLI farming system has the highest level of education, labor, land, and assets and meets most household food needs through market purchases compared with the other two. The RPNLI has the least level of resources over all but has a slightly higher level of education and land resources compared with the RPLI. The clusters formed the basis for further analysis and the results were interpreted in a comparative manner.

Non-parametric Tests

The Kruskal–Wallis test was applied to examine the differences and compare the 3 clusters while the Mann–Whitney U test was used to identify or locate specific cluster relations.

The Kruskal–Wallis test

It is the most widely used non-parametric technique for testing the null hypothesis that several samples have been drawn from the same identical population. The test uses more information than the median test, it is usually more powerful and is preferred when the available data are measured on at least the ordinal scale. A significant Kruskal–Wallis test indicates that at least one sample stochastically dominates one

Table 9.2 Descriptive of cluster variables for the 3 farming systems

Item	RPLI-FS ($n = 52$)	RRLI-FS ($n = 39$)	RPNLI-FS ($n = 29$)	All ($n = 120$)	Sig. of mean diff.
Age	47.31 (11.05)	49.90 (11.10)	53.07 (12.77)	49.54 (11.63)	*b
Education years	7.02 (3.81)	10.28 (3.04)	8.41 (3.98)	8.42 (3.86)	*a, c
Labor hours allocated to farm	1244.42 (418.63)	1463.08 (320.35)	864.26 (465.35)	1223.61 (457.47)	***a, b, c
Labor hours allocated off-farm	121.85 (79.65)	170.46 (75.45)	129.10 (70.09)	139.40 (78.55)	***a, c
Household size	4.19 (1.469)	7.03 (1.94)	7.24 (2.01)	5.85 (2.28)	***a, b
Asset value (₦)	78,177.53 (170,186.02)	115,604.25 (148,144.45)	65,714.00 (105,094.67)	87,329.20 (149,774.23)	***a, b
Farm size (ha)	3.67 (4.94)	15.48 (39.84)	3.98 (3.10)	7.58 (23.45)	***a, c
Household non food (₦)	191,289.70 (102,059.91)	355,382.36 (189,594.53)	459,596.41 (274,882.53)	309,460.61 (214,494.57)	*** a, b
Household food expense (₦)	91,254.23 (71,443.95)	197,300.51 (128,820.44)	65,453.79 (60,954.68)	119,484.17 (106,794.7)	*** a, b

Note: *RPLI_FS* resource poor labor intensive farming system; *RRLI-FS* resource rich labor intensive farming system; *RPNLI-FS* resource poor non-labor intensive farming system

***Significant at 1% based on Kruskal–Wallis test. Values in parenthesis are standard deviation. "a" significant between clusters 1 and 2; "b" significant between clusters 1 and 3; "c" significant between clusters 2 and 3

other sample. The test does not identify where this stochastic dominance occurs or for how many pairs of groups stochastic dominance obtains.

The test statistic is given by

$$K = (N - 1)\frac{\sum_{i=1}^{g} n_i(\bar{r} - r_i)^2}{\sum_{i=1}^{g}\sum_{j=1}^{n_i}\left(r_{ij} - \bar{r}\right)^2} \tag{4}$$

where

n_i the number of observations in group i
r_i the rank (among all observations) of observation j from group i
N is the total number of observations across all groups

$$\bar{r}_i = \frac{\sum_{j=1}^{n_i} r_{ij}}{n_i} \tag{5}$$

$\bar{r}$ is the average of all the r_{ij} (Source: Wayne 1990).

The Mann–Whitney U test

This was used to examine the location of the mean differences between the clusters. It is a non-parametric test of the null hypothesis that two samples come from the same population against an alternative hypothesis, especially that a particular population tends to have larger values than the other.

The statistic is given by:

$$U_1 = R_1 - \frac{n_1(n_1 + 1)}{2} \tag{6}$$

where n_1 is the sample size for sample 1, and R_1 is the sum of the ranks in sample 1 (Corder and Foreman 2014).

Regression Models

The general linear model was used to identify profile green SMEs options and the possible impact on family income. Endogenous switching regression model was used to examine the effect and counter factual of migration on different farming systems.

General Linear Model (GLM) Univariate Analysis

The GLM univariate procedure provides regression analysis and analysis of variance for one dependent variable by one or more factors and/or variables. The factor variables divide the population into groups. It is used to test null hypotheses about the effects of other variables on the means of various groupings of a single dependent variable. It is also used to investigate interactions between factors as well as the effects of individual factors, some of which may be random. Simply put, the GLM univariate procedure is used to model the value of a dependent scale variable based on its relationship to categorical and scale predictors (Rawlings et al. 1998). The functional form is given as:

$$Y = XB + U \quad \text{or} \quad Y = b_0 + bx + u \tag{7}$$

This can be expanded to give the following:

$$\begin{cases} Y_1 = X_{11}\beta_1 + \cdots + X_{1L}\beta_L + \varepsilon_1 \\ \cdots \\ Y_j = X_{j1}\beta_1 + \cdots + X_{jL}\beta_L + \varepsilon_j \end{cases}$$

The matrix is given by:

$$\begin{pmatrix} Y_1 \\ \cdot \\ Y_j \end{pmatrix} = \begin{bmatrix} X_{11} & \cdot & X_{1L} \\ \cdot & \cdot & \cdot \\ X_{j1} & \cdot & X_{jL} \end{bmatrix} \begin{pmatrix} \beta_1 \\ \cdot \\ \beta_L \end{pmatrix} + \begin{pmatrix} \varepsilon_1 \\ \cdot \\ \varepsilon_j \end{pmatrix} \tag{8}$$

where

$Y = \begin{pmatrix} Y_1 \\ \cdot \\ Y_j \end{pmatrix}$ is a matrix of outcome variables (Observed data).

$X = \begin{bmatrix} X_{11} & \cdot & X_{1L} \\ \cdot & \cdot & \cdot \\ X_{j1} & \cdot & X_{jL} \end{bmatrix}$ is a matrix pre-program variables or covariates.

$\beta = \begin{pmatrix} \beta_1 \\ \cdot \\ \beta_L \end{pmatrix}$ is a matrix containing parameters to be estimated.

$\varepsilon = \begin{pmatrix} \varepsilon_1 \\ \cdot \\ \varepsilon_j \end{pmatrix}$ is a matrix containing residuals/errors.

The errors are usually assumed to be uncorrelated across measurements, and follow a multivariate normal distribution.

Endogenous Switching Regression Model

The model was used to examine self-selection in migration as a result of several factors. The general form as described by Lokshin and Sajaia (2004) is based on the behavior of an agent with two regression equations and a criterion function, I_i, that determines which regime the agent faces (see Footnote 1):

$$I_i = 1 \quad \text{if} \quad \gamma Z_i + u_i > 0 \tag{a}$$

$$I_i = 0 \text{ if } \gamma Z_i + u_i \leq 0 \tag{b}$$

$$\text{Regime}_1\colon y_{1i} = \beta_1 X_{1i} + Q_{1i} \text{ if } I_i = 1 \tag{c}$$

$$\text{Regime}_2\colon y_{2i} = \beta_2 X_{2i} + Q_{2i} \text{ if } I_i = 0 \tag{d}$$

Here, y_{ji} are the dependent variables in the continuous equations; X_{1i} and X_{2i} are vectors of weakly exogenous variables; and β_1, β_2, and γ are vectors of parameters. It also assumes that u_i, Q_{1i}, and Q_{2i} have a trivariate normal distribution with mean vector zero and covariance matrix which equals

$$\begin{bmatrix} \sigma_\eta^2 & \sigma_{\eta 1} & \sigma_{\eta 2} \\ \sigma_{1\eta} & \sigma_1^2 & \cdot \\ \sigma_{2\eta} & \cdot & \sigma_2^2 \end{bmatrix} \tag{e}$$

where σ_η^2 is a variance of the error term in the selection equation and σ_1^2 and σ_2^2 are variances of the error terms in the continuous equations. $\sigma_{1\eta}$ is a covariance of u_i and Q_{1i}, and $\sigma_{2\eta}$ is a covariance of u_i and Q_{2i}. The covariance between Q_{1i} and Q_{2i} is not defined, as y_{1i} and y_{2i} are never observed simultaneously. It can be assumed that $\sigma_n^2 = 1$ (γ is estimable only up to a scalar factor). The model is identified by construction through nonlinearities (Lokshin and Sajaia 2004; Di Falco and Veronesi 2014). The endogenous switching regression model is used to produce selection-corrected predictions of counterfactual states in the dependent variables.

$$E(y_{1i}/A_i = 1) = x_{1i}\beta_1 + \sigma_{1\eta}\lambda_{1i} \tag{f1}$$

$$E(y_{2i}/A_i = 0) = x_{2i}\beta_2 + \sigma_{2\eta}\lambda_{2i} \tag{f2}$$

$$E(y_{2i}/A_i = 1) = x_{1i}\beta_2 + \sigma_{2\eta}\lambda_{1i} \tag{f3}$$

$$E(y_{1i}/A_i = 0) = x_{2i}\beta_1 + \sigma_{1\eta}\lambda_{2i} \tag{f4}$$

9.5 Results

9.5.1 Summary of Living Standard

The living standard variables show that the resource base of the farming systems contributes to income-generating abilities, food security status, the degree of indebtedness, and the overall living standard of each cluster. While some variables are generally poor, the RRLI system is better off on most measurement variables while the RPLI ranks second in terms of economic success. The RPNLI is the most vulnerable system, and it has the lowest value of most of the criteria (Table 9.3).

9.5.2 Climate Change Experience and Migration Profile

Climate Shock Experience

Several climate change shocks were reported in the systems, the most notable are heavy rainfall, floods, and pests as a result of change in temperature or high humidity. Cocoa and cassava farmers in particular complained about the resurgence of the black pod disease of cocoa and the cassava mealy bug. The awareness of climate change is higher in the RPNLI system. The frequency of the experience was mainly seasonal and the effect on the different crops cultivated differed. The results show that pest and diseases attributed to the fluctuating temperature and levels of humidity as well as floods as a result of rainfall were the most destructive in the area.

Also, most of the crops upon which they depend were affected. The cropping system is hardly monocultural as such multiple crops on the farms were destroyed. Several "adaptation" strategies were mentioned but a critical assessment would show that these are coping strategies and are not necessarily environmentally sustainable or friendly (Table 9.4). The irony of the situation is that even though 42% of the households have migrants as a result of the shocks related to the farm, none mentioned it as a strategy implying that the climate shocks induced the final decision which they view mainly as an overall economic, livelihood, or development strategy.

Migrants and Migration Profile

More male migrants than female, age-wise, are in the economically active bracket. Migrants' educational level is generally low; RPNLI has the highest proportion of the least educated as well as the largest of the well-educated migrants. This could imply that the system has placed value on education for the migrants. The RPLI has about 8% of migrants being the household head or his spouse. Migrants are employed in different activities, the RRLI has the highest number of those who are self-employed or are trading and could remit substantial amount of money. A number of them are employed in different jobs, some of which are not known to their family members. Typically, the migrants have remained within the state or regions in the city centers such as Owerri and Umuahia, etc., and they are relatively short distanced. The south-west and south-east regions are mainly represented by cities such as Lagos and Port Harcourt; Africa region represented are Cameroon and Benin Republic.

The RPNLI has a relative proportion of migrants distributed across different destinations except out of the country. The decision to have at least a member of the family migrate is made more often in the RPLI and RRLI systems while migrants in the RPNLI were self-motivated, but we see evidence of the family alone or family in conjunction with the individual making a decision on the strategy to take. The

Table 9.3 Summary of living standard analysis of farming systems

Item	RPLI-FS (n = 52)	RRLI-FS (n = 39)	RPNLI-FS (n = 29)	All (n = 120)	Sig. diff. of mean
Family income (FI) (₦)	490,359.70 (892,864.70)	1,189,126.00 (189,940.70)	467,444.70 (563,466.20)	711,921.10 (988,086.20)	***a
Farm income (₦)	316,437.00 (139,378.80)	1,013,188.00 (396,950.80)	353,237.80 (1,223,444.00)	551,774.60 (849,217.70)	***a, c
Total off-farm income (₦)	173,922.70 (798,929.50)	175,938.46 (310,356.43)	114,206.90 (157,814.85)	160,146.50 (557,538.60)	
Off-farm income-HH (₦)	168,384.20 (798,883.90)	143,323.08 (281,338.86)	89,379.31 (137,209.02)	141,146.50 (551,547.60)	
Off-farm income-SP (₦)	5538.46 (39,938.41)	32,615.38 (87,011.38)	24,827.59 (81,003.56)	19,000.00 (69,233.98)	**c
Annual cash balance					
% with negative balance	55.00	31.00	54.00	46.00	
Net cash balance (₦)	240,456.70 (959,759.10)	1,359,468.00 (3,600,776.00)	198,321.60 (1,047,792.00)	593,952.60 (2,253,480.00)	**a, c
Independence-res. owner					
Debt owed (₦)	1029.23 (5367.76)	769.23 (3542.68)	1793.10 (9278.93)	1129.33 (6063.73)	
Rent on land (₦)	2326.92 (11,201.75)	14,251.28 (56,376.92)	3568.97 (17,694.70)	6502.50 (34,231.27)	
Food supply and security					
Food expenditure (₦/year)	91,254.23 (71,443.95)	197,300.51 (128,820.44)	65,453.79 (60,954.67)	119,484.17 (106,794.69)	***a, b
Home consumption (₦)	92,128.13 (252,922.98)	47,286.76 (57,311.59)	170,300.00 (501,203.35)	95,444.34 (290,690.41)	
Supply non-food needs					
Non-food exp. (₦/year)	191,289.70 (102,059.91)	355,382.36 (189,594.53)	459,596.41 (274,882.53)	309,460.60 (214,494.57)	***a, b
Health expense (₦/year)	6697.08 (8719.84)	17,770.97 (19,293.37)	24,229.66 (25,325.09)	14,533.13 (18,855.78)	**a, b
Days lost to illness	66.38 (223.87)	2.72 (5.42)	128.97 (677.24)	60.62 (366.73)	
Education					
Education-head (years)	7.02 (3.81)	10.28 (3.04)	8.41 (3.98)	8.42 (3.86)	***a, c

(continued)

Table 9.3 (continued)

Item	RPLI-FS (n = 52)	RRLI-FS (n = 39)	RPNLI-FS (n = 29)	All (n = 120)	Sig. diff. of mean
Education spouse (years)	9.11 (3.86)	10.13 (2.70)	9.62 (3.60)	7.77 (4.35)	*a
Social security and safety					
Remittance (₦/year)	3103.46 (5874.70)	4004.87 (7539.29)	9060.69 (11,106.01)	4836.08 (8237.19)	***b c
Net remittance (₦/year)	1071.15 (9431.81)	2012.82 (8069.18)	5889.62 (12,099.65)	2541.66 (9855.60)	b
Asset value (₦)	78,177.54 (170,186.02)	115,604.25 (148,144.45)	65,714.00 (105,094.67)	87,329.20 (149,774.23)	***a, b

***Significant at 1%; **significant at 5%; *significant at 10%. All tests are Kruskal–Wallis. Values in parenthesis are standard deviation. "a" significant between clusters 1 and 2; "b" significant between clusters 1 and 3; "c" significant between clusters 2 and 3. $1 = NGN 183

Table 9.4 Climate shock incidence and experience

Items	RPLI (n = 52)	RRLI (n = 39)	RPNLI (n = 29)	Total (n = 120)
Climate change incidence observed				
Nothing	15.30	12.90	3.40	10.50
Heavy rainfall	38.50	30.80	27.50	32.70
Rainfall, flood, pest	13.40	12.80	38.70	21.63
Floods	17.20	15.40	6.80	13.13
Pests	3.80	12.90	6.90	7.87
High temperature	9.60	12.80	10.30	10.90
Erosion	1.90	0.00	6.80	2.90
Frequency of incidence				
Seasonal	11.50	7.70	20.70	12.50
Monthly	1.90	2.60	0.00	1.70
Weekly	1.90	2.60	3.40	2.50
Occasionally	3.80	0.00	3.40	2.50
Adaptation				
Nothing	44.20	56.40	44.80	48.30
Water supply and control	9.60	0.00	0.00	5.80
Chemicals	26.80	28.20	10.30	23.30
Cropping related	7.60	33.40	17.10	35.70

RPNLI has more migrants remitting food, money, and other items to the families left behind. In all systems, at least 50% of the migrants care for their families by sending remittances to them. In the RPNLI, food and money are the most common items remitted. The RPNLI have the best response in terms of remittances and its consistency (Tables 9.4 and 9.5). Apart from receiving, migrant families also send money and food particularly to those migrants who are yet to find a footing. Remittance purposes differ between the systems, within the RPLI, it is used for family up keep and school fees, in the RRLI is mainly for family and household need, while in the RPNLI, it is for farming and family up keep. Further discussions with respondents seem to indicate that there could be a cultural perspective to migration.

The number of migrants per system is significantly different, with the RPNLI having the highest number; they also receive the highest amount of remittances. The net value of remittances differs only between the resource poor systems (Table 9.6).

9.5.3 Climate Change-Induced Migration Components Effects on Living Standard

The climate change-induced migration components were examined against family income and living standard components. Erosion and remittance have negative effects on family income and asset/income component of the farm families but have no significant effects were reflected in the other living standard components. However, strangely, the flood component seems to have a positive effect on family income, economic viability of household head and asset/income components but negative on food security and vulnerability components. The other shock variables behave consistently except for crop losses due to erosion which appears to have a positive effect on food security components. The effects of the components were found to be different from one cluster to the other. The results exhibited imply that climate shocks and migration as a response catalyze some other behaviors in the different systems and as such the overall impact is not overwhelmingly negative (Table 9.7).

9.5.4 The Impact of Climate Change-Induced Migration on Living Standard

The decision to migrate in the systems is mostly by the family or a joint decision with the individual; apart from it being induced by climate shocks, the final act of migrating or not migrating could also be as a result of unobservable differences between households; also, the hypothesis that migrant households were not significantly more productive or better off in terms of living standard needed to be examined. As such an endogenous switching regression model was estimated using full information maximum likelihood method.

Table 9.5 Migrant characteristics and profile

Items	RPLI (n = 52) (%)	RRLI (n = 39) (%)	RPNLI (n = 29) (%)	Total (n = 120) (%)
CC-I migrants	25.00	35.90	51.70	35.00
Sex				
Male	15.40	28.20	44.80	26.70
Female	9.60	5.10	6.90	7.50
Age				
<14	0.00	0.00	3.40	0.80
14–60	24.80	36.00	47.80	33.90
Education				
Primary education	3.80	2.60	20.60	7.50
JSS	7.70	5.10	13.80	8.30
SSCE	13.50	20.50	6.90	14.20
Tertiary	0.00	2.60	10.30	3.40
Position in family				
First	11.50	20.50	42.90	21.80
Household Head/spouse	7.70	0.00	0.00	3.40
Employment status				
Unemployed	1.90	5.20	6.90	4.20
Trading/self-employed	3.80	10.30	6.90	6.60
Teaching/civil service	5.70	0.00	10.30	1.70
Schooling	0.00	2.60	6.90	2.50
Farming	1.90	0.00	0.00	0.80
Unknown employment	9.60	7.70	20.70	11.70
Destination				
Within Abia	13.50	13.20	31.00	17.60
South-east	3.80	0.00	10.30	4.20
South-west	0.00	5.30	6.90	3.40
South–south	1.90	5.30	3.40	3.40
Africa	1.90	7.90	0.00	0.80
Decision to leave				
Family	19.2	18.00	7.30	14.8
Self and family	0.00	7.70	6.90	4.20
Self	3.80	10.30	27.60	11.70

Table 9.6 Interaction between migrants and migrants' families

Items	RPLI (n = 52) (%)	RRLI (n = 39) (%)	RPNLI (n = 29) (%)	Total (n = 120) (%)
Still in touch	25.00	35.90	48.30	34.20
Remittance sending	13.50	20.50	28.60	19.30
Frequency of remittance				
Monthly	7.70	5.10	10.30	7.50
Quarterly	0.00	0.00	6.90	1.70
Bi-monthly	3.80	5.10	3.40	4.20
Yearly	3.80	0.00	6.90	3.30
Item sent by family				
Food	5.70	12.90	13.70	10.10
Money	13.50	20.50	34.40	20.80
Use of remittance received				
Farming and upkeep	0.00	0.00	3.40	0.80
Upkeep	5.80	15.40	17.20	10.80
School fees	1.90	0.00	0.00	0.80
Other household needs	0.00	2.60	0.00	0.80
Net remittance				
Positive	25.00	23.30	47.90	29.70

Table 9.7 Remittance value differences

Item	RPLI-FS (n = 52)	RRLI-FS (n = 39)	RPNLI-FS (n = 29)	All (n = 120)	Sig. diff. of mean
Number of migrants	1	2	2	1	***a, b
Value of migrant remittance	3103.46	4004.87	9060.69	4836.08	***b, c
Value family remittance	1525.00	1653.85	1351.76	1525.01	
Net remittance	1071.15	2012.82	5889.62	2541.66	**b

***Significant at 1%; **significant at 5%; *significant at 10%. All tests are Kruskall–Wallis. Values in parenthesis are standard deviation. "a" significant between clusters 1 and 2; "b" significant between clusters 1 and 3; "c" significant between clusters 2 and 3. $1 = NGN 183

Family income is significantly influenced by farm output in both migrant and non-migrant models (Table 9.8). Other factors which influence it differ in both systems; remittance and household size positively affect it in the migrant households while in the non-migrant households, climate shock variables, other economically viable adults and farm size have a positive effect. The factors which have a negative influence are age in the non-migrant systems and pest and disease incidence in the migrants' system. The fact that climate shock variables have a positive influence on family income suggest that the two groups of people experience climate shocks differently and cope in different ways; more importantly it suggests that climate shocks trigger some economic flows which the non-migrants may be in a position to harness better than the migrants. This could also be the reason behind the choice not to migrate. The statistical significance of the correlation coefficient ($\rho 0$) suggests that there is selection effect; hence, unobserved factors affect both the migration decision and family income. In particular, there is negative selection bias for both non-migrants and migrants' households as $\rho 0$ is negative and significant while $\rho 1$ is not statistically significant.

9.5.5 Counterfactual Analysis: Effect of CC-IM on Living Standard

The study also investigated the effect of having migrated in response to climate change on living standard. In other words, to estimate the treatment effect as already mentioned, unobserved heterogeneity in the propensity to migrate in response to climate change creates a selection bias that should not be ignored. The endogenous switching regression model is used to produce selection-corrected predictions of counterfactual states in the dependent variables. It is used to compare the expected states of migrant households ($a =$ f1) relative to the non-migrants ($b =$ f2), and to investigate the expected states of the dependent variables in the counterfactual hypothetical cases ($c =$ f3) that the migrant households did not migrate, and ($d =$ f4) that the non-migrant households migrated. The conditional expectations for the different dependent variables four cases are defined as follows:

$$E(y_{1i}/A_i = 1) = x_{1i}\beta_1 + \sigma_{1\eta}\lambda_{1i} \tag{f1}$$

$$E(y_{2i}/A_i = 0) = x_{2i}\beta_2 + \sigma_{2\eta}\lambda_{2i} \tag{f2}$$

$$E(y_{2i}/A_i = 1) = x_{1i}\beta_2 + \sigma_{2\eta}\lambda_{1i} \tag{f3}$$

$$E(y_{1i}/A_i = 0) = x_{2i}\beta_1 + \sigma_{1\eta}\lambda_{2i} \tag{f4}$$

Table 9.8 Effect of migration on living standard and its components-all

	Family income			Economic viability H/head			Asset			Income			Vulnerability		
Parameter	*t*	Sig.	Partial eta^2	*t*	Sig.	Partial eta^2	*t*	Sig.	Partial eta^2	t	Sig.	Partial eta^2	*t*	Sig.	Partial eta^2
Intercept	2.678	0.009	0.06	−4.624	0	0.16	−0.638	0.525	0.004	0.259	0.796	0.001	−2.164	0.033	0.04
PLI	2.441	0.016	0.051	2.983	0.004	0.074	1.868	0.064	0.03	−4.405	0	0.148	2.243	0.027	0.043
PNLI	−0.524	0.602	0.002	0.954	0.342	0.008	0.266	0.79	0.001	−0.852	0.396	0.006	3.188	0.002	0.083
Migrants	2.178	0.032	0.041	0.471	0.638	0.002	0.161	0.872	0	−0.973	0.333	0.008	0.542	0.589	0.003
Remittance	−3.641	0	0.106	1.043	0.299	0.01	−4.399	0	0.147	−0.557	0.578	0.003	0.462	0.645	0.002
Incidence	3.506	0.001	0.099	0.482	0.631	0.002	2.153	0.033	0.04	−0.112	0.911	0	−1.589	0.115	0.022
Crop loss	−1.771	0.079	0.027	−1.537	0.127	0.021	−1.863	0.065	0.03	1.069	0.287	0.01	−0.281	0.779	0.001
Sex H/head	0.362	0.718	0.001	3.963	0	0.123	−0.007	0.994	0	1.508	0.134	0.02	0.762	0.448	0.005
R-squared	0.289			0.211			0.213			0.133			0.073		

Cases (f1) and (f2) represent the actual expectations observed in the sample. Cases (f3) and (f4) represent the counterfactual expected outcomes. In addition, following Heckman, the effect of the treatment "to migrate" on the treated (TT) is calculated as the difference between (f1) and (f3),

$$\mathrm{TT} = E(y_{1i}/A_i = 1) - E(y_{2i}/A_i = 1) = x_{1i}(\beta_1 - \beta_2) + (\sigma_{1\eta} - \sigma_{2\eta})\lambda_{1i} \quad \text{(f5)}$$

which represents the effect of climate change-induced migration on the dependent variables of the farm households that actually have migrants in response to climate change. Similarly, the effect of the treatment on the untreated (TU) is calculated for the farm households that actually do not have migrants in response to climate change as the difference between (f4) and (f2),

$$\mathrm{TU} = E(y_{1i}/A_i = 0) - E(y_{2i}/A_i = 0) = x_{2i}(\beta_1 - \beta_2) + (\sigma_{1\eta} - \sigma_{2\eta})\lambda_{2i} \quad \text{(f6)}$$

The results show that migration made migrants' households worse off in terms of family income and living standard variables. Also, it showed that non-migrant households were better off not having migrants; also, if they had migrated, they would have had a lower family income and they would have been more vulnerable (Table 9.9).

9.6 Discussion

The research findings present different perspectives on the challenges and opportunities presented by climate shocks which farm families tend to respond to by seeking off-farm income or asking economically viable members of the household to migrate. The different farming systems identified had significantly different resources and their allocation and use differed as well. But more importantly, it signifies the differences that need to be taken into consideration in the process of introducing an intervention. The findings with respect to the research questions are discussed below (Table 9.10).

9.6.1 Characteristics of Farming Systems

The Resource Poor Labor Intensive system experience rainfall, floods, and pest on all crops and respond through water supply and control practices as well as use of chemicals. The decision to migrate in response to these shocks is made mostly by the families; 25% of the households in the system have at least one CC-I migrant. Remittance, migrant, and crop loss (erosion) components of CC-IM are significant drivers of migration.

Table 9.9 Family income effect

Log likelihood = −1701.83	1	Wald chi^2(13) = 61.08	Prob > chi2 = 0
Family income	CC-IM	Migrant	Non-migrants
Ext. laboratory hours		2934.094 (6876.11)	5514.835 (7967.51)
Flood/erosion incidence		−128,692 (234,821.7)	367,104.5* (195,265.2)
H-head labor farm	−3.5E−05 (0.002718)	1020.39 (1332.162)	775.9735 (984.41)
Farm size Ha	0.00157 (0.016654)	−32,103.1 (24,553.13)	5342.345* (3132.10)
Age (H-head)	−0.02,527 (0.020007)	−5513.09 (12,392.28)	−11,590.7* (6612.21)
Level of education	0.594329 (1.283,832)		
Marital status	0.472697 (1.395,315)		
Net remittance (₦)	1.26E−05 (1.78E−05)		
Total quantity of crops lost	−7.5E−05 (0.00017)		
Pre_1	3.856,337*** (0.775199)		
Off-farm income		312,172 (221,392.2)	282,931.9 (182,420.3)
Pest/disease incidence		−472,325.8** (213,460.4)	−257,702 (179,507.6)
CC-I migrants' remittance (₦)		24.53,732** (9.958,247)	18.46,662 (14.26)
Value of asset (₦)		0.259375 (0.807033)	0.302606 (0.67)
Type of crop affected		16,946.09 (305,134.9)	651,006.2*** (190,485.4)
Farm output		26.95,058*** (7.347,509)	26.29,709*** (9.14)
Household size		117,907.5** (46,290.28)	−17,919.2 (37,926.53)
Other economically viable adults		126,063.9 (201,071.6)	440,123.7** (201,689)
Constant	−1.69,163 (1.341,062)	−35,175.3 (636,647.6)	85,946.42 (439,472.1)
Sigma		303.24	736.18
Rho		−1.5	−4.44***

LR test of indep. eqns.: chi2(2) = 4.83 Prob > chi2 = 0.0892***, **, * represent 1%, 5%, and 10% significance level, respectively

The Resource Rich labor Intensive System: The households in the system describe their experience of climate shocks mostly based on a specific event; rainfall and floods are the most common but the frequency of these experiences is not well described. All the major economic and minor crops are affected by these shocks but only 44% are making effort to adapt to the situation. The most common "adaptation" strategies

Table 9.10 Average expected states of outcome variables

	Decision stage		
Sub-samples	Migrate	Not migrate	Treatments effects
Family income			
Migrant households	640,622.2	724,442.1	−83,819.95***
Non-migrant households	406,045.2	573,712.7	−167,667.56***

***, **, *Represent 1%, 5%, and 10% significance level, respectively

are the use of chemicals and crop-related activities such as repairing damaged crops or replanting. The decision to migrate is made mostly by the family but sometimes in conjunction with the migrant. Flood experience or damage was done as a result of floods leads to migration, with respect to the components, remittance, migrant, and crop loss (erosion) are significant drivers.

The Resource Poor Non-Labor-Intensive system: Compared with the other systems, it has the lowest asset and income. The climate shock experienced by the system is mainly rainfall, floods, and pests; this affects both major and minor crops on the farm. About 45% of those who experience these shocks respond by doing nothing. The most common "adaptation" strategy is the use of chemicals. Over 50% of the households in the system have at least one CC-I migrant. Over 40% of the migrants are holding the first position among children; hence, age and educational level might play a role in the choice of who should migrate. The decision to migrate, however, is taken most of the time by the individual. The most important climate change shocks which drive migration in the system are flood and pest incidence and with respect to the component, it is the migrant characteristic, remittance, and erosion shock.

9.6.2 *Migration and Migrants' Profile in the Farming Systems*

Reuveny (2007) suggests that people leaving in lesser developed countries are likely to leave environmentally degraded environments and that this would be the most probable response to climate change as well. This will not only happen at the national level but within different communities those who have low adaptive capacities are likely to leave. The RPNLI has the highest percentage of households with migrants followed by the RRLI. The large percentage of households with migrants among the RRLI is unexpected, but could be interpreted in light of the fact that migration is a "silent" response to the shocks they experience although over 50% claim to do "nothing." Apart from this, the general pattern follows other research output: the most educated leave, in some cases, the household heads or the spouse but in most cases, the oldest child or oldest male child. Migration is, in most cases, short distance to a more urban or environmentally friendly environment where off-farm income opportunities are higher. It is mainly a rural–urban pattern of migration (Ayinde et al. 2014; Glaeser

2013; IOM 2015). The decision to migrate is seen to be made mainly by the family and in some cases with the migrant in a joint process; however, in the RPNLI, most of the individuals made the decision themselves. Also, remittances received from migrants are used to meet different family needs which are not farm related and as such may not lead to the commercialization of production activities (Kifle 2007; Ajaero and Onokala 2013). Households also encourage migration to ensure that a generation of relatively prosperous members evolves; migration is a strategy to give the next generation a better chance at life. Families may not always be in a position to fund the cost of migration but would give the emotional and psychological support required for the decision to translate into action. The migration pattern follows the NELM theory except that the high percentage of individuals making the decision to migrate in the RPNLI indicates that the decision-making process could be influenced by the socio-economic context of the households and of the individual.

9.6.3 Effects of CC-IM Components on Living Standard

With respect to living standard, the hypothesis was that the overall living standard should increase. But when the components are examined against family income and living standard components, the output suggests unobservable interactions which cause remittances to have a negative effect on family income and the flood incidence component has a positive impact on it. With respect to the economic viability of the household head and asset/income component, it reduces both when the crop loss due to erosion incidence is high. They, however, increase with increased flood incidence. All the components apart from erosion reduce food security while flood and erosion worsen the vulnerability component. As such different aspects of the living standard are affected differently by CC-IM components. Effects of migration and those of climate change on health, livelihoods, and income as a measure of living standard are well documented (Adger et al. 2009; Mertz et al. 1999) but this research presents a different perspective to the effects considering core living standard components.

9.7 Conclusion

The overarching question was to examine how G-SMEs should be implemented in Africa in the presence of increasing migration induced by climate change.

One may safely conclude that farming systems can be identified and defined by characteristics or variables such as the level of education particularly of the household head and spouse; the resource base which they have ownership and control over; assets owned both durable and non-durable which can easily be converted to cash; relations within the household as defined by resource allocation and use decisions and market relations defined by external purchase of resources, inputs and services as well as the sales of own resources such as labor and products or service. These variables summarized as **Education, Resources, Assets, Household and Market relations (ERHAM) define** the degree of independence from resource owners, income generation, living standard, and innovation adoption. The study identified the RPNLI as the most vulnerable FS based on the 8 criteria of living standard. Through the criteria, multiple vulnerable points were identified in the FS; these are the dependence on other resource owners for consumption loans and productive resources such as land; health burden typified by the number of days lost to illness and average amount expended on cures; a bi-directional dependence on remittances by both migrants and their households; the lack of security in access to water, energy, food supply, and the negative annual cash balance experienced by at least 50 percent of the resource poor farming systems.

Migration as a response to climate shocks is more common among males who are usually the household head or eldest child in the home. The receiving location is mainly within the state of the study area or south-east zone; as such migration here may be characterized as being short distanced. Migrants relate with the sending household through remittances, which is bi-directional. This is an indicator that migrants could be used as potential market development strategy for Green SMEs. Different CC shocks are experienced and the way it is experienced also vary as such several coping strategies were developed; most of which are environmentally unfriendly. Ecosystems friendly chemicals, water harvesting/recycling techniques are needed. Holistic Ecosystems Based Adaptation strategies (EBAs) should be developed for all the CC shocks experienced and the farm situation.

CC-IM was defined through its components to avoid ambiguity. The components are migrant characteristics, remittance, flood incidence, and crop loss (driven by erosion); through these, potential sink holes of G-SMEs at farm and family levels were identified. The effects of remittances, crop loss (erosion) on production, and living standard variables indicate that they could harm the growth and development of a G-SME. It is important to note these and have them factored into any Green SMEs innovation strategy to ensure their survival. Also, the potential economic impact of CC-IM was identified—flood incident—which has a positive effect on gross margins signifying a trigger of economic variables which lead to increased prices and hence increased income, most likely through the effect of scarcity.

Further investigations through the ESR model using the full information maximum livelihood approach confirm that unobservable characteristics influence the decision to migrate. The effects on the production, productivity, and living standard variables vary but suggest that migration will not necessarily make migrants' household better off than non-migrant households but it does improve the situation of the migrants that they could have been worse off if no one migrated.

The result strongly indicates that migrants' households are more likely to be the vulnerable households in the different farming systems; as such migration is not a sustainable solution to the problems or risks associated with climate variability or shock. The strategy tested introduced G-SMEs in the absence and presence of migration and it showed that Green SME is a potentially viable alternative for labor use, migration, and income generation, particularly in the most vulnerable farming systems. Policy interventions using G-SMEs as an alternative should take the following points into consideration:

- Crops of economic importance to the households.
- Focus on the entrepreneurial and economically viable FS.
- Should not be labor demanding, should fit into current labor resource capacity.
- Water and energy situation have promising opportunities for success if G-SMEs related to them are developed and introduced.
- Turning the "flood economic" opportunity to an advantage. It is a market driver.

The results provide G-SME spaces and suggest the use of a value chain approach in this wise. Several spaces for G-SMEs included energy (bio fuel) water recycling, organic chemicals within the context of a full ecosystem-based adaptation strategy.

Policy Recommendations
In light of the findings of the study, the following recommendations are made:

- Identify CC shock prone crops of economic importance cassava, cocoa, yam, rice, and develop G-SMEs around them.
- Develop G-SMEs with credit and thrift content particularly for women.
- Pilot phase should focus on the economically viable and entrepreneurial FS—but it should be noted that dependency ratio could be high for such systems).
- Develop several "baskets" of SMEs for different FS with the ability to meet multiple needs within the family–farm–household system.
- Target migrants at popular migrants receiving locations as possible market agents.
- Seasonal introduction of associated products to harness the market drive potential of climate shocks.
- A value chain approach should be considered in the G-SMEs introduction.

Acknowledgements The author wishes to appreciate Mr. Victor Ejiba's assistance in collecting the data used for the study. The United Nations University Institute for Natural Resources in Africa provided the funding to visit the Institute at the University of Ghana, Legon, to analyze the dataset and write up the research chapter. The conducive work environment and friendly atmosphere which was given to me is very much appreciated.

References

Adger WN, Hallie E, Winkels A (2009) Nested and tele-connected vulnerabilities to environmental change. Front Eco Env 7(3):150–157

Ajaero CK, Onokala PC (2013) The effects of rural-urban migration on rural communities of southeastern Nigeria. Int J Pop Res 1–10

Akinsanmi OA (2005) Gender and food security of rural families in Imo state, south-east Nigeria. Margraf Weikersheim, Stuttgart
Ayinde JO, Torimiro DO, Koledoye GF (2014) Youth migration and agricultural production: analysis of farming communities of Osun state. Nig J Agric. https://doi.org/10.4314/njae.v18i1.11
Bro R, Smilde AK (2014) Principal component analysis. Anal Methods 6:2812–2831. https://doi.org/10.1039/c3ay41907j
Corder GW, Foreman DI (2014) Nonparametric statistics: a step-by-step approach. Wiley, New York
De Haas H (2007) International migration, national development and the role of governments: the case of Nigeria. In: Adepoju A, van Naerssen T, Zoomers A (eds) International migration and national development in Sub-Saharan Africa: viewpoints and policy initiatives in the country of origin. Brill Publishers, Leiden
Di Falco S, Veronesi M (2014) Managing environmental risk in presence of climate change: the role of adaptation in the Nile basin of Ethiopia. Env and Res Eco 57:553–577
Doppler W (2000) Setting the frame: the environmental perspectives in rural and farming systems analyses. In: Doppler W, Koutsouris A (eds) Rural and farming systems analysis environmental perspectives. Margraf Verlag, Weikersheim, pp p1–10
Doppler W (2002) Farming and rural systems approaches published lecture material. Hohenheim University, Stuttgart
Edame GE, Ekpeyong BA, Fonta WM (2011) Climate change, food security and agricultural productivity in Africa: issues and policy directions. Int J Hum Soc Sci 1(21):205–223
Fletcher G, Clark MS (1997) Blackwell handbook of social psychology: interpersonal processes. Blackwell, London
Glaeser EL (2013) A world of cities: the causes and consequences of urbanization in poorer countries. NBER working paper 19745. http://www.nber.org/papers/w19745
Hoddinott J (1994) A model of migration and remittances applied to western Kenya. Oxf Econ Pap 46:459–476
Janssen MA, Ostrom E (2006) Empirically based, agent-based models. Econ Soc 11(2):1–13
Jolly S, Reeves H (2005) Gender and migration. Bridge, Institute of Development Studies, UK
Kifle T (2007) Do remittances encourage investment in education? Evidence from Eritrea. J Afr Stud 4(1)
Lokshin M, Sajaia (2004) Maximum likelihood estimation of endogenous switching regression model. Stata J 4(3):282–289
Machiori L, Maystadt J, Schumacher I (2012) The impact of weather anomalies on migration in sub-saharan Africa. J Env Eco Manag 63(3):355–374
Mensah-Bonsu A, Burger K (2008) A model of migration: getting the permission of the households. www.researchgate.net/publication/226939548A_
Mertz O, Mbow C, Reenberg A et al (1999) Adaptation strategies and climate vulnerability in the sudano-sahelian region of West Africa Atmos. Sci Let 12:104–108. https://doi.org/10.1002/asl.314
Nwajiuba CU, Ejiogu OA (2008) Population and migration nexus: evidence from communities of varying population densities in South East Nigeria. In: Proceedings of the Oxford business and economics conference. Oxford University, Oxford, 22–24 June 2008
Olajide OA (2013) Male migrant's perceptions of agricultural problems and benefits of rural-urban migration in Oyo state Nigeria. Int J Agric Sci 3(2):85–90
Ravenstein E (1889) The laws of migration. J R Stat Soc 52:241–301
Rawlings JO, Pantula SG, Dickey DA (1998) Applied regression analysis. Statistics. Springer, Berlin
Reuveny R (2007) Climate change-induced migration and violent conflict. Polit Geogr 26(6):656–673
Rouse R (1991) Mexican migration and the social space post modernism. Diaspora J Trans Stud 1(1):8–23
Sinha BRK (2005) Human migration: concepts and approaches. Geo Bull 3(4):403–414

Stark O (1991) The migration of labor. Blackwell, Oxford
Stark O, Bloom DE (1985) New economics of labor migration. AEA papers and proceedings, pp 173–178
Taylor JE (1999) The new economics of labor migration and the role of remittances in the migration process. Int Mig 37(1):63–88
Wayne WD (1990) Kruskall Wallis one-way analysis of variance by ranks. Appl Non-Parametr Stat, 2nd edn. PWS-Kent, Boston, pp 226–234
World Migration Report (2015) Migrants and cities-new partnerships to manage mobility. International Organization for Migration

Chapter 10
An Assessment of Factors Shaping Green Growth Uptake in the Forest Sector at Rural Community Level in South Africa

Chidiebere Ofoegbu

10.1 Introduction

Most rural communities in African countries are typically underdeveloped in terms of infrastructure and government services (Dlamini 2014; Maponya and Mpandeli 2013). Employment opportunities are limited, alternative economic activities are scarce, and consequently, the people are highly reliant on forests and natural resources for subsistence and income (Asare et al. 2013; Shackleton 2004). Firewood, building poles, medicinal plants, and edible fruits are some of the forest products commonly used by rural households for livelihood sustenance (DWAF 2005a; Shackleton 2004). In the same vein, in South Africa, DWAF (2005a) estimated that over 80% of rural households use fuelwood as their primary source of energy. The primary health care of about 27 million rural poor households in South Africa is heavily dependent on medicinal forest products (DAFF 2012). Apart from direct-use livelihood contributions of forests and forest products to most rural households (DAFF 2010), forests are also key components of ecotourism on which the economy of many rural communities relies (Asare et al. 2013).

Despite the importance of forests, current mode of forest use and management in most rural communities in South Africa makes it difficult for the communities to translate forest resources wealth into sustained economic growth and livelihood resilience to climate change challenges (Shackleton and Shackleton 2004; Shackleton and Campbell 2001). In most communities, considerable unsustainable forest use and management practices are on the rise, particularly in forests situated within and at the periphery of rural settlements (Matose 2008). As a result, the growing rate of deforestation and forest degradation has become a major concern for the enhancement of rural livelihood sustainability in most communities across Africa (Dercon 2014; Holmes-Watts and Watts 2008). This situation is expected to be exacerbated by

C. Ofoegbu (✉)
University of Pretoria, 5-15 Plant Sciences Complex, Pretoria, South Africa
e-mail: ofoegbu.c@gmail.com

C. Atewamba and D. Yong Ngondjeb (eds.), *Inclusive Green Growth*, Advances in African Economic, Social and Political Development,
https://doi.org/10.1007/978-3-030-44180-7_10

climate change (IPCC [Intergovernmental Panel on Climate Change] 2007). It is projected that climate change impacts will diminish the capacity of forests to provide goods and services that are vital to rural livelihood sustenance (Davis et al. 2010), and community development (David 2005).

In South Africa, concerns about climate change and the need to maximize benefits from forests for sustainable rural development are driving national interest in the "green growth" or "green economy" concept (Bishop et al. 2015; WEF 2013; Musyoki 2012). Green economy is defined as "an economy that results in improved human well-being and social equity, while significantly reducing environmental risks and ecological scarcities" (UNEP 2011; Musyoki 2012). South Africa recognizes green growth as a pathway to sustainable development (DEA 2011). Moreover, South Africa is a signatory to several international conventions that seek to promote transition to a green growth pathway (Smit and Musango 2015; Montmasson-Clair 2012). In the same vein, the forest sector is viewed as a critical link in the transition to a green growth pathway at rural community level in the South African green growth landscape (Musyoki 2012; Resnick et al. 2012; Slunge and Loayza 2012).

However, despite growing recognition of the importance and economic function of the forest sector in facilitating transition to "green growth" and improving resilience of rural livelihoods to climate change, there are concerns that the rhetoric for transitioning to a green growth pathway may not translate into practical action at either the community or sectoral level (Dercon 2014; Slunge and Loayza 2012). Much of the discussion on "green growth" has remained relatively vague in terms of specifics, including the forest sector and rural communities (Dercon 2014). The question of how various green growth strategies and green investments interact with current mode of forest use and management at rural community level in South Africa is rarely asked (UNECE 2009). Consequently, factors that might facilitate or hinder transition to green growth in the forest sector at rural community level remain largely unknown.

This study was designed to address this challenge by assessing current mode of forest use and management at rural community level in South Africa with a view to locating space for forest-based green growth initiatives development and factors that might facilitate or hinder uptake of such initiatives. Forest dependent rural communities of the Vhembe district were purposely selected as a case study.

The specific objective and associated research questions are:

1. How does forest use and management contribute to employment, income, equity, and social inclusiveness in rural community development?
2. What forest sector/forest products offer the best opportunity for transitioning to forest green growth initiatives in the community?
3. Are there any factors that hinder or facilitates the transitioning to a green growth pathway via initiatives based on these forest products?

10.2 Literature Review

The term "green economy" first appeared in the scientific discourse in the report "Blueprint for a Green Economy" about two decades ago (Hynes 2012). Since then interest in transitioning to a green economy has evolved and intensified (UNITAR 2013). Several reports, e.g. the EMG Report, the UNEP Green Economy Report and the report of the UN Secretary General's High-level Panel on Global Sustainability, are pointing to green economy as a new economic paradigm—one in which economic growth is not delivered at the expense of growing environmental risks, ecological scarcities, and social disparities (PEP 2013).

At the visionary level, UNEP (2011) considers the green economy as: "*An economy that results in improved human well-being and social equity, while significantly reducing environmental risks and ecological scarcities*" (UNITAR 2013). At the operational level, the green economy is seen as economic growth in income and employment driven by investments that: reduce carbon emissions and pollution; enhance energy and resource efficiency; and prevent the loss of biodiversity and ecosystem services. In the same vein, the World Resources Institute (WRI) views a green economy as an alternative growth and development pathway; one that can generate growth and improvements in people's lives in ways consistent with sustainable development, i.e. sustaining and advancing economic, environmental and social well-being (WRI 2012). Similarly, UN DESA reviewed various definitions of green economy, and concludes that various existing definitions of a green economy are generally consistent, having sustainable development as their ultimate objective and aims to reconcile economic development and environmental sustainability, without ignoring social aspects (UN DESA 2012).

Meanwhile, the forest sector is one of the identified key priority sectors in the green growth landscape of South Africa (UNEP 2011b, 2013). In the same vein, the South Africa's national climate change response strategy recognized that the forest sector is strategic for climate change mitigation and adaptation strategy, and transition to a lower-carbon economy (DEA 2011; Musyoki 2012). Interest in forest green growth in South Africa's rural community context is primarily driven by the need to address the multiple challenges of climate change, forest degradation, and rural economic development (Kaggwa et al. 2013; Musyoki 2012).

In this regard, OECD (2012) envisaged that the forest sector will deliver maximum green growth benefits to rural economy through forest initiatives that promote economic growth, social inclusiveness, and environmental sustainability. In the same vein, African Development Bank recommended that green growth will yield the greatest impact on Africa's rural economy within the two focal areas of: (1) efficient/sustainable management of natural assets and (2) building resilience of livelihoods (AfDB 2012). UNEP (2013) and FAO (2009) envisaged that the forest sector will play an increasingly important role in jobs creation, enhancement of rural livelihood sustenance, and resilience to climate change impact.

The forest sector is thus well positioned to deliver green growth benefits to rural communities in South Africa (UNEP 2011b). Opportunities for transitioning to a

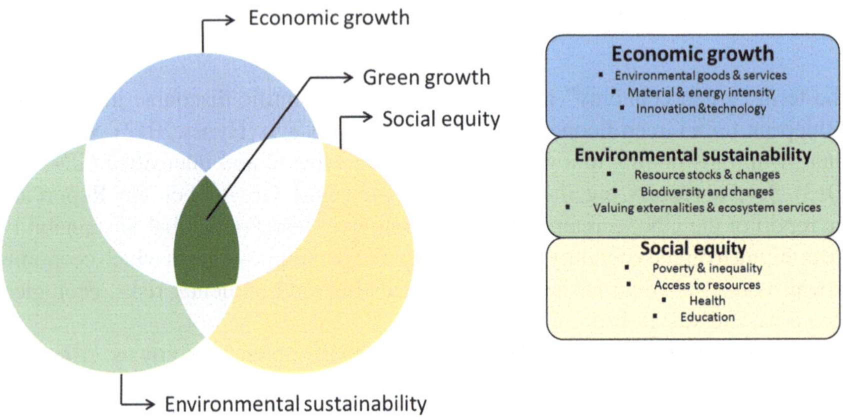

Fig. 10.1 Green growth concept

green growth pathway in the forest sector arise in many ways and forms, namely through the growth in markets for existing products, technical innovation in creating new products and market opportunities, increased efficiencies in product utilization, improved forest management practices, and improved business arrangements for small-scale business enterprises (FAO 2009; UNECE 2009). However, these need to be based on equity and socially inclusive principles that integrate rural host communities meaningfully (WWF 2013). Forest-based green growth initiatives generally have sustainable development as the ultimate objective and integrate with measures to reconcile economic development and environmental sustainability, without ignoring the social aspects (UN DESA 2012) (see Fig. 10.1).

At the operational level, forest green growth is to be driven by investments that: reduce carbon emissions and forest degradation; enhance effectiveness and efficiency of forest resource utilization; and prevent the loss of biodiversity and ecosystem services (WRI 2012; OECD 2012; UNEP 2011b). These notwithstanding, there are different perspectives and agendas shaping the path to green growth (WWF 2013). Consequently, transitioning to green growth is very challenging for many developing countries (Slunge and Loayza 2012).

10.3 Methodology

10.3.1 Description of Study Area

The study was conducted in Vhembe District Municipality, Limpopo Province of the Republic of South Africa as shown in Fig. 10.2. The district shares international borders with Zimbabwe and Botswana in the north and north-west, respectively. The district has a total population of 119,988,654.4% of them being females. Vhembe

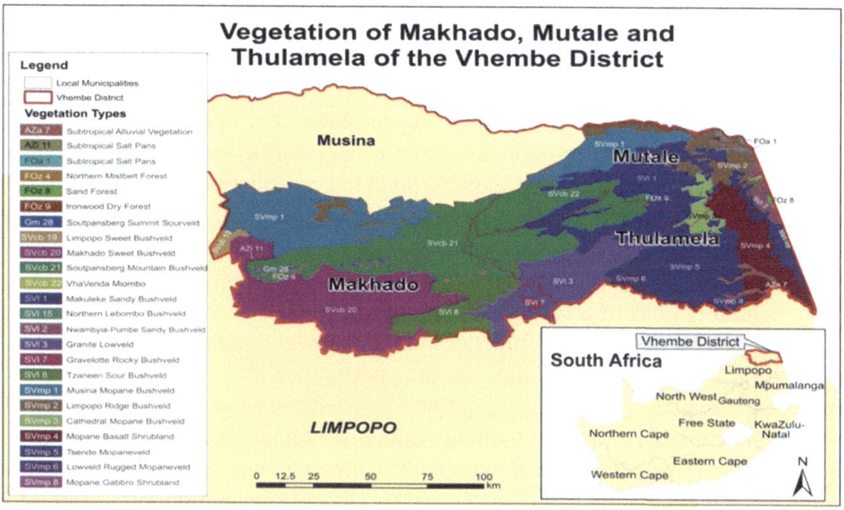

Fig. 10.2 Vegetation map of Vhembe district of South Africa (Ofoegbu et al. 2016)

has the second lowest access to infrastructure amongst districts in the province. The main languages spoken are Tshivenda (69%) and Xitsonga (27%) (CoGTA 2012). In order to closely examine the role that different forest types, social, economic, and infrastructural contexts play in household forest use and management and its implication for forest green growth initiative uptake, three distinct research areas were selected in Vhembe district. The selected areas are in these three municipalities: Makhado, Mutale, and Thulamela. The landscape of selected communities in Thulamela, Makhado, and Mutale has typical woodland, savanna, and semi-arid zones characteristics, respectively (Rosmarin 2013).

In each of the selected municipality, we further selected a replicate of seven rural communities. This gave a combined total of 21 rural communities, which were then surveyed for this study. A total of 366 households were selected from the 21 rural communities using stratified proportionate random sampling procedures. For questionnaire administration, each selected household was subjected to a non-probability purposive sampling technique whereby we selected a respondent from the household that (a) was over 20 years of age and (b) has lived for more than five years in the community.

10.3.2 Measurable Variables: Key Questions and Indicators

The study draws on several sources of information. Primary data were gathered through household questionnaire survey conducted in forest-based rural communities in Vhembe district. Secondary data included the Municipalities demographic and socio-economic document by The South African Demarcation Board, and the South African National census (2011) document, conducted by Statistics South Africa (StatsSA 2012).

The questionnaire contains series of questions on opportunity for green entrepreneurship in forest resource use and management in the community, level of social inclusiveness in forest management, potential contribution of innovative forest management to green growth in the community, and ways of promoting equity in forestry propelled development in the community. In general, the questionnaires were structured in a way that allowed deduction to be made on how forest management in each community links to green growth objectives.

Forest use and management practices in the study communities were examined with a view to identifying specific, viable opportunities in forest resource use and management that each community can harness to achieve economic, environmental, and social well-being and help stimulate resilient sustainable development. In addition, opportunity to yield the following co-benefits: (i) employment; (ii) resilience to climate change; (iii) social inclusion; and (iv) income generation were examined.

10.3.3 Data Analysis

Data from the questionnaire survey were subjected to weighting adjustment in order to correct for possible problems of either over- or under-representation of variables (Bethlehem 2015). The sample was weighed against the actual population to arrive at a weighted sample. The weighted data were then subjected to statistical analysis using the Statistical Package for the Social Sciences (SPSS) (Levesque 2007).

Discrete variables were summarized by the frequency of each code within the questionnaire and summary statistics computed for all numeric variables. Categorical data on the opinion of respondents were analysed based on individual responses. Descriptive statistics and chi-square test were used to analyse: households' dependence on formal and informal forest sector, and opportunities and challenges for social inclusiveness in forest management. The Pearson chi-square was used where the expected cell frequency was ≥ 5, while maximum-likelihood (M-L) chi-square was used when the expected cell frequencies were lower than five (Clewer and Scarisbrick 2001). However, in order to identify variables that were independent predictors of participation in forest management, we used logistic regressions from where the estimated odds ratios (y) were derived to ascertain the effect of the predictors on respondents' participation in forest management. Odds ratios were used to measure the magnitude of strength of association or non-independence between two binary data values. A p-value of $p < 0.05$ represented statistical significance in hypothesis testing and 95% confidence intervals were used to describe the estimation of unknown parameters (Clewer and Scarisbrick 2001).

10.3.4 Specification of the Logistic Regression Model

The target modelled variable was participation in forest management. We took each indicator as a binary outcome and used logistic regression to model a number of explanatory variables including employment status (yes 1; no = 0), farming skill (yes 1; no = 0), animal husbandry skill (yes 1; no = 0), carpentry skill (yes 1; no = 0), years of residency (≤38) (yes 1; no = 0), years of residency (39–52) (yes 1; no = 0), years of residency (53–65) (yes 1; no = 0), years of residency (66+) (yes 1; no = 0), age of respondent (≤38) (yes 1; no = 0), age of respondent (39–52) (yes 1; no = 0), age of respondent (53–65) (yes 1; no = 0), age of respondent (66+) (yes 1; no = 0), and academic qualification (yes 1; no = 0). The chi-square test at $a = 0.05$ significance level was used to assess the goodness of fit of the models.

10.4 Results and Discussion

10.4.1 Forest Use and Household Income

Households' dependence on forest-based income was examined in terms of formal and informal forest sectors' contribution to household income with a view to locating space for forest green growth initiatives development and factors that might facilitate or hinder uptake of such initiatives. We used the result from the reconnaissance survey to identify key formal forest sector in the area (which are tourism, sawmill, tree plantations, furniture and wood artisan/carving), and key forest products traded in the informal sector.

10.4.1.1 Formal Sector

Despite the high abundance of forests including plantations in the area, formal forest sector's contribution to household income was generally low and varied significantly ($P = 0.000$) per sector and between municipalities (Table 10.1). In Thulamela, tree plantation (60.4%), sawmill (56%), and tourism (51%) sectors were the three top-contributing sectors to household income. In Makhado and Mutale, formal forest sector's contribution to household was lower with all surveyed sectors making less than 42% contribution to household income.

From the result, forest types appear to play a significant role in the contribution of the formal forest sector to household income. Thus, in Thulamela, communities with typical woodland-forest type coupled with high abundance of tree plantations, the formal forest sector was observed to make more contribution to household income.

Table 10.1 Contribution of selected formal forest sectors to households' income

Forest sector	Contribution rating	Proportion of respondents (%) in		
		Makhado (n = 156)	Mutale (n = 110)	Thulamela (n = 100)
Tourism	High	12.7[a]	31.2[b]	45.0[b]
	Medium	0.6[a]	2.8[a]	4.0[a]
	Low	4.5[a]	1.8[a]	2.0[a]
	No contribution	82.2[a]	64.2[b]	49.0[b]
Sawmill	High	15.9[a]	33.9[b]	50.0[b]
	Medium	3.2[a]	3.7[a]	3.0[a]
	Low	4.5[a]	3.7[a]	3.0[a]
	No contribution	76.4[a]	58.7[b]	44.0[b]
Tree plantation	High	26.3[a]	38.2[a, b]	53.5[b]
	Medium	6.4[a]	4.5[a]	5.9[a]
	Low	4.5[a]	2.7[a]	1.0[a]
	No contribution	62.8[b]	54.5[a]	39.6[b]
Furniture	High	12.1[a]	25.5[b]	48.0[c]
	Medium	3.8[a]	2.7[a]	3.0[a]
	Low	2.5[a]	1.8[a]	1.0[a]
	No contribution	81.5[a]	70.0[a]	48.0[b]

NB Each superscript letter denotes a subset of "Makhado, Mutale and Thulamela" categories whose column proportions do not differ significantly from each other at the 0.05 level

10.4.1.2 Informal Sector

However, despite the low contribution of the formal forest sector to job creation and household income in the study area, there is widespread informal trade in forest products in the study communities. This informal forest sector was found to be a significant contributor to job creation and households' income in the study area (Table 10.2).

Apart from charcoal, all the forest products surveyed were observed to make significant contributions to households' subsistence and income. Firewood and timber were the most commonly traded wood products in the study area and also the highest contributor to household income. Firewood contribution to household income was observed to be significantly different ($p = 0.000$) within the municipalities. The majority of respondents in Makhado (97.5%), Mutale (97.2%), and Thulamela (68%) rated the firewood contribution to their household income as high. The Bonferroni test, however, showed that firewood contribution to household income did not differ significantly between Mutale and Makhado, but showed significant difference between Thulamela, and Makhado/or Mutale. Unlike in other parts of Africa, charcoal and mushroom were not popularly traded for income by households in the area. The majority of respondents in Makhado (81.2%), Mutale (71.4%), and Thulamela

Table 10.2 Informal forest product trade contribution to job creation and household income

Forest product	Informal forest product trade	Proportion of respondents (%) in		
		Makhado (156)	Mutale (n = 110)	Thulamela (n = 100)
Firewood	Yes	100^{a}	100^{a}	79^{b}
	No	0^{a}	0^{a}	21^{b}
Wild fruits and food	Yes	94.3^{a}	100^{b}	73^{c}
	No	5.7^{a}	0^{b}	27^{c}
Timber/construction wood	Yes	83.3^{a}	45.9^{b}	42^{b}
	No	16.7^{a}	54.1^{b}	58^{b}
Charcoal	Yes	17.9$^{a, b}$	29.4^{b}	15^{a}
	No	82.1$^{a, b}$	70.6^{b}	85^{a}
Thatch grass	Yes	75.5^{a}	85.3^{b}	67^{a}
	No	29.5^{a}	14.7^{b}	33^{a}
Wild vegetables	Yes	91.7^{a}	94.5^{a}	62^{b}
	No	8.3^{a}	5.5^{a}	38^{b}
Mushroom	Yes	82.2^{a}	42.2^{b}	42^{b}
	No	17.8^{a}	57.8^{b}	58^{b}
Honey	Yes	86^{a}	51.4^{b}	41^{b}
	No	14^{a}	48.6^{b}	59^{b}
Medicinal plants	Yes	62.8^{a}	90.8^{b}	40^{c}
	No	37.2^{a}	9.2^{b}	60^{c}
Fodder	Yes	75.6^{a}	94.5^{b}	49.5^{c}
	No	24.4^{a}	5.5^{b}	50.5^{c}
Weaving fibre	Yes	68.2^{a}	45.9^{b}	40^{b}
	No	31.8^{a}	54.1^{b}	60^{b}
Bush meat/edible insects	Yes	84.1^{a}	98.2^{b}	66^{c}
	No	15.9^{a}	1.9^{b}	34^{c}

NB Each superscript letter denotes a subset of "Makhado, Mutale and Thulamela" categories whose column proportions do not differ significantly from each other at the 0.05 level

(84%) were not involved in charcoal trade. This was due mainly to the fact that culturally the people prefer firewood and rarely use charcoal for cooking. Likewise, majority of respondents in Mutale (58%) and Thulamela (57%) were not involved in the sale of mushrooms (Table 10.2).

Bush meat/wild edible insect particularly mopani worms were widely consumed in the study communities. The consumption of this forest product was highest in Mutale (92.7%), followed by Makhado (71.3%), and lowest in Thulamela (63%). The Pearson chi-square test showed a significant difference ($P = 0.000$) in the rate of consumption across the municipalities. The Bonferroni test, however, showed that there was no significant difference between the rate of consumption of this product in Makhado and Thulamela.

Informal trade in forest products is mostly due to poor development of markets and structure regulating production, harvest and trade of these forest products. Managing these issues will create opportunity for maximizing benefits from production and trading of these forest products.

10.4.1.3 Prospects of the Formal and Informal Forest Sectors to Support Green Growth

As shown in Sect. 4.1.1, current contribution of the formal forest sector to household income and employment opportunity in the study communities is significantly lower than the informal sector. Furthermore, contribution of the formal forest sector is highly dependent on availability of thriving tree plantations and associated forest industries. Expanding the forest resource base of a community is thus a required precondition for the development of a vibrant formal forest sector in a community and by implication a precondition for the development of a formal forest sector based green growth initiatives. However, expanding the forest resource base of a community may only deliver green growth benefits in the long run. This is because afforestation is a long-term project. Thus, the formal forest sector does not offer good prospects for transitioning to green growth in the short term in the study communities.

Moreover, not all rural communities in South Africa, like the case of Mutale, have the favourable environment for forest expansion e.g. afforestation or tree plantation development (Chamberlain et al. 2005). The informal forest sector in this regard offers a good option for immediate uptake of green growth initiatives and delivery of green growth benefits to rural host communities. More so, in rural communities where there are favourable environment for tree plantation development, like the case of Makhado, the informal forest sector was observed to make significant contributions to households' income and employment opportunity. This is largely due to the fact that tree plantations in such communities are highly degraded and are unable to support vibrant forest industries. Thus, given the prevailing condition, the informal sector is best positioned for immediate uptake of green growth initiatives in the forest sector at rural community level in South Africa. This postulation is supported by the findings in Table 10.2 where the informal forest sector was observed to make more contribution to households' income in Mutale and Makhado and the communities are characterized as arid vegetation and degraded forests, respectively. Indeed the informal forest sector is crucial to rural development and livelihood sustainability and transitioning to green growth at rural community level in South Africa (Montmasson-Clair 2012; Musyoki 2012).

10.4.2 Linking Local Level Forest Use and Management to Green Growth Pathways

In this section, we offer a critical discussion of current modalities of forest use and management in the informal sector at rural community level in South Africa with a view to teasing out critical factors that will shape transitioning to green growth. This discussion forms the basis for a more in-depth discussion of how to facilitate transitioning to green growth in the forest sector at rural community level in South Africa.

10.4.2.1 Forest Use in the Informal Sector and Link to Green Growth

As shown in Table 10.2, the informal sector is the most significant contributor to employment creation and household income in most of the study communities. The sector similarly offers significant business opportunities for many of the rural dwellers who operate as small-scale forest entrepreneurs. This form of employment/or livelihood strategy provides important livelihood security for most rural households, who often are amongst the poorest and most marginalized groups in South Africa. In the same vein, the South African Department of Agriculture, Forestry and Fisheries estimated that up to 100,000 households are engaged in small-scale trade in forest products (DAFF 2012; DWAF 2005b).

In response to the green growth transition agenda, we attempt to identify most commonly traded products per community by municipalities. This is to help identify forest products with best prospects for transitioning to green growth pathways in the study area (Table 10.3).

All identified forest products per municipality in Table 10.3 showed potential for promotion of forest-based green entrepreneurship in the study communities. These products can be effective in facilitating transitioning to a green growth pathway in the informal forest sector at the community level. They showed good prospects for immediate and widespread impact on job creation, income generation, and uptake of forest green growth initiatives in the study communities. However, not all of these products are widely traded across the study communities, and by implication

Table 10.3 Most commonly traded forest product per municipality

Municipality	Forest product
Makhado	Firewood, wild fruits and food, timber/construction wood, thatch grass, wild vegetables, mushroom, honey, medicinal plants, fodder, weaving fibre, bush meat/edible insects
Mutale	Firewood, wild fruits and food, thatch grass, wild vegetables, honey, medicinal plants, fodder, bush meat/edible insects
Thulamela	Firewood, wild fruits and food, thatch grass, wild vegetables, bush meat/edible insects

their suitability for green growth initiative uptake varies. There are many constraints, some of which are universal to all informally traded forest products (Shackleton and Campbell 2001), and some specific to each product that may restrict the products suitability for inclusion in green growth initiatives (Shackleton and Shackleton 2004). In this regard, firewood and edible insects' production and trade offer the best prospect for forest green growth initiatives across all surveyed communities. Specifically for Makhado, construction wood and honey offer the second-best option. While the second-best options for Mutale are wild fruits and food and medicinal plants, the second-best option is wild vegetables for Thulamela municipality. These products are widely traded in the municipalities and can be produced in the municipalities with comparative advantage.

The most obvious constraints to transitioning to a green growth pathway in the informal forest sector at rural community level are the current mode of forest use and management. Forests in most rural communities of South Africa are treated as open access resources with little or no regulation over their use and management (DAFF 2010). Consequently, unsustainable forest use and management practices leading to forest degradation are on the increase in most rural communities across South Africa (Berliner 2005). This condition is constraining forests' ability to contribute sustainably to household livelihood security (Berliner 2005; Dovie 2003), and may impede transitioning to a green growth pathway in the forest sector at rural community level (Berliner 2005). Other issues that may inhibit transitioning to green growth in the forest sector at rural community level are discussed in Sect. 4.2.2.

10.4.2.2 Regulatory Factors for Green Growth Uptake in the Informal Forest Sector

It is unlikely that rural forest entrepreneurs will be able to grow and position themselves for meaningful participation in a green growth setting without at least some external intervention (Resnick et al. 2012). There are many constraints, some of which are universal to all informally traded forest products (Shackleton and Campbell 2001), and some specific to each product that may restrict the ability of rural participants in the informal forest sectors to benefit maximally from a green growth initiative. Notable is the fact that most rural participants in the informal forest sector do not have the technical skills, knowledge, and resources to develop their businesses and position themselves for meaningful participation in green growth initiatives (Shackleton and Shackleton 2004).

Additionally, availability of the various forest products traded informally in the study communities is not uniform. There is considerable differentiation in the seasonality or all-year-round availability of these products (DWAF 2005a). Some products are available all year round, while some are available on seasonal basis (DAFF 2010; DWAF 2005b). This will have significant impact on the operations and profitability of any green growth initiatives based on such products (UNEP 2013; Musyoki 2012). For example, products such as wild fruits and edible insects are widely traded across the study communities and present good prospect for green growth initiatives uptake.

However, the seasonality that characterizes their availability may pose a significant constraint to scaling up and profitability of these initiatives. Another major limitation is the issue of access to credit facilities. As supported by Shackleton and Shackleton (2004), most rural forest entrepreneurs do not have access to credit facilities, which is crucial if they wish to expand their businesses for meaningful participation in a green growth initiative.

In addition, the challenge of structuring and organizing the numerous informal forest entrepreneurs in the study communities for collective benefit in a forest green growth initiative will need to be properly handled in order to maximize green growth benefits via informal forest sector in the area. Organizing and restructuring rural forest entrepreneurs are indeed an important prerequisite for successful implementation of forest green growth initiatives. As shown by the case of Dukubuku wood carvers and reforestation project funded by UNDP and GEF in South Africa. The project organized wood carvers in a single cooperative. This helped facilitate sustainable use, conservation, and restoration of the Dukuduku forest while also improving the socio-economic status of the rural host community through income generation by selling non-timber wood products (UNDP South Africa 2015). This is supported by DWAF (2005b) assertion, "rural forest entrepreneurs require better organization to collectively bargain and lobby for their interest and have access to resources and training support to improve their productivity and participate in value adding production opportunities". Facilitating forest green growth initiative uptake through reorganization of informal forest sectors in the study communities will also require improving efficiency of product exploitation and handling throughout the value chain (AfDB 2012; DAFF 2012).

Improving efficiency of forest product exploitation and handling along the value chain have direct links to availability of products in sufficient quantity to meet increasing production demand. As supported by Shackleton and Shackleton (2004), forest product sourcing via destructive harvesting method particularly of slow-growing species will pose a major challenge in terms of sustainability to the dependent forest industry. Furthermore, technical issue related to regulation of rates of product exploitation from forests in the community can have implication for green growth initiatives. According to Geldenhuys (2004), long-term biologically sustainable supply of forest products in the right quantities is a prerequisite for enterprise development. Forest product supply and sustainability is thus a major challenge that needs to be properly handled in stimulating rural forest entrepreneurs' participation in green growth initiatives. Other constraints such as efficient methods of processing and market development may significantly hinder development of forest green growth initiatives based on the informal forest sector in the study communities (DWAF 2005a).

Overcoming these constraints that might hold back delivery of green growth benefits via informal forest sector to study communities is crucial to harnessing the benefits of green growth for sustainable development at rural community level in South Africa. As observed by DWAF (2005a), creating enabling mechanisms can help unlock the potential of the informal forest sector in fostering a transition to green growth in the forest sector particularly at rural community level.

Perhaps of greater significance for ensuring sustainability of forest green growth initiatives than technical constraints are the issue of equitability and social inclusiveness in access to forests and sharing of benefits from forest ventures. These can have substantial impact on efforts to secure long-term resource supplies from the forest and sustainable forest management in the communities (Shackleton 2004). This can have implication for private investors that may have interest in forest green growth initiatives in the community.

10.4.2.3 Issues of Equity and Social Inclusiveness in Forest Use and Management and Implication for Green Growth Uptake

Although it is recognized that forest green growth initiatives such as forest enterprise development can play an effective role in sustainable forest use and management, and delivery of green growth initiatives to the study communities (FAO 2009), issues relating to equity and social inclusiveness are identified to be crucial to profitability and sustainability of such initiatives (UN DESA 2012). As pointed out by Geldenhuys (2002), equitable benefit sharing or distribution of benefits from forest enterprise, and rights to access and benefits from forest resources are crucial to profitability and sustainability of forest enterprise and forest management initiatives. In the same vein, Shackleton and Shackleton (2004) observed that in places where local people were well organized, effectiveness of forest development initiatives were ensured and local participants enjoyed greater benefits from the initiative.

We consequently examined equity and social inclusiveness in current state of forest use and management and how these can be enhanced via forest green growth initiatives. We began by examining equity in access to forest and forest resources in the study communities. We found that there were no political or sociocultural barriers to: access to forests or forest resources; or entrant or pursuant of effective forest enterprise business. We, however, noted that there are some socio-economic issues that limit people's access to forests. The results in Fig. 10.3 showed that socio-economic issues in the form of permit issuance and travel distance are the key factors that limiting people's access to forests. However, over 80% of respondents in all the study communities do not experience any difficulty in accessing forest resources in their locality.

We further examined current state of social inclusiveness in forest management in the communities. We found that there is poor community inclusion in forest management in the area. Result in Fig. 10.4 showed that over 70% of all respondents were not in any way involved in forest management in their community.

The observed poor status of community inclusion in forest management in the study communities is a major concern for the delivery of green growth benefits. As noted by Turyahabwe et al. (2006), improved participation of local people in forest use and management, and forest enterprise development is essential for sustainability and profitability of forest-based development initiatives. Thus, government and local authority needs to actively encourage all segments of the community to participate unhindered in forests and forest enterprise management and development.

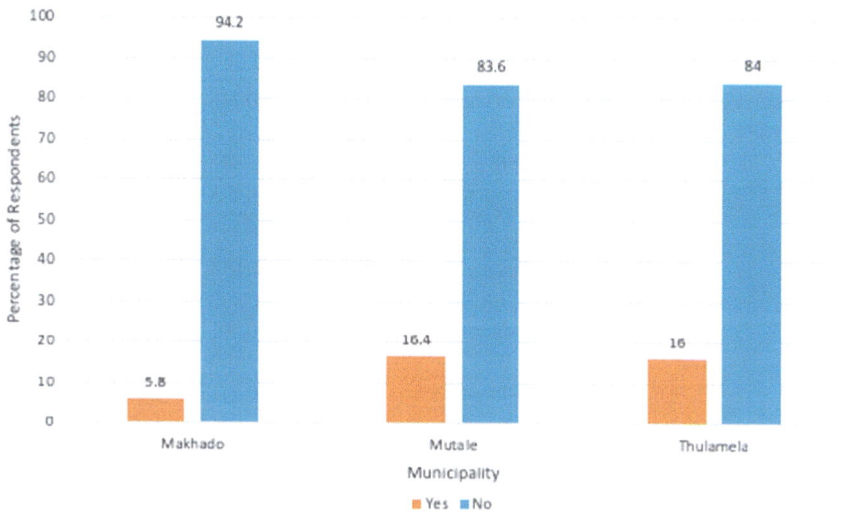

Fig. 10.3 Perceived constraints to accessing forest products in the study communities

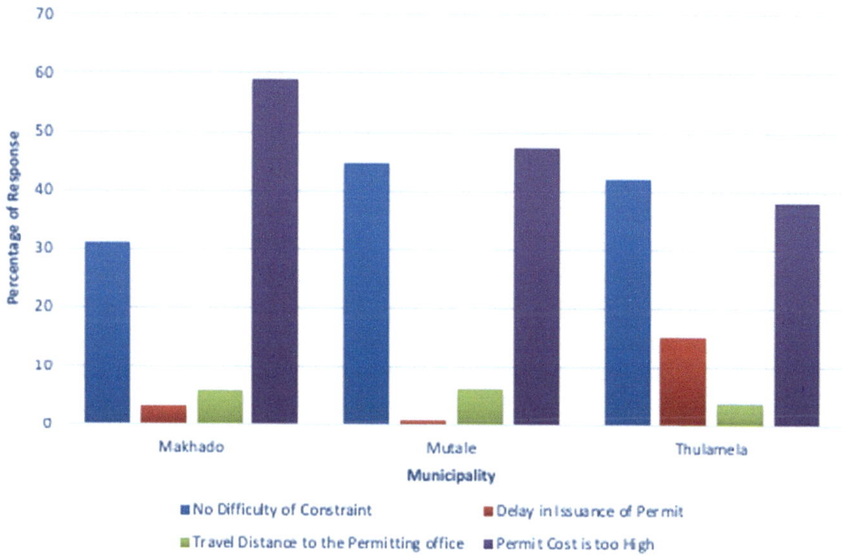

Fig. 10.4 Respondent's participation in forest management in Vhembe district

We furthermore investigated factors that drive participation in forest use and management as a way of determining strategy to improve social inclusiveness in forest management. We used logistic regression to examine socio-economic variables that drives people's participation in forest management (see Table 10.4).

Table 10.4 Factors influencing respondents' participation in forest management

Dependent variable	Independent variable	Odds ratio	Lower	Upper	*P* Value
Participation in forest management	Employment status (employed)	0.633	0.246	1.630	0.343
	Farming skill	3.662	1.493	8.980	0.005*
	Animal husbandry skill	1.037	0.495	2.175	0.923
	Carpentry skill	2.431	0.455	12.995	0.299
	Years of residency (≤38)	0.339	0.078	1.466	0.148
	Years of residency (39–52)	1.651	0.301	9.047	0.564
	Years of residency (53–65)	0.514	0.131	2.020	0.341
	Years of residency (66+)	0.302	0.082	1.109	0.071
	Age of respondent (≤38)	0.572	0.132	2.491	0.457
	Age of respondent (39–52)	0.475	0.135	1.677	0.247
	Age of respondent (53–65)	0.619	0.194	1.976	0.418
	Age of respondent (66+)	1.003	0.311	3.234	0.995
	Formal education	0.776	0.343	1.757	0.544

* significant at 0.005

Results in Table 10.4 showed that participation in forest management in the study communities was significantly influenced by possession of farming skills. This suggests that people with skills to farm, to plant and raise trees are more likely to participate in forest enterprise business and forest green growth initiatives. This links to issue of capacity development. The challenge of weak capacity is a widespread phenomenon in Vhembe district and has been cited as a significant challenge to sustainable development in the district (CoGTA 2012). In the same vein, weak capacity at rural community level has serious implication for promotion of sustainable forest utilization and management in the study communities. As observed by Nelson and Agrawal (2008), capacity development at rural community level has positive correlation with sustainability and effectiveness of community-based forest initiative and forest green growth initiatives.

Furthermore, we assessed training need of the people for capacity improvement towards facilitation of social inclusiveness in forest management by asking the respondents to indicate the type of training that would most interest them in order to improve their capacity to participate meaningfully in forest management (Table 10.5).

Table 10.5 Respondents required skills and knowledge for active participation in forest management

Training	Proportion of respondents (%) in		
	Makhado (n = 156)	Mutale (n = 110)	Thulamela (n = 100)
Sustainable forest resource harvest practices	33.3	35.5	14.9
Training on best practice in forest regeneration	18.6	23.6	23.8
Business opportunity in non-timber forest resources such as honey, thatch grass	11.5	2.7	11.9
Training on agroforestry practice	12.8	7.3	21.8
None	23.7	30.9	27.7

The majority of respondents expressed the need for in-depth training on several types of forest management practices so as to improve their capacity to participate meaningfully in integrated sustainable forest and forest enterprise management (Table 10.5). Respondents in Makhado (33.3) and Mutale (35.5) were more interested in training on sustainable forest resource harvest practices, while respondents in Thulamela (23.8) were more interested in training on best practice in forest regeneration. Provision of these skills will definitely play an active role in facilitating people's active participation in forest rehabilitation and sustainable forest management programmes. Interestingly, there was also high interest in training for business skills, NTFPs, and forest product marketing.

However, it is important to point out that capacity enhancement does not necessarily have to be through formal education. As shown by our result in Table 10.4, possession of formal education does not have any significant impact on people's participation in forest management initiatives. The onus is therefore on government to provide the people with relevant skills for capacity enhancement as a way of facilitating social inclusiveness in forest use and management, and forest green growth initiatives in the communities. Providing people with needed skills is therefore necessary and essential for uptake of forest-based green growth initiatives in the communities (Holmes-Watts and Watts 2008; Turyahabwe et al. 2006). As observed by Musyoki (2012), rural households' capacity enhancement is a key tool for promotion of green entrepreneurship and decoupling of household livelihood activity from destructive forest use practices in the rural areas of Vhembe district (Musyoki 2012).

10.4.3 Forest Management Approaches and Intervention Criteria for Connecting the Informal Forest Sector with Green Growth Initiatives

The findings of this study have shown the strategic need for capacity enhancement and promotion of equity and social inclusiveness in fostering uptake of green growth initiatives in the forest sector at rural community level in South Africa. However, there is equally the need for proper administration of sustainable forest management principles to ensure optimal delivery of green growth benefits. For instance, while it is important to promote equitable access to forests and forest products for stimulation of forest green growth initiative uptake, it is, however, very important that there is management guide to regulate it. For example, when regulation for access to and harvest of forest products is not enforced, the people tend to engage in illegal activities. In our case study, we observed that the people often engage in harvesting of live trees (Fig. 10.5) for firewood which is against the regulation guiding forest product exploitation in the community. The people were supposed to harvest only dead trees.

Although the informal forest sector is an important component of the social, economic, and political settings of rural communities in South Africa (Smit and Musango, 2015), most of current production activities in the informal sector are not

Fig. 10.5 Live branches harvested for firewood

in compliance with the principle of sustainable forest management and in most cases involve exploitative and illegal activities can be a source of both deforestation and forest degradation (Smit and Musango 2015). The onus is thus on forest management authority to come up with a strategy to help tackle these challenges.

Current forest management arrangements do not adequately cater for the need of the informal forest sector to sustainably harness products from the forest for enterprise development. While annual sustainable yield of products, e.g. timber (traded in the formal sector) from most forests, are well established (Chamberlain et al. 2005), annual sustainable yield of products, e.g. medicinal plants (traded in the informal sector), are unknown (Makunga et al. 2008; Geldenhuys 2004). There is therefore the need for a more comprehensive approach to forest management at rural community level in order to cater for the informal sector needs.

These issues cannot be ignored, and policies and management strategies that tie green growth and informal forest sector together at rural community level should aim not only to maximize environmentally protective or restorative activities, but also to minimize environmentally destructive and exploitative activities and ensue sustainably supply of forest products for rural livelihoods (Smit and Musango 2015).

The implementation of forest management strategy that will effectively connect the informal forest sector with green growth initiatives requires an understanding of the informal forest sector that extends beyond assumptions that informal activities are disorganized and chaotic in form or practice (Smit and Musango 2015). A hybrid forest management approach is needed that is open to such duality; embraces both "the traditional and the modern, the small scale and the big scale, the informal and the formal"; addresses power imbalances; and protects the vulnerable (Smit and Musango 2015). For this reason, we advocate an approach that recognizes complexity and contextual realities, when planning for the informal sector in forest management.

We opined that it is important to distinguish between management objectives for products traded "formally" and management objectives for products traded "informally" in instituting forest management guide and practices at rural community level. This is essential to accommodate community rules, norms, and unique characteristics. This process will entail an in-depth participatory approach thereby facilitating adoption of inclusive forest management strategy in the community. Indeed successful transition to a green growth pathway in the informal forest sector at rural community level will require inclusion of and strengthening of the absorptive capacity of the local forest management authority.

10.5 Conclusion and Recommendations

There are myriad of opportunities to stimulate transitioning to a green growth pathway in the South African forest sector particularly at rural community level. This study has identified and prioritized forest sectors and forest products suitable for green growth initiatives uptake at rural community level in South Africa while also contributing to management of climate change and sustainable forest management.

With majority of the population in the study communities engaged in the informal forest sector for subsistence and income, the informal forest sector provides an appropriate foundation for transitioning to green growth pathway for sustainable forest management and sustainable development in the area. While additional comprehensive research may be warranted, empirical evidence from this study showed that promotion of forest enterprise based on firewood and edible insects (e.g. Mopani worms) offers good prospects for profitable forest green growth initiatives in the study communities. However, sustainability and profitability of such enterprise will depend largely on implementation of sustainable forest management practices that promotes equitable and socially inclusive access to forests and responsible forest exploitation practices. This will require enhancement of the people's capacity to participate in responsible forest use and management.

The green growth initiative indeed can be a powerful means to turn the tide in favour of forests and help rural communities realize full benefits from their forest resource base in making progress towards a sustainable climate resilient future. However, much needs to be done to spur development of policies and business strategies that place forest ecosystems' goods and services at the centre of future economic development and climate change mitigation and adaptation at rural community level. These will entail resolving issues related to:

- Organizing the informal forest sector particularly the numerous small and medium scale forestry enterprises in the study communities for collective bargaining and lobbying.
- Provisioning of relevant and well-targeted technical assistance to these entrepreneurs on modalities for forest business and forest management as a way of ensuring sustainability of product supply.

References

AfDB (2012) Facilitating green growth in Africa: perspectives from the African Development Bank. Discussion paper on green growth. African Development Bank, Tunis

Asare RA, Kyei A, Mason JJ (2013) The community resource management area mechanism: a strategy to manage African forest resources for REDD+. Phil Trans R Soc B 368:1–9

Berliner D (2005) Systematic conservation planning for the forest biome of South Africa: approach, methods and results of the selection of priority forests for conservation action. Department of WaterAffairs and Forestry (DWAF), Pretoria

Bethlehem J (2015) Applied survey methods: a statistical perspective. Retrieved 11 Mar 2015, from Weighting adjustment. http://www.applied-survey-methods.com/weight.html

Bishop R, Eis J, Harrison P, Shukla P, Awafo E A, Herweijer C, Hourcade JC (2015) Assessing and communicating benefits of green growth. In Green Growth Best Practice Initiative (ed) Green growth in practice: lessons from country experiences. The Green Growth Best Practice (GGBP), London, pp 79–102

Chamberlain D, Essop H, Hougaard C, Malherbe S, Walker R (2005) Part I: the contribution, costs and development opportunities of the forestry, timber, pulp and paper industries in South Africa. Final report—29 June 2005. Genesis Analytics (Pty) Ltd., South Africa

Clewer AG, Scarisbrick DH (2001) Practical statistics and experimental design for plant and crop science. Wiley, West Sussex, pp 1–22

DAFF (2010) Policy principles and guidelines for control of development affecting natural forests. Department of Agriculture, Forestry and Fisheries, Pretoria

DAFF (2012) Integrated growth and development plan 2012. Department of Agriculture, Forestry and Fisheries, Pretoria

David LS (2005) Integrating climate change adaptation into forest management. For Chronicle 81(5):691–695

Davis C, Stevens N, Archer E, Sinden LA, Nkambule C (2010) Engaging stakeholders in global change risk and vulnerability planning: a case study of the Kruger to Canyons Biosphere Region. In: Science real and relevant conference. CSIR, Pretoria, pp 1–18

Department of Environmental Affairs: DEA (2011) National climate change response white paper. Private Bag X447, Pretoria 0001. https://www.environment.gov.za/sites/default/files/docs/climate_change_governance.pdf

Department of Cooperative Governance and Traditional Affairs: CoGTA (2012) Vhembe district municipality profile. Department of Cooperative Governance and Traditional Affairs, Pretoria

Dercon S (2014) Is green growth good for the poor? The World Bank Research Observer by Oxford University Press, Oxford. http://wbro.oxfordjournals.org/

Dlamini CS (2014) African forests, people and climate change project: forest and climate change policies. Strategies and programmes in the SADC and COMESA regions. Afr For Forum Work Pap Ser 2(17), 39 pp

Dovie DB (2003) Rural economy and livelihoods from the non-timber forest products trade. Compromising sustainability in southern Africa? Int J Sustain Dev World Ecol 10:247–262

DWAF (2005a) Draft key issue paper on forestry and poverty in South Africa. Department of Water Affairs and Forestry, Pretoria

DWAF (2005b) Key issue paper on forestry enterprise development. Department of Water Affairs and Forestry, Pretoria

FAO (2009) Green jobs. An Int J For For Ind Unasylva 60(3). ISSN 0041–6436. ftp://ftp.fao.org/docrep/fao/012/i1025e/i1025e01.pdf. 28 Nov 2013

Geldenhuys CJ (2004). Meeting the demand of Ocotea bullata bark. Implications for the conservation of high value and medicinal tree species. In: Lawes MJ, Eeley HAC, Shackleton CM, Geach BGS (eds) Indigenous forests and woodlands in South Africa: policy, people and practice. University of KwaZulu-Natal Press, Scottsville, pp 517–550

Geldenhuys CJ (2002) Concepts and approach towards development of sustainable resource use of non-timber forest products: the example of bark harvesting for traditional medicine. In: Seydack AHW, Vorster T, Vermeulen WJ, Van der merve IJ (eds) Multiple use management of natural forests and savanna woodlands: policy refinements and scientific progress. In: Proceedings of natural forests and savanna woodlands symposium III, 6–9 May 2002, Berg-en-Dal, Kruger National Park, pp 91–104

Holmes-Watts T, Watts S (2008) Legal frameworks for and the practice of participatory natural resources management in South Africa. For Policy Econ 10:435–443

Hynes W (2012) Green growth and developing countries: a summary for policy makers. The Organization for Economic Co-operation and Development (OECD), Paris

IPCC (2007) Africa. Climate Change 2007: impacts, adaptation and vulnerability. In: Parry L, Canziani OF, Palutikof JP, van der Linden PJ, Hanson CE (eds), Contribution of Working Group II to the fourth assessment report of the Intergovernmental Panel on Climate Change. Cambridge University Press, Cambridge UK, PP 433–467

Kaggwa M, Mutanga SS, Nhamo G, Simelane T (2013) South Africa's green economy transition: implications for reorienting the economy towards a low-carbon growth trajectory. South African Institute of International Affairs Occasional Paper 168, Johannesburg

Levesque R (2007) SPSS programming and data management: a guide for SPSS and SAS users, 4th edn. SPSS Inc., Chicago. ISBN 1-56827-390-8

Makunga NP, Philander LE, Smith M (2008) Current perspectives on an emerging formal natural products sector in South Africa. J Ethnopharmacol 119:365–375
Maponya P, Mpandeli S (2013) Perception of farmers on climate change and adaptation in Limpopo Province of South Africa. J Hum Ecol 42(3):283–288
Matose F (2008) Towards community-based forest management in Southern Africa: do reforms lead to improved livelihoods for the poor? Workshop on forest governance & decentralization in Africa. University of the Western Cape, Durban, pp 1–9
Montmasson-Clair G (2012) Green economy policy framework and employment opportunity: a South African case study. Working Paper Series 2012-02-Trade and Industrial Policy Strategies, Pretoria
Musyoki A (2012) The emerging policy for green economy and social development in Limpopo, South Africa. United Nations Research Institute for Social Development, Geneva
Nelson F, Agrawal A (2008) Patronage or participation? Community-based natural resource management reform in Sub-Saharan Africa. Dev Change 39(4):557–585
Ofoegbu C, Chirwa PW, Francis J, Babalola FD (2016) Perception-based analysis of climate change effect on forest-based livelihood: the case of Vhembe District in South Africa. Jàmbá J Disaster Risk Stud 8(1):a271. http://dx.doi.org/10.4102/jamba.v8i1.271
OECD (2012) Green growth and developing countries: a summary for policy makers. Organization for Economic Co-operation and Development, Pretoria
Poverty-Environment Partnership (PEP) (2013). Building an inclusive green economy for all–opportunities and challenges for overcoming poverty and inequality
Resnick D, Tarp F, Thurlow J (2012) The political economy of green growth: cases from Southern Africa. Pub Admin Dev 32:215–228. https://doi.org/10.1002/pad.1619
Rosmarin J (2013) Vhembe biosphere reserve. Retrieved 1 Aug 2014, from Environmental Affairs. www.environment.gov.za/?q=content/projects_programmes/manand_thebiosphere_reserves/list/vhembe
Shackleton C, Shackleton S (2004) The importance of non-timber forest products in rural livelihood security and as safety nets: a review of evidence from South Africa. S Afr J Sci 100:658–664
Shackleton M (2004) Assessment of the livelihoods importance of forestry, forests and forest products in South Africa. Rhodes University, Grahamstown
Shackleton SE, Campbell BM (2001) Devolution in natural resource management: institutional arrangements and power shifts. A synthesis of case studies from southern Africa. SADC Wildlife Sector Natural Resource Management Programme, Lilongwe & WWF (Southern Africa), Harare
Slunge D, Loayza F (2012) Greening growth through strategic environmental assessment of sector reforms. Public Admin Dev 32:245–261. https://doi.org/10.1002/pad.1623
Smit S, Musango JK (2015) Towards connecting green economy with informal economy in South Africa: a review and way forward. Ecol Econ 116:154–159
Statistics South Africa: STAT SA (2012) Census 2011, Statistical release–P0301.4/Statistics South Africa. Pretoria
Turyahabwe N, Geldenhuys CJ, Watts S, Banana AY (2006) Technical and institutional capacity in local organizations to manage decentralized forest resources in Uganda. South Afr For J 208(1):63–78. https://doi.org/10.2989/10295920609505263
UNECE (2009) The forest sector in the green economy. Geneva timber and forest discussion paper 54. http://sustainabledevelopment.un.org/content/documents/807DP-54.pdf. 20 Nov 2013
UNDP South Africa (2015) Sustainable use and greening in one: seeing forest for the trees. Retrieved 4 Dec 2015, from UNDP South Arica, Environment and Energy, 15 Sept. http://www.za.undp.org/content/south_africa/en/home/ourwork/environmentandenergy/successstories/sustainable-use-and-greening-in-one—seeing-forest-for-the-tree.html
United Nations Department of Economic and Social Affairs (UN DESA) (2012) A guidebook to the green economy issue 1: Green economy, green growth, and low-carbon development-history, definitions and a guide to recent publications. UN DESA, New York
United Nations Institute for Training and Research (UNITAR) (2013) Introduction to a green economy: concepts and applications. E-learning course. Oct–Dec 2013

UNEP (2013) Green economy scoping study: South African green economy modeling report (SAGEM)–focus on natural resource management, agriculture, transport and energy sectors
United Nations Environment Programme (UNEP) (2011) Towards a green economy: pathways to sustainable development and poverty eradication. UNEP, Nairobi
UNEP (2011b) Forests: investing in natural capital, towards a green economy. http://www.unep.org/greeneconomy/Portals/88/documents/ger/GER_5_Forests.pdf. 12 Nov 2013
World Economic Forum (2013) The green investment report: the ways and means to unlock private finance for green growth. WEF, Geneva
World Resources Institute (WRI) (2012) Q&A: what is a green economy?. WRI, Washington
WWF (2013) Linking green economies and sustainable livelihoods for rural wellbeing in Asia. WWF Asia-Pacific SD4C, Jakarta

Chapter 11
An Assessment of the Impact of Barriers on Eco-Innovation Within Small and Medium-Sized Manufacturing Firms in the Peripheral Regions of Lagos

Maruf Sanni

11.1 Introduction

It is now widely accepted that innovative small and medium enterprises (SMEs) are capable of driving the green economy and facilitating a transition to low-carbon development. They play an important role in reinvigorating the economy and creating employment during economic crises. Together with the microenterprises, they provide close to 80% of employment and other related income in Nigeria (SMEDAN/NBS 2012). They also provide value addition to critical raw materials as well as contributing to export earnings in the country (SEDIN/SMEDAN 2013). Therefore, SMEs are crucial to the development of the economies of developing countries such as Nigeria.

In Nigeria, it is also typical of SMEs to operate in clusters. Some popular SMEs clusters include the leather cluster in Kano, the tie and dye cluster in Osogbo and Abeokuta, the Otigba information and communication technology cluster in Lagos, the automotive cluster in Nnewi, and the leather, feather, and fashion cluster in Aba (Oyelaran-Oyeyinka 2007). Another defining characteristic of SMEs is that the sole proprietor or the entrepreneur is critical to the success or failure of innovation activities or efforts of the firms (Adegbite et al. 2007; Kickul and Gundry 2002; Cosh and Hughes 2000). This is so because of the veracity of how quick decisions with regard to innovation efforts can be made especially if the entrepreneur is equipped with the right sources of knowledge. This is different when compared with big firms where the manager of the firm would have to cut through a long chain of bureaucracy before critical decisions on innovation efforts can be made. However, this is not without its attendant risk for the SMEs as a rash decision by the sole proprietor or the entrepreneur could spell doom for the firm.

M. Sanni (✉)
National Centre for Technology Management, Federal Ministry of Science and Technology, Obafemi Awolowo University, Ile Ife PMB 012, Osun State, Nigeria
e-mail: maruf.sanni@nacetem.gov.ng

C. Atewamba and D. Yong Ngondjeb (eds.), *Inclusive Green Growth*, Advances in African Economic, Social and Political Development,
https://doi.org/10.1007/978-3-030-44180-7_11

Small firms gain a competitive advantage over big firms in servicing dispersed local markets (Sanusi 2003). In addition to this, SMEs have also been recognized as the link between the community and the larger global economy (Henderson 2002). However, commercialization of innovative goods and services by SMEs (when compared with big firms) is hampered by many obstacles ranging from inadequate cash flow, low technological capability, and insufficient material resources, and as such, they tend to have less propensity to eco-innovate (Bos-Brouwers 2010; Carrillo-Hermosilla et al. 2009). Meanwhile, eco-innovation has been defined as the production, assimilation, or exploitation of a product, production process, service or management or business method that is novel to the organization (developing or adopting it) and which results (throughout its life cycle) in a reduction of environmental risk, pollution, and other negative impacts of resources use (including energy use) compared to relevant alternatives (Kemp and Pearson 2007). Many of the challenges contributing to low propensities to eco-innovate become even more critical when such firms are located in peripheral regions of big business hubs. This is more so because the location of a business enterprise can set different conditions for operating a business in terms of access to public and technical services. For instance, issues such as inadequate skills, logistical challenges, communication and market inaccessibility, and poor infrastructural facilities are known to characterize peripheral regions when compared with a central location (SEDIN/SMEDAN 2013; Gatrell 2001) or a large hub of business clusters.

Very few studies have examined barriers to eco-innovation, especially among small and medium-sized manufacturing firms in regions surrounding a big hub of business clusters like that of Lagos state in Nigeria. Understanding the barriers is key to the development of appropriate eco-innovation policies. At the same time, it has been suggested that more often than not, firms that are located in peripheral regions may require specific policy interventions which might be different from those required by firms located in central business districts (Cooke et al. 1997). It should be noted at this juncture that peripheral firms as conceptualized in this study are those firms located outside a big business hub, but are able to transact business with such hub on a daily basis. Lagos is chosen in this study as the main commercial hub with which other cities in the country transact business. This is because the city represents the commercial nerve center of the country. It also plays a critical role in the dynamics of both internal and external trade in the country, especially in the manufacturing sector. For instance, the city accounts for over 60% of industrial and commercial activities in the country and harbors almost 50% of Nigeria's skilled workers (Nwagwu and Oni 2015). It has been estimated that the metropolitan Lagos accounts for over 53% of manufacturing employment in Nigeria. Lagos also contributes the largest share of the 7% of national gross domestic product (GDP) coming from the manufacturing sector. In 2013, manufacturing was estimated to have contributed $35 billion to the national economy (Leke et al. 2014).

There is a dearth of information on eco-innovation from peripheral SMEs, their significance, as well as the benefits of resource efficiency, pollution reduction, and sustainable production. This study therefore sets out to examine barriers that could limit the capability of small and medium-sized manufacturing firms located in the peripheral regions of Lagos state in Nigeria, to eco-innovate. This study intends to contribute to the knowledge base and suggest to policy makers ways that could promote the development of eco-innovation among small and medium-sized manufacturing firms located in peripheral regions.

The remaining part of the chapter is organized as follows. Section 11.2 reviews the literature and lays out the theoretical framework for the analysis of barriers to eco-innovation among SMEs. Section 11.3 presents the methodology and empirical model to estimate the effects of barriers on eco-innovation. Section 11.4 analyzes and discusses the results. Section 11.5 provides the conclusion and some policy recommendations.

11.2 Review of Barriers to Eco-Innovation

Eco-innovation has been described as a fuzzy concept used in many disciplines and as such requires both theoretical and methodological explanations (Andersen 2010). The concept is at the interface of the environmental and economic objectives of firms trying to promote green competitiveness as an important aspect of environmental technological development. Many analytical methods have been proffered to investigate the process of environmental technological change. Some of them include corporate environmental management theory (Welford 1996; Steger 1996), induced innovation theory (Ruttan 2000), actor-network analysis (Luiten and Blok 2004; Luiten 2001), environmental economics (Jaffe et al. 2002), organization studies (Khanna et al. 2006), industrial economics (Cole et al. 2005; Kerr and Newell 2001), and evolutionary economics (Unruh 2000; Foxon et al. 2005, etc.). In the meantime, the application of different theories to eco-innovation processes and management demands a logical theoretical framework.

With regard to this study, the evolutionary economics approach suggests an appropriate framework with which barriers to eco-innovation can be analyzed and relevant policies can be recommended (Dosi et al. 1988; Freeman and Perez 1988; Freeman 1992; Arthur 1994; Metcalfe 1995; Nelson and Winter 2002). When compared with neoclassical economics, evolutionary economics approach is more appropriate to analyze transition and learning processes. It assumes bounded rationality and rules of thumb rather than optimization (Rennings 1998). The evolutionary approach uses biological terms of selection and variation in describing the innovation process. For instance, man-made environmental problems or external forces are known to put selective pressure on society to create new technologies such as eco-innovation. Adoption of eco-innovation as a strategy to improve the competitiveness of firms may involve long-term radical technological change. In this context, it has been suggested that ecological irreversibility is one of the main drivers of eco-innovation. The evolutionary perspective gives detailed explanations on technological change and competition, and socio-technical systems, critically examining pathways and different types of innovations (Nill and Kemp 2009).

Central to evolutionary approaches is the belief that technological change is a path-dependent process that comes from the interaction between supply and demand factors, social groups, and other economic forces (del Río et al. 2010). Evolutionary theory has proved to be a robust approach in understanding the complex dynamics of the world around us. Evolutionary economics, with underline core principles in evolutionary theory is fast becoming a more relevant alternative method to mainstream economics (Nelson and Winter 1982). It has been used for economic analysis of technological change and innovation. Some scholars have argued that it provides a more realistic theory of these phenomena than mainstream economic methods (Verspagen 2009). In an evolutionary process, novelty and selection interact to advance change. Evolutionary economic theory also sheds light on the potential of the market to go green within the context of socioeconomic dynamics. This is based on the assumption that economic process is not static, rationality is bounded, information is lacking, and markets are fundamentally imperfect (Andersen 2010). Based on this background, the study creates a framework with which to analyze barriers to eco-innovation among SMEs in the peripheral regions. This is illustrated in Fig. 11.1.

Although there are many studies on the determinants of eco-innovation, very few scholars have systematically examined barriers to eco-innovation (Marin et al. 2015). Other scholars such as Mohnen and Röller (2005), Smallbone and North (1999), March et al. (2002), and Segarra-Blasco et al. (2008) classified barriers to innovation into costs, knowledge and markets as well as broad categories of internal and external barriers. Some of the few studies on barriers to eco-innovation investigated how failures of the system of innovation could create barriers such as inability to provide

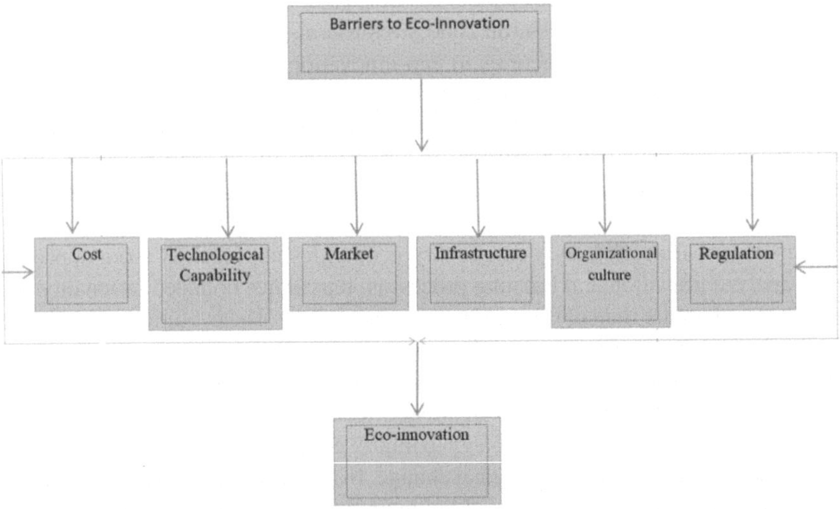

Fig. 11.1 Conceptual framework for barriers limiting eco-innovation among SMEs in peripheral regions. *Source* Author's illustration

infrastructural facilities and a favorable investment climate, inconsistency in government policies, lack of continuity in governance, and weak institutional structures (Foxon and Pearson 2008). However, these studies did not take into consideration barriers within firms as most of the failures enumerated above are systemic failures. Some other studies on barriers to eco-innovation examined firm-level barriers (del Río et al. 2010), while others investigated barriers to leadership in environmental corporate strategies (Runhaar et al. 2008).

It has also been identified that the techno-economic characteristics of eco-innovation could also be a barrier to its implementation. For instance, firms usually find it difficult to introduce eco-innovations when it is too costly or the implementation does not fit into the existing production process (del Río et al. 2010). In all, it should be noted that it is challenging to separate these barriers from one another as they often influence and reinforce one another. For instance, different categories of eco-innovations are likely to be influenced by different set of barriers. At the same time, these barriers could be sector-specific. Taking this further, the intensity of barriers on eco-innovation may also be location-specific. Eco-innovative firms located in a highly urbanized environment may be shielded from certain barriers when compared with firms in peripheral regions. This is so because evidence has suggested that majority of peripheral regions are not conducive for sustainable business. Such regions are usually devoid of strategic actors, support institutions for innovation, and technological change (Todtling and Trippl 2005; Isaksen 2001). In addition, they also lack innovation platforms that are supported by resources and networking forums, as well as training and knowledge support systems (Doloreux and Dionne 2008).

There are certain levels of eco-innovativeness within SMEs (Noci and Verganti 1999) but to a varying degree when compared with big firms. More often than not, when SMEs decide to eco-innovate, they do this to increase economic performance and the quality of goods and services (Revell et al. 2010; Suh et al. 2005). Ultimately, such actions do have positive effects on the environment and the national economy (Schaltegger and Synnestvedt 2002; Dyllick and Hockerts 2002). Studies have shown that eco-innovative SMEs operate differently from big firms (Klewitz and Hansen 2011; Moore and Spence 2006; Tilley 2000). They are also diverse with regard to their size and sector specificities (Hillary 2006). Some of the distinctive characteristics put forward by many scholars pertain to the lack of bureaucracy in the channel of communication, fewer hands at the management and organization structures, and owner–manager system of business operations (Bos-Brouwers 2010). These distinctive features allow SMEs to make quicker decisions and respond to the constantly changing nature of green markets (Jenkins 2006). Eco-innovative SMEs tend to face unique challenges that relate to time constraints, lack of skilled personnel, insufficient technological know-how, and inaccessibility to financial resources (Azzone and Noci 1998; del Brío and Junquera 2003; Spence 1999; Bos-Brouwers 2010). These issues and many more differentiate SMEs from large firms in terms of sustainability. In fact, these factors are responsible for why many SMEs do not invest or implement eco-innovation (Noci and Verganti 1999). In other words, they serve as barriers to the introduction of eco-innovation among the majority of SMEs. From the foregoing, I propose the following hypothesis:

H_{1a}: Lack of technological capabilities affects the ability of the SMEs to introduce eco-innovation

In general, barriers to innovation can be broadly classified into two. The first category termed endogenous barriers relate to those obstacles that are found within the firm while the exogenous barriers are found outside the firm's sphere of influence (Piatier 1984). Some of the endogenous barriers include lack of sufficient skilled labor, inaccessibility to financial resources, lack of access to the market, lack of access to appropriate information and communication technology (ICT), and lack of capacity to generate intellectual property rights (Rush and Bessant 1992). Other internal barriers such as the attitude of the manager to risk bearing and employee's inability to appreciate innovation (Hadjimanolis 1999) have also been identified to influence the innovative performance of firms. Assink (2006) also categorized internal barriers into four groups: adoption, rigid mindset, risk, and nascent barriers. The first category relates to the large chain of bureaucracy and organizational culture that hamper introduction of innovation. These have to do with the tendency of the firm to rely on the success of past innovation efforts. The second group of internal barriers identified by the scholars are those focused on lack of competency to search, find, use, and absorb new opportunities for competitive advantage. Risk barriers also constitute strong impediments to innovation efforts; they constrain firms to depend on routine and experience, as they fear the introduction of radical innovations to the market. Meanwhile, nascent barriers deal with the inability of the top managers to be able to chart an innovative management strategy for the firm (ibid.). Issues such as these led to the proposition of the following hypothesis:

H_{1b}: Organizational culture is highly associated with the ability of the SMEs to introduce eco-innovation

Other barriers such as constricting government regulations, rigid bureaucratic systems, insufficient infrastructural facilities, and inaccessibility to business partners are termed exogenous barriers. These barriers are usually beyond the control of firms. In furtherance of this classification, Hadjimanolis (1999) divided the exogenous barriers into supply, demand, and business environment. When enterprises find it difficult to access technological information, raw materials, and finance for the innovation process, they are said to be experiencing supply barriers. However, demand barriers occur as a result of how customers perceive the risk associated with an innovation, as well as the influence of domestic or foreign market limitations. Business environment barriers concern various government regulations, antitrust measures, and policy actions that impede innovation processes. There are also situations where a weak national innovation system can hamper innovative collaboration that could lead to introduction of competitive goods and services. Issues such as these are usually related to the institutional and economic context of a particular country (Holzl and Janger 2012). Based on the foregoing, the following hypothesis was formulated:

H_{1c}: Government's regulation of the business environment affects the level of eco-innovation among SMEs in peripheral regions

Another major challenge affecting the socioeconomic and technological development in Nigeria is electricity supply. It has been estimated that only 40% of the population in Nigeria draws electricity from the national grid. This energy situation also puts immense pressure on the industrial sector. For instance, the Manufacturers Association of Nigeria (MAN) has claimed that its members spend over N1.8 billion (US $11, 340 million) per week on running electricity generating sets to provide energy for manufacturing operations (NERC 2011). As a result of this, 40% of production costs are attributed to energy costs. Such costs are usually spread across the entire production process resulting in high cost of goods and services in the country. The MAN has since advocated for the regular supply of about 2000 MW of electricity to the industrial sector in order to be able to function optimally and produce at more competitive prices (Ojo 2009). In view of this, the sustainability of the industrial sector could be at risk unless other sources of sustainable energy or energy-efficient strategies are exploited to reduce energy costs. In addition, the bad state of major roads linking the commercial nerve center and other areas has made the transportation of goods and services difficult and costly. This has even become more challenging for those firms located in peripheral regions. Based on these issues, the study hypothesizes that

H_{1d}: Insufficient infrastructure affects the ability of SMEs to produce goods or services that reduce negative impacts on the environment

The issue of inadequate supply of electricity and other factors such as the rapid depletion of raw materials, the need to reduce energy costs and carbon footprints have given many SMEs the impetus to find alternative ways to produce in ways that reduce their negative impacts on the environment, as well as promoting savings in the production processes. Unfortunately, besides the general obstacles that plague the innovation efforts of SMEs, eco-innovative SMEs are also affected by high costs of innovation. The effect of this barrier is usually higher for SMEs when compared to big firms because more often than not SMEs find it difficult to access loans to finance innovation efforts that can allow them to scale up and operate more efficiently. Based on this, the study puts up the following hypothesis:

H_{1e}: The high cost of innovation negatively impacts on the capability of SMEs to eco-innovate

Based on the proposition that there could be a positive association between innovative performance and the intensity of barriers experienced by firms, D'Este et al. (2012) proposed that two categories of innovation barriers should be differentiated: "revealed" and "deterring" barriers. They referred to revealed barriers as those difficulties that firms encounter when they engage in innovation activities. While going through such difficulties, firms are more aware of the intricacies of such difficulties and learn in the process of trying to overcome the challenges. Meanwhile, deterring barriers are those that actually prevent firms from implementing or introducing innovation. In view of this, many small and medium-sized firms usually looked outside for knowledge in order to reduce the impact of such barriers. This paper tests the veracity of this assumption by putting up the following hypothesis:

H_2: Deeper search for external knowledge affects eco-innovation among SMEs

Firm size is one of the most important variables in determining the drivers of eco-innovation. Many scholars have also used it as a control variable in econometric analysis in eco-innovation studies. There seems to be a positive association between eco-innovation and big firms as they usually attract attention from the civil society and other environmental non-governmental organizations (Kesidou and Demirel 2012; Kammerer 2009). Big firms have highly organized research and development (R&D) sections (Kesidou and Demirel 2012) and have better access to loan facilities and skilled labor (Kammerer 2009; Walz 2011; Rave et al. 2011). As such, they have a greater propensity to invest and implement eco-innovation strategies. In the meantime, SMEs are usually characterized by fewer employees, which can be a barrier to the implementation of eco-innovation. Moreover, SMEs often lack financial, technological capabilities, and human resources. SMEs therefore face an uphill task to address the complexity of eco-innovation. In light of the issues raised above, there is a need for critical investigation of the barriers to eco-innovation among SMEs as they face different conditions from those experienced by big firms. Therefore, the following hypothesis is proposed:

H_3: The intensity of barrier on eco-innovation is a function of firm size

11.3 Methodology

11.3.1 Data Collection

There are many variables, indicators, or proxies that had been used to measure eco-innovation in the literature (Arundel and Kemp 2009; Kemp 2010). Some of the important proxies that have been used include environmental patents or environmental investments (Jaffe and Palmer 1997; Brunnermeier and Cohen 2003; Nameroff et al. 2004). These measures have been criticized for the tendency to under- or over-estimate eco-innovation, particularly when it comes to the case of incremental eco-innovation (De Marchi and Grandinetti 2012). Another commonly used indicator is the environmental R&D expenditure. The challenge with the use of this indicator is that it only focuses on the input side of innovation activities and does not consider the output side (Mazzanti and Zoboli 2006). In order to avoid some of these shortcomings, the study adopted the method used by Horbach et al. (2013), Borghesi et al. (2012), De Marchi (2012), De Marchi and Grandinetti (2012), and Horbach (2008) in the measurement of eco-innovation.

For the purpose of this study, the Nigerian Community Innovation Surveys (NCIS) for the 2005–2007 and 2008–2010 data are used. The survey implementation and procedures are based on the "Guidelines for Collecting and Interpreting Innovation Data (2005)" jointly developed by the OECD and the Eurostat popularly referred to as the Oslo Manual (OECD-EUROSTAT 2005), and these were adapted for the

New Partnership for Africa's Development (NEPAD) innovation initiatives. The two waves of the NCIS survey represent two repeated cross-sectional data. The survey used the Industrial Classification of all Economic Activities (ISIC revision 3.1) for the sectoral classification of economic activities. Manufacturing firms fell into divisions between 15 and 37. The sampling of data took place using a multistage systematic random sampling technique. In general, the NCIS covers a wide range of indicators on innovation activities such as innovation expenditure, objectives/effects of innovation, public funding, innovation collaboration, source of information for innovation, and factors that hamper innovation activity, intellectual property rights, and government support for innovation. The survey ensured that all firms that responded in the first edition were also given the same set of questionnaires to complete in the second edition. Eight hundred and ninety manufacturing firms responded to the survey of which 113 and 411 were non-innovating and micro firms, respectively. The final sample had 366 after non-innovating, and all the microenterprises had been removed.

On firm size, manufacturing firms were categorized as follows: Those with employee size between 1 and 9 were categorized as micro, between 10 and 49 employees were classified as small, those between 50 and 199 were categorized as a medium, while those with 200 and above were classified as large. From this initial categorization, the study constructed a variable "firm size," which took a value of 1 for small and medium-sized firms below 200 employees, and a value of 0 for big firms with more than 200 employees. Given the fact that Lagos is the biggest business hub in the country, this study considered it a commercial business hub, and firms located within it conducted a lot of business transactions with firms from other parts of the country. As such, these locations of the firms (i.e., outside the commercial hub) were in this study considered as the peripheral regions. One of the important variables (used in the empirical model as the dependent variable, *Eco-innovation*, see Table 11.1) is related to the level of eco-innovation within the firm. It is based on the data on the outcome/effects of the innovations introduced by the firm that "reduced environmental impacts or improved health and safety." Firms were allowed to choose four options, reporting if the outcome of innovation was nonexistent, low, medium, or high, in the periods 2005–2007 and 2008–2010. The dependent variable takes the value of 3 if the enterprise reported high effects, a value of 2 for medium effects, 1 for low effects, and 0 for no effects of the outcome of innovation.

11.3.2 Data Analysis

The analysis is divided broadly into two parts. The first part uses descriptive statistics to give a general overview of the characteristics of the firms in the peripheral regions of Lagos. This is complemented by use of the correlation analysis to examine the relationship between the introduction of eco-innovation and the various factors that are likely to have an influence on its implementation among the SMEs in the peripheral regions. The second part which concerns the econometric analysis is carried out using an ordered probit regression technique to investigate the impact of

Table 11.1 Variable descriptions

	Variable name	Variable description
Dependent variable	Eco-innovation	Innovations that reduced environmental impacts with 3 "high effects," 2 "medium effects," 1 "low effects," 0 "no effects"
Barriers	Cost barrier	3 "high" impact, 2 "medium" impact, 1 "low", 0 "no impact"
	Knowledge barrier	3 "high" impact, 2 "medium" impact, 1 "low", 0 "no impact"
	Market barrier	3 "high" impact; 2 "medium" impact, 1 "low", 0 "no impact"
	Regulation barrier	3 "high" impact, 2 "medium" impact, 1 "low", 0 "no impact"
	Infrastructure barrier	3 "high" impact; 2 "medium" impact, 1 "low", 0 "no impact"
	Organizational culture barrier	3 "high" impact, 2 "medium" impact, 1 "low", 0 "no impact"
External knowledge sources	Customers	1 "highly relevant," 2 "medium relevance," 1 "low relevance," 0 "no relevance"
	Competitors	1 "highly relevant," 2 "medium relevance," 1 "low relevance," 0 "no relevance"
	Public research institutes	1 "highly relevant," 2 "medium relevance," 1 "low relevance," 0 "no relevance"
	Professional and industry associations	1 "highly relevant," 2 "medium relevance," 1 "low relevance," 0 "no relevance"
	Suppliers of equipment and materials	1 "highly relevant," 2 "medium relevance," 1 "low relevance," 0 "no relevance"
Other variables	Sector	Manufacturing sector classified by the 2-digit ISIC
	Firm size	Employee between 10 and 49 (1 "small"); 50 and above (2 "large")
Controls	Age	Logarithm of firm's age
	Part of a conglomerate	1 enterprise is part of a group; 0 otherwise
	Product market	"1" for firms that sell their goods only within the country, or "0" for firms with foreign markets

the factors on the firms' capabilities to introduce eco-innovation. This econometric technique allows for exploration of the relative influence of explanatory variables on firms' propensity to eco-innovate. This is carried out to find empirical evidence to support the theoretical propositions on the impact of innovation barriers on the level of eco-innovativeness among the small and medium-sized manufacturing firms in the peripheral regions of Lagos.

11.3.2.1 The Empirical Model

The study opted for the ordinal probit regression model because it is assumed that there is a latent continuous metric underlying the ordinal responses observed in the data. In an ordered probit model, an underlying score is estimated as a linear function of the independent variables and a set of cut points. The probability of observing outcome i corresponds to the probability that the estimated linear function, plus random error, is within the range of the cut points estimated for the outcome:

$$\Pr\left(\text{outcome}_j = i\right) = \Pr\left(k_i - 1 < \beta_1 x_{1j} + \beta_2 x_{2j} + \cdots + \beta_k x_{kj} + uj \leq k_i\right)$$

uj is assumed to be normally distributed. In either case, we estimate the coefficients $\beta_1, \beta_2, \ldots, \beta_k$ together with the cut points $k_1, k_2, \ldots, k_I - 1$, where I is the number of possible outcomes. k_0 is taken as $-\infty$, and k_I is taken as $+\infty$. All of this is a direct generalization of the ordinary two-outcome probit model.
where

x_1 Cost barrier
x_2 Technological capability barrier
x_3 Market barrier
x_4 Infrastructure barrier
x_5 Organization barrier
x_6 Regulatory barrier
x_7 External knowledge
x_8 Sector
x_9 Firm size
x_{10} Firm's age
x_{11} Product market
x_{12} Part of a conglomerate.

The variables are defined in Table 11.1. In a study such as this, researchers usually tend to estimate the slope parameter(s) and the threshold parameters. However, since the ordered probit model is nonlinear, exact algebraic expressions for their parameters do not exist. Instead, to compute the parameters, iterative estimation methods are required. In this study, the maximum likelihood estimation method is adopted. This is based on its efficiency in estimating ordered probit model slope parameters and thresholds (Winship and Mare 1984). Multicollinearity among the variables was tested using correlation analysis (see Table 11.6), while the assumptions of the ordered probit regression model were assessed by conducting tests of parallel lines (see Table 11.7).

11.4 Results and Discussion

11.4.1 Characteristics of Firms in the Peripheral Regions

Many variables such as age of the firm, ownership structure, market within which the firms transact business are known to affect the level of innovativeness. These factors can serve as obstacles to innovation activities at the firm level (Radwan and Pellegrini 2010; Cohen 2010). Based on this, the study controlled for the following: age of the firm "*firm's age,*" whether a firm is part of a group of companies or not "*part of a conglomerate,*" and if they sold their goods in the local or international market "*product market*" (see Table 11.3). Firms with lots of experience in implementing innovation activities are likely to have an advantage over young start-ups in the same sector. There are also reasons to believe that manufacturing firms that are part of a global group of firms are likely to have access to more resources than "stand-alone" firms. At the same time, firms that engage in international trade are likely to have developed certain innovation strategies that could make them more competitive. The breakdown of the sub-sectors of the manufacturing firms under study is shown in Table 11.2.

Table 11.2 Sub-sectors of the manufacturing firms

Sub-sector	Number of firms	Percentage
Food and beverages	76	20.8
Textiles and wearing apparel	24	6.6
Leather tanning and dressing, etc.	12	3.3
Wood and paper products, etc.	25	6.8
Publishing, reproduction of recorded media	27	7.4
Chemicals and chemical products	58	15.9
Rubber and plastic products	16	4.4
Non-metallic mineral products	27	7.4
Basic metals and recycling	19	5.2
Fabricated metal, etc.	30	8.2
Machinery, equipment, and vehicles	20	5.5
Furniture	31	8.5
Other	1	0.3
Total	366	100.0

Table 11.3 Characteristics of firms in the peripheral regions

	Total no of firms	Peripheral (%)
Eco-innovative firms (n = 366)		
Low	126	35.8
Medium	82	21.6
High	158	42.6
Firm size (n = 366)		
Small	123	39.4
Medium	147	37.6
Large	96	23.0
Product innovation (*n* = 366)	275	72.0
Process innovation (*n* = 366)	327	89.0
Organization innovation (*n* = 339)	229	67.3
Marketing innovation (*n* = 366)	189	56.5
Business conglomerate (*n* = 366)	91	22.3
Internal R&D department (*n* = 366)	176	42.2
Continuous R&D (*n* = 175)	84	43.7
Occasional R&D (*n* = 175)	91	56.3
External search for knowledge (*n* = 366)	111	26.6

The observation emerging from Table 11.3 reveals that 64.2% of firms in the peripheral regions of Lagos are eco-innovative as they produce goods and services with medium to high positive effects on the environment. This could be a pointer to the fact that if provided with the right business environment, manufacturing firms in the peripheral regions could be as eco-innovative as that of those in the big business hub. This result also suggests that it is possible for firms to be eco-innovative irrespective of their locations. With regard to firm size, many firms (77%) in the peripheral regions were in the small and medium categories. Table 11.3 also shows that firms in the peripheral regions undertook different types of innovation as follows: product innovation (72.0%), process innovation (89.0%), organization innovation (67.3%), and marketing innovation (56.5%). However, few studies have shown which of the categories of innovation are common and which one makes the most impact on the firm's performance (Oke et al. 2004; Kanter 1985). Many small and medium-sized manufacturing firms usually attempt to innovate so as they increase their turnover

(Bala-Subrahmanya 2005; Oke et al. 2004). It is not surprising that few manufacturing firms in peripheral regions were part of a conglomerate (22.3%).

Because of the technical nature of eco-innovation and the relevance of R&D in the development and introduction of environmental innovation, the study examined how R&D is being incorporated into innovation activities. Results in Table 11.3 show that only 42.2% of firms in the peripheral regions carry out in-house R&D. Further exploration of these statistics reveals that 43.7% of firms in the peripheral region engage in continuous R&D. Interestingly, however, more firms among those located in the peripheral regions (56.3%) are likely to engage in occasional R&D. It must, however, be noted that most firms in developing countries like Nigeria do not have sufficient in-house capabilities to carry out R&D, and as such most of the innovation is usually incremental in nature. This is reflected in the number of firms that claimed to have carried out in-house R&D (see Table 11.3). In this context, some firms in developing countries do acquire foreign technologies which they learn and adapt to suit local conditions (Aggarwal 2000). More often than not, the majority of small and medium-sized firms lack the resources to compete and invest in training and R&D as compared to large firms (Goedhuys 2007). As a result of this, many studies have shown that collaboration with external sources of knowledge increases the chances of SMEs successfully introducing innovation (Rothwell and Dodgson 1991). This even becomes more pertinent when they are faced with lots of barriers that limit them from innovating. Thus, collaboration with external sources of knowledge is an important aspect of the technological learning process (Figueiredo 2002, 2003). Unfortunately, relatively few manufacturing firms in the peripheral regions of Lagos (26.6%) were found to collaborate with external sources of knowledge.

The study also considered barriers to eco-innovation. These were identified by asking the manufacturing firms to respond to questions that are related to a set of six barrier constructs, namely high cost of innovation, lack of technological capability, inaccessibility to market, lack of infrastructure, rigid organizational culture, and regulation (see Table 11.4). Firms were asked to state how important were these factors in hampering their innovation activities or projects or influencing a decision not to innovate during 2005–2007 and 2008–2010 periods. Firms were allowed to choose four options, reporting if the barrier factor was nonexistent, low, medium, or high. Each barrier construct took the value of 3 if the firm reported high degree

Table 11.4 Barriers to eco-innovation among the SMEs in peripheral regions of Lagos

	Firms	Mean rank	Std. deviation	Minimum	Maximum
Cost barrier	209	2.03	0.898	0	3
Infrastructure barrier	209	1.95	0.884	0	3
Regulatory barrier	217	1.59	1.237	0	3
Market barrier	210	1.52	0.824	0	3
Technological capability barrier	208	1.21	0.805	0	3
Organization barrier	200	1.02	0.959	0	3

Table 11.5 Sources of external knowledge among the SMEs in the peripheral regions

Sources of external knowledge	Mean rank	Very important	Moderately important	Important	Not important
Customers	1.55	26.7	29.5	15.7	28.1
Suppliers	1.49	27.2	26.3	14.7	31.8
Competitors	1.19	20.8	21.8	13.4	44.0
Industry associations	0.98	12.4	22.6	15.2	49.8
Consultants and private laboratories	0.67	9.2	12.0	15.2	63.6
Public R&D institutes	0.46	4.1	10.6	12.0	73.3
Knowledge institutions	0.45	5.1	10.6	8.8	75.6

of importance, a value of 2 for medium degree of importance, 1 for low degree of importance, and 0 for the nonexistence of the barrier.

With regard to external knowledge access, the study considered sources of knowledge such as suppliers of equipment, materials, components or software; clients or customers; competitors or other enterprises; consultants, commercial laboratories, private R&D institutes; universities or other higher education institutions; government or public research institutes; and professional and industry associations (see Table 11.5). For each external knowledge source, the firms were given four options to choose from, namely highly important, of medium importance, of low value to the innovation process, or not used at all, in the periods 2005–2007 and 2008–2010. In addition, each source of knowledge construct took the value of 3 if the enterprise reported high degree of importance, a value of 2 for medium degree of importance, 1 for low degree of importance, and 0 for nonexistence of the source.

11.4.2 Relationship Between Eco-Innovation and Barriers Among the SMEs in the Peripheral Regions

The section examines the relationship between the introduction of eco-innovation and the barriers that limit its implementation among SMEs in the peripheral regions of Lagos. Table 11.6 presents the results from the correlation analysis of the relationship between the variables. Cursory look at the results (Table 11.6) shows that the level of eco-innovativeness among the SMEs in the peripheral regions is significantly associated with barriers related to innovation costs ($r = 0.14$; $p < 0.05$), access to technological capability ($r = 0.17$; $p < 0.05$), market ($r = 0.21$; $p < 0.01$), insufficient infrastructure ($r = 0.17$; $p < 0.05$), and regulation by the government ($r = 0.65$; $p <$

Table 11.6 Relationships between eco-innovation and barriers among the SMEs in the peripheral regions

		Eco-innovation	A2	A3	A4	A5	A6	A7	A8	A9	A10	A11
	Eco-innovation	1										
A2	Cost barrier	0.14*	1									
A3	Technological barrier	0.17*	0.49**	1								
A4	Market barrier	0.21**	0.38**	0.51**	1							
A5	Infrastructure barrier	0.17*	0.39**	0.46**	0.42**	1						
A6	Organization barrier	0.11	0.12	0.34**	0.36**	0.27**	1					
A7	Regulatory barrier	0.65**	0.10	0.03	0.31**	0.04	0.13	1				
A8	External knowledge	0.25**	0.10	0.24**	0.23**	0.02	0.05	0.30**	1			
A9	Firm's age	0.03	−0.11	−0.03	−0.03	0.01	0.06	0.06	−0.03	1		
A10	Product market	0.01	0.07	0.09	−0.09	−0.02	−0.02	−0.04	0.02	−0.09	1	
A11	Part of a conglomerate	−0.01	−0.13	−0.09	−0.02	0.02	−0.10	−0.04	0.10	−0.07	−0.04	1

$^{*}p < 0.05$, $^{**}p < 0.01$

0.01). The use of external knowledge was also found to be positively associated ($r = 0.25$; $p < 0.01$) with eco-innovation among the SMEs. The fact that all the barriers are positively associated with the level of eco-innovativeness could be an indication that the more the propensity to eco-innovate, the higher the intensity of the barriers is. This result, seemingly counterintuitive, has also been reported by Galia and Legros (2004) where they hinted that firms with high levels of innovativeness are usually confronted with more innovation obstacles than firms that are less innovative. Some authors have even suggested that how firms perceive innovation barriers could be an indication of how successful they are overcoming such obstacles (Tourigny and Le 2004; Baldwin and Lin 2002). Those that see barriers as opportunities often tend to realize more successful innovations.

11.4.3 Assessment of the Impact of Barriers on Eco-Innovation Among the SMEs in the Peripheral Regions

This section investigates the extent to which several categories of barriers impact on the level of eco-innovativeness of SMEs in the peripheral regions of Lagos metropolis. The results of the ordinal probit regression analysis (Table 11.7) show that technological capability and regulatory barriers came up as the most significant obstacles that affected the propensity of the SMEs to introduce eco-innovation. In addition to this, the result of the analysis also shows that those firms that search deeply for external knowledge are more likely to increase their chances of being eco-innovative. Another important finding from this analysis is that the sector in which an SME belonged to mattered.

The results in Table 11.7 have several implications for eco-innovation among SMEs in peripheral regions. The results suggest that firms tend to eco-innovate more when they are facing more obstacles. These results corroborate studies that have shown that innovating firms (as it is the case in this study) tend to have higher levels of barriers when compared with non-innovating firms (Mohnen and Rosa 1999; Baldwin and Lin 2002). This is due to the fact that innovating firms usually invest more and derive benefits from high-risk innovative ventures. However, as they risk more, they also incur higher costs in areas such as technological capability development, staff training, acquisition of R&D and infrastructural facilities, among others (Baldwin 1996).

Another explanation for the finding could be obtained from the phenomenon of learning-by-doing model of technology adoption. Innovative firms are known to be capable of engaging in a process of technological learning to overcome innovation barriers. They also tend to engage and interact more with knowledge-based entities (Archibugi and Pietrobelli 2003; Triguero et al. 2013). Innovative firms are usually confronted with many challenges that are associated with innovative strategies. More often than not, firms do not completely understand the ramification of

Table 11.7 Results of the ordinal probit regression model

	Estimate	Std. error	Exp(B)
Cost barrier	−0.07	0.13	0.93
Technological capability barrier	0.31*	0.16	1.36
Market barrier	−0.15	0.15	0.86
Infrastructure barrier	0.20	0.13	1.22
Organization barrier	−0.06	0.11	0.94
Regulatory barrier	0.83**	0.10	2.28
External knowledge	0.70**	0.29	2.02
Sector	0.04**	0.02	1.04
Firm's size	0.02	0.26	1.02
Firm's age	−0.03	0.23	0.97
Product market	−0.26	0.28	0.77
Part of a conglomerate	−0.12	0.26	0.89
No. of observation	164		
Nagelkerke R^2	0.53		
Chi-square	108.58**		
Test of parallel lines	$X^2 = 12.11$ ($p > 0.05$)		

*$p < 0.05$, **$p < 0.01$

the problems they face until they actually engage in the innovation process. As such, non-innovating firms may not have a full grasp of the challenges of innovation strategies as compared to innovating firms. A proof of this can be found in a study conducted by Baldwin and Lin (2002) where they reported that the rating of obstacles by non-innovating firms is usually lower when compared with innovating firms. Their study showed that it is by actively engaging in innovative strategies that firms learn about the magnitude of the challenges they have to overcome. As innovative firms confront more challenges while introducing complex innovative strategies (such as eco-innovation), technological learning takes place, more understanding of the obstacles is achieved, problem-solving competencies are developed, and barriers are eventually surmounted. In other words, non-innovating firms may not necessarily pass through such a process because obstacles are not truly experienced until firms actually face them.

The results of the analysis in Table 11.7 also show that the regulatory barrier is positively associated with a high level of eco-innovativeness. As a matter of fact, the result suggests that SMEs that had encountered regulatory barriers are twice more likely to eco-innovate than those that had not experienced the barrier. This result is in line with a study conducted on the paper and pulp industry in Nigeria where it was reported that national regulations such as guidelines and standards for the mitigation and control of pollution, for environmental impact assessment, and for liquid waste effluent treatment plants were very crucial for the adoption of eco-innovation in production processes (Adelegan et al. 2010). It should be noted that one of the peculiarities of eco-innovation is its association with government policies (Milliman and Prince 1989; Brunnermeier and Cohen 2003). Some of these policies address environmental issues such as natural resource depletion, energy sources, and supply. Most of these policies manifest in the form of economic instruments such as taxation and quality standards. As a matter of fact, some scholars believe that well-crafted environmental policies have the potential to spur eco-innovation giving opportunities to manufacturing firms to be competitive in the green business environment (Porter and van der Linde 1995). Based on the results of this study, we accept the hypothesis that states that government's regulation of the business environment affects the level of eco-innovation among SMEs in peripheral regions.

The results in Table 11.7 also show strong support for the view that the sector in which an SME belongs to influences its level of eco-innovation. The implication of this finding for SMEs is that the impact of barriers varies according to the sector in which a particular firm belongs to. As such, sectoral differences can limit or enhance eco-innovation performance in the medium to the long run. Thus, it is important to examine how sectoral differences affect the level of eco-innovativeness in order to elicit useful industry or sector-specific information. Such information is pertinent for the clarification of inter-industry differences, which helps to explain specific characteristics of innovation processes with regard to knowledge sources and learning, and how such processes are motivated (Dosi 1982; Oltra and Saint Jean 2009).

As important as knowledge is to innovation activities, it tends to be specific to the sector in which a firm belongs to (Malerba and Orsenigo 1996, 1997). For instance, although sector aggregation may be useful in understanding export competitiveness of a country, capturing the nuances of interactive learning process, environmental regulatory framework, and innovation activities at firm level may be difficult with such an approach (Costantini and Mazzanti 2012). Therefore, analyzing sectoral differences can assist in improving the robustness of empirical results on innovation activities at firm level. Information from such analysis allows the understanding of the specificity of innovation barriers within a particular manufacturing sub-sector. As much as government regulations and policies have significant impacts on the introduction of eco-innovation, appreciating the disparities in different types of policies across different sectors of manufacturing industries is equally important (Ashford et al. 1985).

The finding on the effect of deep search for external knowledge (see Table 11.7) shows that firms that search deeply for external knowledge are twice likely to be more eco-innovative than their counterparts. The implication of this result is that the intensity of knowledge search is very important in helping SMEs in peripheral regions to eco-innovate. Coming from the results of this study, firms search for knowledge beyond their boundaries as part of their eco-innovation strategies. Evidence has shown that eco-innovative activities require more external sources of knowledge than "conventional innovations" (Horbach et al. 2013). Many of the knowledge inputs for eco-innovation come from various sources (Rennings and Rammer 2009). Unfortunately, lack of technological and absorptive capacities as well as market failures makes internalization of such knowledge challenging (Pisano 1990; Mowery et al. 1998). In view of this, firms can interact or collaborate with external knowledge sources which might help to reduce the effects of some of the innovation barriers by combining resources and exchanging useful information (Cassiman and Veugelers 2006).

It should be noted that technological and managerial capabilities within firms are important in the utilization and improvement of technical knowledge gathered from external sources (Triguero et al. 2013). In other words, even though the acquisition of external knowledge may be a good strategy for innovation, the utilization of such knowledge could be difficult without the necessary technological or managerial capabilities. From the foregoing, it is clear that a certain level of collaboration with external sources of knowledge such as customers, suppliers, competitors, and knowledge institutions such as public research institutes and universities is a necessary condition for eco-innovation (De Marchi and Grandinetti 2013). Therefore, the hypothesis put forward in this study which states that deeper search for external knowledge affects the level of eco-innovation among SMEs in peripheral regions is accepted. In all, these analyses have shown that technological capabilities and the regulatory environment are pertinent to the introduction of eco-innovation by SMEs in peripheral regions of Lagos metropolis. It must, however, be stated that characteristics of firms such as the deep search for external knowledge and the sector/s they belong to also affect the level of eco-innovation.

11.5 Conclusion and Policy Recommendations

This study investigated innovation barriers that could limit the capability of small and medium-sized manufacturing firms in peripheral regions of Lagos metropolis to eco-innovate. Borrowing from evolutionary economics and innovation theories, the study categorized barriers to innovation into six types as follows: cost, technological capability, market, infrastructure, organizational culture, and regulatory barriers. Factors such as the deep search for external knowledge and the sector of firms were included in the analysis as part of the characteristics of firms that could affect the level of eco-innovation among the firms.

In terms of barriers to eco-innovation, the initial exploratory results from the correlation analysis suggest that barriers such as the high cost of innovation, insufficient technological capability, inaccessibility to market, insufficient infrastructure, and unfavorable regulation by the government affect the introduction of eco-innovation by SMEs in the peripheral regions of Lagos metropolis. However, further analysis using an advanced analytical technique (the ordinal probit regression) suggests that most of these aforementioned barriers are important but not sufficient to drive or limit the introduction of eco-innovation among the SMEs. The results revealed that eco-innovation among SMEs had a significant positive association with the technological capability barrier and the regulatory barrier. At the same time, the firm's characteristics such as the deep search for external knowledge and the sector of the firm were also found to significantly affect the introduction of eco-innovation among the firms.

It is interesting to find out that there is a positive association between certain barriers and the level of eco-innovativeness among the SMEs. Although this might seem counterintuitive, these results provide empirical support for findings from other studies, for example, Galia and Legros (2004) and Baldwin and Lin (2002), where they found that innovative firms are usually faced with more obstacles than firms that are not innovative. These studies also suggested that the more the innovative firms become, the more obstacles they confront, and the more learning they experience. In line with these thoughts, some scholars (e.g., Tourigny and Le 2004; Baldwin and Lin 2002) have suggested that the perception of obstacles by firms could also be an indication of how successful they surmount barriers. Those that consider obstacles as opportunities are likely to be more successful than those that perceive them as barriers. In other words, the more obstacles firms encounter, the more solutions they create, and the more innovation they introduce to the market. For instance, many firms in Africa have found ways to overcome many innovation barriers, by looking inward and creating networks that serve as sources of technical expertise and support structures for credit facilities and market access (Oyelaran-Oyeyinka 2007; Barr 2002; Fafchamps 2001).

This study has made important contributions to sustainability theory, as well as the strategic eco-innovation management and policy making for eco-innovation, particularly for SMEs in peripheral regions. With regard to theory, the study substantiates that innovation is a learning process, and perception of innovation barriers matters. The study also suggests that technological capability and the regulatory framework are critical factors in the introduction of eco-innovation among SMEs in peripheral regions. The study also throws more light on the importance of learning-by-doing as an avenue to generate knowledge to eco-innovate. In terms of strategic management of eco-innovation in peripheral regions, SMEs should focus their attention on the learning-by-doing model of technology adoption, as well as seeking to interact with various external sources of knowledge to complement their in-house innovation strategies. This becomes more important and clearer when SMEs understand that innovation is a social process, and they need to engage with external actors in order to enhance their competitiveness in the emerging green market. To improve eco-innovation policies, law makers should be aware that a substantial level of eco-innovation takes place in peripheral regions, and as such, care should be taken to nurture and strengthen SMEs that engage in such innovative activities. Considering the fact that the sector/s in which SMEs belong to matter in the introduction of eco-innovation, policy makers should avoid making blanket eco-innovation policies for all the manufacturing sub-sectors in the country.

With regard to the limitation of the study, it is important to mention at this juncture that the concept of peripheral regions as operationalized in this study may not truly represent the ideal peripheral region. However, the study gave a succinct description of the dynamics of barriers to eco-innovation between a big business hub and surrounding firms that transact business with it on a daily basis. Meanwhile, it will be interesting to investigate the extent to which barriers that are specific to a particular sector influence eco-innovation among SMEs within the context of developing countries.

Acknowledgements Financial support from the United Nations University Institute for Natural Resources in Africa (UNU-INRA), Ghana, is graciously acknowledged. The findings and conclusions expressed in this chapter are those of the author and should not necessarily be attributed to the institute.

References

Adegbite SA, Ilori MO, Irefin IA, Abereijo IO, Aderemi HOS (2007) Evaluation of the impact of entrepreneurial characteristics on the performance of small scale manufacturing industries in Nigeria. J Asia Entrep Sustain. Viewed 10 April 2016, from http://findarticles.com/p/articles/mi_qa5499/is_200705/ai_n21302203/print

Adelegan JA, Carlsson B, Perelli S (2010) Eco-innovation and corporate performance: the African experience. Paper presented at the Berlin conference on the human dimensions of global environmental change, Berlin, 8–9 Oct 2010

Aggarwal A (2000) Deregulation, technology imports and in-house R&D efforts: an analysis of the Indian experience. Res Policy 29(9):1081–1093

Andersen MM (2010) On the faces and phases of eco-innovation: on the dynamics of the greening of the economy. Paper to be presented at the summer conference 2010 on "opening up innovation: strategy, organization and technology", Imperial College, London Business School, 16–18 June 2010

Archibugi D, Pietrobelli C (2003) The globalisation of technology and its implications for developing countries: windows of opportunity or further burden? Technol Forecast Soc Chang 70(9):861–883

Arthur B (1994) Increasing returns and path dependence in the economy. University of Michigan Press

Arundel A, Kemp R (2009) Measuring eco-innovation. UNU-MERIT working papers

Ashford NA, Ayres C, Stone RF (1985) Using regulation to change the market for innovation. Harv Environ Law Rev 9:419–466

Assink M (2006) Inhibitors of disruptive innovation capability: a conceptual model. Eur J Innov Manag 9:215–233

Azzone G, Noci G (1998) Seeing ecology and 'green' innovations as a source of change. J Organ Change Manag 11(2):94–111

Bala-Subrahmanya MH (2005) Technological innovations in Indian small enterprises: dimensions, intensity and implications. Int J Technol Manage 30(1/2):188–204

Baldwin J (1996) Innovation: the key to success in small firms. In: de la Mothe J, Paquette G (eds) Evolutionary economics and the new international political economy. Pinter, London, pp 238–256

Baldwin J, Lin Z (2002) Impediments to advanced technology adoption for Canadian manufacturers. Res Policy 31(1):1–18

Barr AM (2002) The functional diversity and spillover effects of social capital. J Afr Econ 11(1):90–113

Borghesi S, Cainelli G, Mazzanti M (2012) The European emission trading scheme and environmental innovation diffusion: empirical analyses using Italian CIS data. Quaderno Università degli Studi di Ferrara
Bos-Brouwers H (2010) Corporate sustainability and innovation in SMEs: evidence of themes and activities in practice. Bus Strategy Environ 19:417–435
Brunnermeier SB, Cohen MA (2003) Determinants of environmental innovation in US manufacturing industries. J Environ Econ Manag 45(2):278–293
Carrillo-Hermosilla J, del Gonzalez PR, Konnola T (2009) What is eco-innovation? In: Eco-innovation. Palgrave Macmillan, UK, pp 6–27
Cassiman B, Veugelers R (2006) In search of complementarity in innovation strategy: internal R&D and external knowledge acquisition. Manag Sci 52(1):68–82
Cohen WM (2010) Fifty years of empirical studies of innovative activity and performance. In: Handbook of the economics of innovation, vol 1, pp 129–213
Cole M, Elliott R, Shimamoto K (2005) Industrial characteristics, environmental regulations and air pollution: an analysis of the UK manufacturing sector. JEEM 50:121–143
Cooke P, Uranga MG, Etxebarria G (1997) Regional innovation systems: institutional and organizational dimensions. Res Policy 26(4):475–491
Cosh A, Hughes A (2000) British enterprise in transition: growth, innovation and public policy in the small and medium-sized enterprise sector 1994–1999. University of Cambridge
Costantini V, Mazzanti M (2012) On the green and innovative side of trade competitiveness? The impact of environmental policies and innovation on EU exports. Res Policy 41(1):132–153
D'Este P, Iammarino S, Savona M, von Tunzelmann N (2012) What hampers innovation? Revealed barriers versus deterring barriers. Res Policy 41(2):482–488
De Marchi V (2012) Environmental innovation and R&D cooperation: empirical evidence from Spanish manufacturing firms. Res Policy 41:614–623
De Marchi V, Grandinetti R (2012) Who are the green innovators? An empirical analysis of firm's level factors driving environmental innovation adoption. Paper presented at the DRUID conference, Copenhagen Business School, Copenhagen, 19–21 June 2012
De Marchi V, Grandinetti R (2013) Knowledge strategies for environmental innovations: the case of Italian manufacturing firms. J Knowl Manag 17(4):569–582
del Brío J, Junquera B (2003) A review of the literature on environmental innovation management in SMEs: implications for public policies. Technovation 23(12):939–948
del Río P, Carrillo-Hermosilla J, Könnölä T (2010) Policy strategies to promote eco-innovation. J Ind Ecol 14(4):54–557
Doloreux D, Dionne S (2008) Is regional innovation system development possible in peripheral regions? Some evidence from the case of La Pocatiere, Canada. Entrepr Reg Dev 20:259–283
Dosi G (1982) Technological paradigms and technological trajectories. Res Policy 11:147–162
Dosi G, Freeman C, Nelson R, Silverberg G, Soete L (eds) (1988) Technical change and economic theory. Pinter, London
Dyllick T, Hockerts K (2002) Beyond the business case for corporate sustainability. Bus Strategy Environ 11:130–141
Fafchamps M (2001) Networks, communities and markets in Sub-Saharan Africa: implications for firm growth and investment. J Afr Econ 10(Suppl. 2):109–142
Figueiredo P (2002) Learning processes features and technological capability accumulation: explaining inter-firm differences. Technovation 22(11):685–698
Figueiredo PN (2003) Learning, capability accumulation and firms differences: evidence from latecomer steel. Ind Corp Chang 12(3):607–643
Foxon T, Pearson P (2008) Overcoming barriers to innovation and diffusion of cleaner technologies: some features of a sustainable innovation policy regime. J Clean Prod 16(1):148–161
Foxon TJ, Pearson P, Makuch Z, Mata M (2005) Policy driver sand barriers for sustainable innovation: final report. Imperial College, London. Viewed 10 Apr 2016, from http://www.sustainabletechnologies.ac.uk/Projects/policy.htm

Freeman C (1992) The economics of hope: essays on technical change, economic growth and the environment. Pinter, London
Freeman C, Perez C (1988) Structural crises of adjustment, business cycles and investment behaviour. In: Dosi G et al (eds) Technical change and economic theory. Pinter, London
Galia F, Legros D (2004) Complementarities between obstacles to innovation: evidence from France. Res Policy 33(8):1185–1199
Gatrell JD (2001) Integrated dependence: knowledge-based industries in peripheral regions. Econ Dev Rev 17(3):63
Goedhuys M (2007) Learning, product innovation, and firm heterogeneity in developing countries; evidence from Tanzania. Ind Corp Chang 16(2):269–292
Hadjimanolis A (1999) Barriers to innovation for SMEs in a small less developed country (Cyprus). Technovation 19:561–570
Henderson J (2002) Building the rural economy with high-growth entrepreneurs. Econ Rev 87(3):45–70
Hillary R (2006) Using network approaches to engage small and medium-sized enterprises in environmental management systems. In: Marinova D, Annandale D, Phillimore J (eds) The international handbook on environmental technology management. Edward Elgar, Cheltenham, pp 241–250
Holzl W, Janger J (2012) Innovation barriers across firms and countries. WIFO working papers, no. 426
Horbach J (2008) Determinants of environmental innovation—new evidence from German panel data sources. Res Policy 37:163–173
Horbach J, Oltra V, Belin J (2013) Determinants and specificities of eco-innovations compared to other innovations—an econometric analysis for the French and German industry based on the community innovation survey. Ind Innov 20(6):523–543
Isaksen A (2001) Building regional innovation systems: is endogenous industrial development possible in the global economy? Can J Reg Sci 24:101–120
Jaffe A, Palmer K (1997) Environmental regulation and innovation: a panel data study. Rev Econ Stat 79:610–619
Jaffe AB, Newell RG, Stavins RN (2002) Environmental policy and technological change. Environ Resource Econ 22(1):41–70
Jenkins H (2006) Small business champions for corporate social responsibility. J Bus Ethics 67:241–256
Kammerer D (2009) The effects of customer benefit and regulation on environmental product innovation: empirical evidence from appliance manufacturers in Germany. Ecol Econ 68(8):2285–2295
Kanter RM (1985) Supporting innovation and venture development in established companies. J Bus Ventur 1(1):47–61
Kemp R (2010) Eco-innovation: definition, measurement and open research issue. Econ Polit 27(3):397–420
Kemp R, Pearson P (2007) Final report MEI project about measuring eco-innovation. UM Merit, Maastricht
Kerr S, Newell R (2001) Policy-induced technology adoption: evidence from the US lead phasedown. J Ind Econ 51(3):317–343
Kesidou E, Demirel P (2012) On the drivers of eco-innovations: empirical evidence from the UK. Res Policy 41(5):862–870
Khanna M, Deltas G, Harrington D (2006) Adoption of pollution prevention techniques: the role of management systems, demand-side factors and complementary assets. Environ Resour Econ World Congr Kyoto
Kickul J, Gundry L (2002) Prospecting for strategic advantage: the proactive entrepreneurial personality and small firm innovation. J Small Bus Manage 40(2):85–97

Klewitz J, Hansen EG (2011) Sustainability-oriented innovation in SMEs: a systematic literature review of existing practices and actors involved. In: Huizingh E, Torkelli M, Conn S (eds) Proceedings of the XXII ISPIM conference, Hamburg, Germany, 12–15 June 2011

Leke A, Fiorini R, Dobbs R, Thompson F, Suleiman A, Wright D (2014) Nigeria's renewal: delivering inclusive growth in Africa's largest economy. McKinsey Global Institute

Luiten E (2001) Beyond energy efficiency. Actors, networks and government intervention in the development of industrial process technologies. PhD thesis, Utrecht, the Netherlands

Luiten E, Blok K (2004) Stimulating R&D of industrial energy-efficient technology. Policy lessons—impulse technology. Energy Policy 32:1087–1108

Malerba F, Orsenigo L (1996) Schumpeterian patterns of innovation. Camb J Econ 19(1):47–65

Malerba F, Orsenigo L (1997) Technological regimes and sectoral patterns of innovative activities. Ind Corp Change 6:83–117

March I, Ganasekaran A, Lloria B (2002) Product development process in Spanish SMEs: an empirical research. Technovation 22:301–312

Marin G, Marzucchi A, Zoboli R (2015) SMEs and barriers to eco-innovation in the EU: exploring different firm profiles. J Evol Econ 25:671–705

Mazzanti M, Zoboli R (2006) Examining the factors influencing environmental innovations. Nota di Lavoro 20. The Fondazione Eni Enrico Mattei (FEEM) note di Lavoro series index

Metcalfe JS (1995) Technology systems and technology policy in an evolutionary framework. Camb J Econ 19S:25–46

Milliman SR, Prince R (1989) Firm incentives to promote technological change in pollution control. J Environ Econ Manag 17:247–265

Mohnen P, Röller L (2005) Complementarities in innovation policy. Eur Econ Rev 49:1431–1450

Mohnen P, Rosa J (1999) Barriers to innovation in services industries in Canada. Cat. No. 88F0017-MIE99007. Statistics Canada, Ottawa

Moore G, Spence L (2006) Small and medium-sized enterprises and corporate social responsibility: identifying the knowledge gaps. J Bus Ethics 67(3):219–226

Mowery DC, Oxley JE, Silverman BS (1998) Technological overlap and interfirm cooperation: implications for the resource-based view of the firm. Res Policy 27(5):507–523

Nameroff TJ, Garant RJ, Albert MB (2004) Adoption of green chemistry: an analysis based on US patents. Res Policy 33(6):959–974

Nelson RS, Winter S (1982) An evolutionary theory of economic change. Harvard University Press, Cambridge

Nelson RR, Winter SG (2002) Evolutionary theorizing in economics. J Econ Perspect 16(2):23–46

NERC (2011) ECN ECoN and FG to incur N177 bn in electricity subsidy. Nigerian Electricity Regulatory Commission, Abuja, Nigeria

Nill J, Kemp R (2009) Evolutionary approaches for sustainable innovation policies: from niche to paradigm? Res Policy 38(4):668–680

Noci G, Verganti R (1999) Managing 'green' product innovation in small firms. R&D Management 29(1):3–15

Nwagwu I, Oni T (2015) Lagos and its potentials for economic growth. Heinrich Böll Stiftung, Abuja

OECD-EUROSTAT (2005) Oslo manual: guidelines for collecting and interpreting innovation data, 3rd edn. OECD, Paris

Ojo E (2009) Manufacturers need 2,000 MW of electricity to stay afloat. Business Day Nigeria, Lagos, Nigeria

Oke A, Burke G, Myers A (2004) Innovation types and their impact on performance in UK SMEs. In: 11th international product development conference, Dublin, Ireland

Oltra V, Saint Jean M (2009) Sectoral systems of environmental innovation: an application to the French automotive industry. Technol Forecast Soc Chang 76(4):567–583

Oyelaran-Oyeyinka B (2007) SME: issues, challenges and prospects. Paper presented at the FSS 2020 international conference, Nigeria, June 2007

Piatier A (1984) Barriers to innovation. Frances Pinter, London

Pisano GP (1990) The R&D boundaries of the firm: an empirical analysis. Adm Sci Q 35(1):153–176
Porter ME, van der Linde C (1995) Green and competitive: ending the stalemate. Harv Bus Rev 120–133
Radwan I, Pellegrini G (2010) Knowledge, productivity, and innovation in Nigeria: creating a new economy. World Bank Publications
Rave T, Goetzke F, Larch M (2011) The determinants of environmental innovations and patenting: Germany reconsidered. Ifo working paper no. 97
Rennings K (1998) Towards a theory and policy of eco-innovation—neoclassical and (co-) evolutionary perspectives. ZEW discussion papers no. 98-24
Rennings K, Rammer C (2009) Increasing energy and resource efficiency through innovation. An explorative analysis using innovation survey data. Czech J Econ Financ 59(1):442–459
Revell A, Stokes D, Chen H (2010) Small businesses and the environment: turning over a new leaf? Bus Strategy Environ 19(5):273–289
Rothwell R, Dodgson M (1991) External linkages and innovation in small and medium-sized enterprises. R&D Manag 21(2):125–138
Runhaar H, Tigchelaar C, Vermeulen WJ (2008) Environmental leaders: making a difference a typology of environmental leaders and recommendations for a differentiated policy approach. Bus Strategy Environ 17(3):160–178
Rush H, Bessant J (1992) Revolution in three-quarter time: lessons from the diffusion of advanced manufacturing technologies. Technol Anal Strateg Manag 4(1):3–19
Ruttan V (2000) Technology, growth and environment: an induced innovation perspective. University of Minnesota, St Paul (Minnesota)
Sanusi JO (2003) Overview of government's efforts in the development of SMEs and the emergence of small and medium industries equity investment scheme (SMIEIS). Paper presented at the national summit on SMIEIS organized by the bankers' committee and Lagos chambers of commerce and industry (LCCI), Lagos, 10 June 2003
Schaltegger S, Synnestvedt T (2002) The link between 'green' and economic success: environmental management as the crucial trigger between environmental and economic performance. J Environ Manage 65:339–346
SEDIN, SMEDAN (2013) Enterprise baseline survey (EBS) 2012. Pro-Poor Growth and Promotion of Employment (SEDIN) Programme and Small & Medium Enterprises Development Agency of Nigeria (SMEDAN), Abuja, Nigeria
Segarra-Blasco A, Garcia-Quevedo J, Teruel-Carrizosa M (2008) Barriers to innovation and public policy in Catalonia. Int Entrep Manag J 4(4):431–451
Smallbone D, North D (1999) Innovation and new technology in rural small and medium-sized enterprises: some policy issues. Environ Plann C Gov Policy 17(5):549–566
SMEDAN/NBS (2012) Survey report on micro, small and medium enterprises in Nigeria. Abuja, Nigeria
Spence L (1999) Does size matter? The state of the art in small business ethics. Bus Ethics Eur Rev 8(3):163–174
Steger U (1996) Managerial issues in closing the loop. Bus Strategy Environ 5:252–268
Suh S, Lee K, Sangsun H (2005) Eco-efficiency for pollution prevention in small to medium-sized enterprises. J Ind Ecol 9(4):223–240
Tilley F (2000) Small firm environmental ethics: how deep do they go? Bus Ethics Eur Rev 9(1):31–41
Todtling F, Trippl M (2005) One size fits all? Towards a differentiated policy approach with respect to regional innovation systems. Res Policy 34:1203–1219
Tourigny D, Le CD (2004) Impediments to innovation faced by Canadian manufacturing firms. Econ Innov New Technol 13(3):217–250
Triguero A, Moreno-Mondéjar L, Davia MA (2013) Drivers of different types of eco-innovation in European SMEs. Ecol Econ 92:25–33
Unruh G (2000) Understanding carbon lock-in. Energy Policy 28(12):817–830

Verspagen B (2009) The use of modeling tools for policy in evolutionary environments. Technol Forecast Soc Chang 76:453–461

Walz R (2011) Opportunities and specific challenges for eco-innovations. In: UN conference promoting eco-innovation: policies and opportunities, Tel Aviv

Welford R (1996) Corporate environmental management: systems and strategies. Earthscan, London

Winship C, Mare RD (1984) Regression models with ordinal variables. Am Sociol Rev 512–525

Chapter 12
Inclusive Green Agricultural Business Model Innovation for Rural Africa: A Conceptual Framework

Yirviel Janvier Métouolé Méda and Calvin Atewamba

12.1 Introduction

The concept of green economy (GE) has re-emerged as a response to the continuing environmental and economic development challenges, especially the recent global crises (climate, energy, food and financial crises) (Allen and Clouth 2012). According to its most used definition, GE is one that results in improved human well-being and social equity, while significantly reducing environmental risks and ecological scarcities (UNEP 2011). Initially interpreted as the intersection between environment and economy, the concept has broadened to include the social dimension, which led to the concept of "*inclusive green economy*" (Allen and Clouth 2012; UNEP 2011). One of the reasons for this broadening of the concept is the need of an inclusive economic growth in order to reduce poverty and improve well-being in developing countries, along with the need to improve environmental management for tackling resource scarcities and climate change (Beltramello et al. 2013). Thus, the implications of transitioning to GE are different in developing countries from those in developed countries where the focus is on a structural change towards a resource-efficient and knowledge-based economy (Beltramello et al. 2013).

In this GE transition, businesses are increasingly found at the centre of public and academic debates. The reason for this is that they are often identified as both the cause of many emerging environmental challenges and the solution to them. The search for ways to achieving GE transition has then given rise to the concept of green business model innovation (GBMI) (FORA 2011). The main idea is to move from traditional business models (TBMs), which characterize the current economy seen as

Y. J. Métouolé Méda (✉)
Professional University Institute, University of Dédougou, Dédougou, Burkina Faso
e-mail: janvier.meda2@gmail.com

C. Atewamba
United Nations University Institute for Natural Resources in Africa, Accra, Ghana
e-mail: atewamba@gmail.com

C. Atewamba and D. Yong Ngondjeb (eds.), *Inclusive Green Growth*, Advances in African Economic, Social and Political Development,
https://doi.org/10.1007/978-3-030-44180-7_12

unsustainable, to green business models (GBMs) that are intended to characterize a GE. This is a challenge that also differs not only across regions, but across economic activities of which agriculture turns to be of biggest concern (UNEP 2011; FAO et al. 2012). Yet, most of the emerging literature on GBMI is still at the general level of conceptualization, with case studies limited to the Nordics and OECD countries (see Beltramello et al. 2013; Bisgaard et al. 2012; Henriksen et al. 2012a, b, c, d). Very little attention has been paid to GBMI in the context of Africa where agriculture is the most important economic sector and whose development is the most challenged in the world.

The development challenges that Africa faces and their connection to the agriculture–environment nexus are well known. Of primary concerns are socio-economic issues related to the high food insecurity and poverty among smallholder farmers resulting from low agricultural productivity and income. About 22.9% of Africans were underfed in 2012 compared to 14.9% in the developing world (FAO et al. 2012). About 50% of the world undernourished people and 75% of malnourished children are smallholder farmers in Africa (UNEP 2011). 70% of the poorest Africans are employed in agriculture and tend to have lower incomes, because of low prices, unfair business practices and lack of transportation, storage and processing infrastructure (UNEP 2011). Agricultural yields reached only 27% of their potentials in Africa compared to 35% for Latin America and the Caribbean, and almost 90% for East Asia. The main reasons for the low agricultural productivity include the failure to adopt the "green revolution" technologies, soil erosion, water constraints, poor agricultural services, crop losses, adverse climate effects, etc. The situation is even likely to worsen in the future with the increasing scarcity of land, water, mineral resources coupled with climate change (IPCC 2014). Yet, unsustainable farming practices are among the primary causes of these environmental challenges that affect food security in the continent.

Agriculture accounts for about 35% of severely degraded land worldwide (UNEP 2011). It also contributes up to 90% to deforestation in developing countries (Benhin 2006). This deforestation is a major cause of the increased accumulation of carbon dioxide in the atmosphere, while fossil fuel use, livestock, wetland rice cultivation, manure management, bush fires and burning of agricultural residues are sources of carbon dioxide and methane emissions (UNEP 2011). In addition, conventional intensive farming systems (e.g. cotton, cocoa, rice in Africa) pollute soils, water and food due to inefficient use of synthetic agrochemicals and lead to health hazards. There is an increasing public awareness over the unsustainability of the current agricultural systems in general and in Africa in particular, which compromises their development in the long term. Accordingly, alternative approaches are being proposed in order to address not only the environmental issues but also the socio-economic ones. Thus, the concept of green agriculture (GA) has emerged together with GE. It refers to the use of farming practices and technologies that simultaneously (i) maintain or increase farm productivity and profitability, ensure the provision of food, create rural jobs, reduce poverty, enhance food security and livelihoods of local communities, improve nutrition and health, while (ii) reducing the negative externalities and gradually leading to positive ones (UNEP 2011, 2013). Koohafkan et al. (2012) define GA as including the use of environmentally friendly and socially equitable technologies and methods, in a world with increasing scarcity of agricultural resources (arable land, petroleum, water and nitrogen), and within a scenario of a rapidly changing climate, social unrest and economic uncertainty.

In rural areas where agriculture is the main economic activity, GE transition is almost synonymous with GA transition. Keeping in mind the issues discussed above and the definition of GA, achieving this GA transition in Africa implies a paradigm shift from the usual agribusiness practices. The underlying agricultural business models need to consider not only the environmental challenges but also the structural socio-economic problems in the continent. This requires radical innovations not only from an environmental perspective but also from a social perspective. Therefore, the green business model innovation (GBMI) as conceptualized so far may be limited in scope since it focuses on environmental issues, and may need to be adjusted for critical social issues. A broader framework that addresses social issues is the sustainable business model innovation (SBMI) perspective that even existed before the GBMI approach (Bocken et al. 2014; Lüdeke-freund 2009). However, sustainability is too broad especially with respect to social issues that are so many and that cannot be all addressed in business at the same time. Some guidance needs to be provided on pressing issues to consider like those that farmers face in Africa.

This paper therefore examines the idea of green-oriented business model innovation in agriculture for rural Africa. The main contribution is the development of a framework of an inclusive green agribusiness model innovation (IGABMI) based on the literature review, depicting how traditional agribusiness models (TABMs) in rural Africa can transform into inclusive green agribusiness models (IGABMs). This new and more operational framework matches better with the concept of inclusive green economy considered as the means for achieving sustainable development. The rest of the paper is organized as follows. Section 12.2 describes the method used to gather the relevant literature. The GBMI framework is discussed in Sect. 12.3. Section 12.4 exposes the components of an IGABMI. Potential drivers and barriers of IGABMI in rural Africa, as well as the role of policy, are discussed in Sect. 12.5. Section 12.6 illustrates the IGABMI framework using a value chain approach. Section 12.7 proposes a framework for assessing and monitoring achievements towards IGABM through a performance index for inclusive green agricultural business; Section 12.8 concludes the study.

12.2 Methods and Literature Used

The study is based on a review of the literature. Desk search started with various scientific databases. An e-mail alert was also created in Google Scholar in order to have regular update on this literature. From this process, relevant scientific works found for the study were related mainly to concepts such as a business model (BM), business model innovation (BMI), social business model innovation (SBMI), green business model innovation (GBMI), green innovation (GI), green economy (GE), green agriculture (GA) and other related concepts. However, a very few literatures on some concepts of primary interest such as GBMI, inclusive business model innovation (IBMI) and agricultural business model (ABM) was found. A simple Google Search was then performed to find relevant reports, including those from which the concepts of GBMI and IBMI originated. The most relevant of the reports constituted the basis for the review of the GBMI, IBMI and ABM as concepts on their own, while the peer-reviewed papers previously identified served to review the sub-concepts that compose GBMI, IBMI, ABM and IGABMI.

12.3 The Green Business Model Innovation Framework and Research Gap

Green business model innovation (GBMI) as a concept looks very complex at first sight, with no generally agreed definition. This is because the concept is still at its very early stage of development. Moreover, it is built on other concepts with less obvious definitions: green innovation, business model, business model innovation. An overview of these concepts is therefore important in order to understand what GBMI means and what it implies for agriculture in rural Africa.

12.3.1 Green Innovation

Innovation is recognized as a major driver of business success, economic development and social progress (Porter and van der Linde 1995). In the search for ways to addressing environmental issues, the concept of green innovation (or eco-innovation) has emerged and is still seen as the key driver of the GE transition (Beltramello et al. 2013; Sarkar 2013; UNEP 2011). From a general perspective, innovation can be defined as "*the implementation of a new or significantly improved product (good or service), or process, a new marketing method, or a new organizational method in business practices, workplace organization or external relations*" (OECD and Eurostat 2005). With an environmental focus, green innovation or eco-innovation is defined as "*the introduction of any new or significantly improved product (good or service), process, organizational change or marketing solutions that reduce the use of natural resources (including materials, energy, water and land) and decrease the release of harmful substances across the whole life cycle*" (Joller 2012).

An innovation in general, or a green innovation in particular, can be incremental; i.e. it involves a minor change in the system such as the introduction of new products; it can also be radical (transformative or systemic) involving a complex change such as a shift in the business model (Bisgaard et al. 2012; Beltramello et al. 2013). Radical green innovations are seen as having the potential to drive the GE transition in a faster way (Beltramello et al. 2013).

12.3.2 Business Model and Business Model Innovation

There is no single definition for business model (BM) (Benson-Rea et al. 2013). Nevertheless, there are two definitions that are popular: BM is "*the rationale of how an organization creates, delivers and captures value*" (Osterwalder and Pigneur 2009); BM is "*a systemic perspective on how to 'do business', encompassing boundary-spanning activities (performed by a focal firm or others) and focusing on value creation as well as on value capture*" (Zott et al. 2011). These two definitions are converging, in that value creation is at the heart of any business model, and can take different forms including economic, social and other forms (Osterwalder and Pigneur 2009; Zott et al. 2011). The conceptualization of value in the BM beyond its economic or financial understanding has been informed by previous works on BM that can create social value including reduction of poverty, human suffering and pollution (Seelos and Mair 2005, 2007; Thompson and MacMillan 2010). Companies capture value (economic) by seizing new business opportunities, new markets and new revenue streams (Zott et al. 2011). This is best described in the business model canvas (Fig. 12.1). The BM canvas generally involves nine building blocks showing the logic of how a company intends to make money: strategic decisions on customer segmentation, products and services to offer (value propositions), business and research partners to engage with, resources to create and channels to deliver value, as well as the underlying cost structure and revenue streams to ensure economic viability of business (Osterwalder and Pigneur 2009).

Apart from its role in value creation, BM has been linked to innovation and firm performance. Thus, two related ideas can be identified from the literature. First, BM is an important driver of innovation (including product, organizational, marketing and technological or process innovation). Intuitively, it is through BM that companies develop and commercialize innovative ideas, technologies, and goods and services (Zott et al. 2011). Thus, one can find expressions like *BM "to capture value from technological innovation*" (Teece 2010) or *BM "for eco-innovation*" (Beltramello et al. 2013; Jing and Jiang 2013).

The second idea is that the BM itself can also be innovated. Thus, BM is a new subject of innovation along with the usual subjects (technology or process, product, marketing, organizational structure) (Zott et al. 2011). This is typically the idea of business model innovation (BMI), coming from the fact that value creation mechanisms are often beyond the scope of technological, product, marketing or organizational innovation (Schaltegger et al. 2011; Girotra and Netessine 2013).

Technological innovation might not guarantee firm success, and value creation might be compromised if the capacity to create new business models does not exist (Teece 2010; Zott et al. 2011). As industries and markets change, there is a need to innovate in the business model in order to stay competitive by rethinking value proposition to the customers, seeking new ways to capture new markets, generating new types of producer–consumer relationships and restructuring activities, resources and partnerships (Jing and Jiang 2013). Innovation becomes business model innovation when two or more elements of a business model are reinvented to deliver value

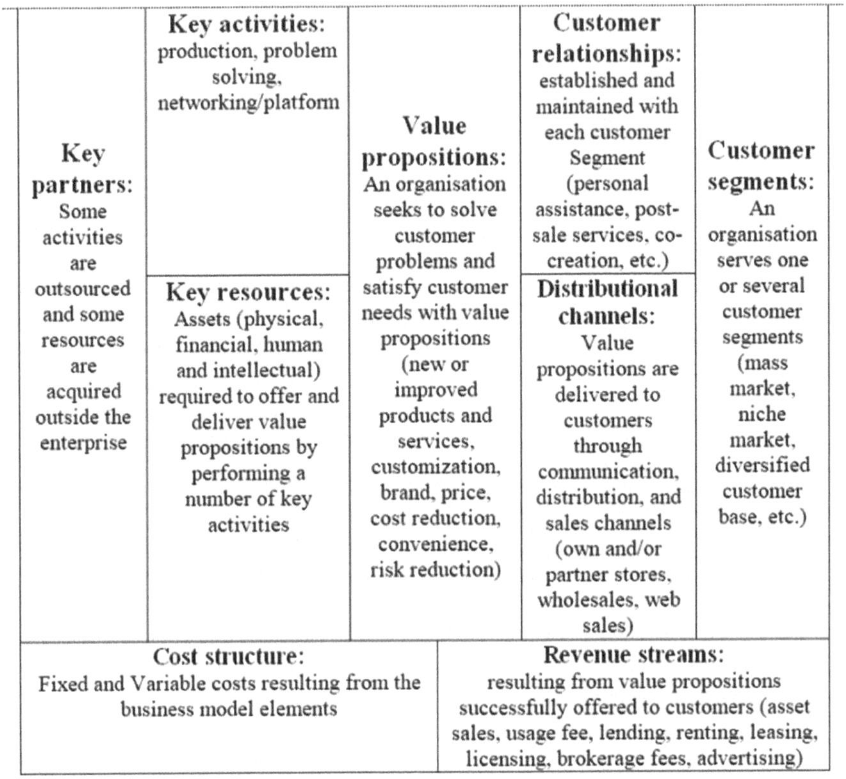

Fig. 12.1 Nine blocks of the business model canvas according to Osterwalder and Pigneur (2009)

in a new way (Joller 2012; Schaltegger et al. 2011). BMI can be understood as an improvement in the building blocks of the BM canvas (Henriksen et al. 2012a).

12.3.3 Green Business Model Innovation

From the previous developments, the concept of green business model innovation (GBMI) can now be understood as a green innovation (or eco-innovation) in the business model. As such, GBMI is a particular case of BMI, i.e. a BMI towards green aspects or in a green manner. When this kind of innovation happens, the resulting new BM can then be referred to as green business model (GBM). Attempts to conceptualize GBMI have been emerging since 2010, namely studies for some organizations such as Nordic Innovation Centre (Henriksen et al. 2012a, b), OECD (Beltramello et al. 2013), European Commission (Bilsen et al. 2013) as well as academic papers (Joller 2012; Jing and Jiang 2013).

Thus, the term "green business models" (GBMs) has been first used by the Danish Enterprise and Construction Authority (FORA) through the project *Green Business Models in the Nordic Region* completed for Nordic Innovation (FORA 2010). FORA (2010) defined GBMs as business models which support the development of products and services (systems) with environmental benefits, reduce resource use/waste and which are economically viable; they aim to create economic and environmental win-win benefits for both the supplier and the customer (FORA 2010).

In the proposal of a second project undertaken on *Green Business Model Innovation* (FORA 2011), GBMI has been defined as "*a new way of doing business and refers to non-technological green innovation in and between companies which change the core businesses from selling a product to selling a (full) service and at the same time retaining ownership of the product and responsibility for its functionality*" (FORA 2011). The emphasis on the "non-technological" aspect, like in the green economy report (UNEP 2011), is certainly because BMI goes beyond the technological innovation as previously discussed. Adopting also this FORA's (2011) definition and the BM canvas, Jing and Jiang (2013) emphasized that GBMs integrate sustainability into the process of operation management. They proposed a framework of GBM including four aspects: the core logic (green value creation and realization); the two transformations (business strategy and operation management); the triple bottom line (economic, social and environmental benefits); and the four interfaces (market offering, enterprise interface, customer interface and profit model).

Using the term "*eco-innovation in business model*", Joller (2012) defined it as "*the new logic/way/rationale [of] how an organization offers its customers value (and creates sustainable revenue streams), while reducing the use of natural resources and decreasing the release of harmful substances across the whole life cycle*". From the implementation of the Nordic Innovation second project, the following definition has been adopted, built on the BM canvas approach, as well as the BMI definition and typology presented above: "*Green Business Model Innovation is when a business changes part(s) of its business model and thereby captures economic value as well as reduces the ecological footprint in a life-cycle perspective*" (Bisgaard et al. 2012). Indeed, despite the initial definition adopted in the project proposal, establishing a clear definition of the GBMI concept for international use was one of the aims of this project (FORA 2011). For the authors, the more parts of a BM are changed and have a green effect, and the more profoundly a green change is taking place within the elements of the business model, the greener is the BMI (Bisgaard et al. 2012). This definition has also been used in the OECD study (Beltramello et al. 2013). As for the European Commission study, Bilsen et al. (2013) used interchangeably the terms "*innovative/green business models*" and "*green/innovative business models*" and defined them as "*the use of innovative technological or non-technological solutions and/or cooperation with other market actors, leading to increased economic or environmental efficiency, resulting in the achievement of a business model more decoupled from environmental costs*".

In addition to the conceptual framework proposed, the related studies have identified various types of GBMs that exist in the Nordic Region and OECD countries and from which other companies can learn to green their business (Bisgaard et al. 2012;

Bilsen et al. 2013 ; Beltramello et al. 2013). These can be grouped into four categories: classic GBMs; incentive models; life-cycle and coordination benefit models; and information and communication technology (ICT) solutions. Classic GBMs are mainly based on the use of improved products or processes. Incentive models are based on how a company incentivizes its consumers so that part or the entire value chain is greener. Life-cycle models apply most to companies, which have a focus on greening the entire value chain and benefit from coordination. However, coordination benefit models also include business models that exploit the proximity of agents. ICTs provide a wide range of solutions for energy and resource use control (video/teleconference or telephone meetings, e-mail, online shopping).

12.3.4 Lessons from GBMI and Research Gap

From the review of the GBMI framework, the question is whether this framework provides rooms for applications in smallholder agriculture that typically characterizes rural Africa. As GBMI has been framed from a general perspective, it could be applied to any economic sector including agriculture. However, the emerging literature on the concept of BM since the mid-1990s and that of GBM since 2010 is giving more attention to the services and manufacturing sectors than agriculture. Thus, they are not comprehensive enough for addressing the challenges that farmers face in Africa. Following this framework, the negative environmental externalities and the related social costs can be addressed through a green agricultural business model innovation (GABMI). However, this will not necessarily make smallholder farmers better off unless they benefit from an equitable sharing of the economic value created and captured through this business model.

The broader framework of sustainable business model innovation (SBMI) or business model innovation for sustainability (BMIS) could address the gaps in the GBMI conceptualization. Sustainable business model (SBM) has been defined as "*a business model that creates competitive advantage through superior customer value and contributes to a sustainable development of the company and society*" (Lüdeke-freund 2009). SBM captures economic, social and environmental value for a wide range of stakeholders, which is consistent with the triple bottom line perspective of GBM proposed by Jing and Jiang (2013). Bocken et al. (2014) defined BMIS as "*innovations that create significant positive and/or significantly reduced negative impacts for the environment and/or society, through changes in the way the organization and its value network create, deliver value and capture value (i.e. create economic value) or change their value propositions*". However, sustainability is too broad especially with respect to social issues that are so many and that cannot be all addressed in business unless some guidance is provided on how to address pressing issues like those that farmers face in Africa. With a special focus on these issues, an inclusive green agricultural business model innovation (IGABMI) is then proposed as a more operational framework with sound theoretical foundations.

12.4 Inclusive Green Agricultural Business Models for Smallholder Farmers

An inclusive green business model innovation (IGBMI) implies an innovation in the BM not only from an environmental perspective but also from a social (inclusive) perspective. An inclusive green agricultural business model innovation (IGABMI) is then the result of both inclusive innovation and green innovation in an agribusiness model (ABM).

12.4.1 Inclusive or Social Business Model Innovation

The social business model (SBM) perspective has emerged from the social entrepreneurship literature starting with the "*bottom of the pyramid*" (BOP) theory. Introduced in the business literature since the early 2000s (Prahalad and Hart 2002), the BOP approach emerged from the failure of the conventional BM to address the needs of the poor, the so-called bottom of the pyramid in the distribution of the world income. Basically, this theory sees the poor people as potential customers (market opportunity) whom companies can serve profitably and alleviate poverty (social value creation) if they learn to rethink their existing strategies and business models. Though the initial BOP approach intended to include the poor in the business model (Prahalad and Hart 2002; Prahalad 2010), it has been criticized for seeing the poor only as consumers rather than producers, while their income is actually lower than what the theory assumes (Karnani 2007; Seelos and Mair 2007). This income could be improved by involving them in more productive activities.

To extend the BOP view, the inclusive business model (IBM) was proposed and first formalized by the United Nations Development Programme (UNDP 2008). IBMs are commercially viable models that benefit low-income communities by including them in a company's value chain on the demand side as clients and consumers, and/or on the supply side as producers, entrepreneurs or employees in a sustainable way (UNDP 2008). They range from multinational corporations to large domestic companies, cooperatives, small and medium enterprises (SMEs), or even not-for-profit organizations that use business principles or social business approaches to achieve their mission (UNDP 2010).

Although the UNDP's definition includes the sustainability aspect that may address equity issues, there is a need to explicitly emphasize equitable sharing of the financial value created and captured along the value chain. The issue is that sustainability is too broad so that its operationalization requires a clarification of the aspects to focus on. It is argued here that those aspects should include at least an equitable sharing of the financial value. Inclusive business model innovation (IBMI) can then be defined as the new logic/way/rationale of how a business offers value to low-income communities by including them in a company's value chain on the demand side as clients and consumers, and/or on the supply side as producers, entrepreneurs or employees in a sustainable way, while creating sustainable revenue streams and their equitable sharing.

From a review of various case studies (Seelos and Mair 2007; Yunus et al. 2010; Dahan et al. 2010), a key message for developing a social or inclusive business model is the importance of partnership between organizations to leverage resources and capabilities. Indeed, though social business models are argued to be self-sufficient (Yunus et al. 2010), these case studies show that support from government, NGOs and other development agencies can be critical to their success.

12.4.2 Inclusive Agricultural Business Models

There is very little effort in providing a conceptual framework for agricultural business models. A recent work by Amador and Vera-Cruz (2013) is one of the rare conceptualizations of what an agricultural business model is. Compared to the general concept of business models discussed previously, the main contribution of this work has been to emphasize the government as key elements of a business model especially in the agricultural sector. Nevertheless, various agricultural business models exist in practice. These have been comprehensively discussed in the literature, especially with respect to their inclusiveness of smallholder farmers. While some authors identified the various arrangements that exist for agricultural investments, others used a "driver model" perspective to describe how smallholder farmers are organized and linked to agricultural markets. These two approaches of agricultural business models are overlapping.

Arrangements for inclusive agricultural investments

Agricultural business models discussed following this approach include government-owned agribusiness, contract farming, management contracts, tenant farming and sharecropping, joint ventures, farmer-owned businesses, upstream and downstream business linkages, large-scale agriculture (Vermeulen and Cotula 2010; Cotula and Leonard 2010). Vermeulen and Cotula (2010) reviewed alternative ways for restructuring agricultural investments in lower- and middle-income countries. Using their inclusiveness criteria, they identified four promising agricultural business models: contract farming schemes, joint ventures, management contracts, and upstream and downstream business links, i.e. value chain relationships. According to the authors, these are based on collaborative arrangements between large-scale investors and local small-scale farmers and communities.

Contract farming model describes pre-agreed supply agreements between farmers (or groups of farmers) and buyers of agricultural outputs. Usually, local farmers agree to grow and deliver agricultural produce for specified quantity and quality at an agreed date. In exchange, the company provides upfront inputs, such as credit, seeds, fertilizers, pesticides and technical advice, all of which may be charged against the final purchase price, and agrees to buy the produce supplied, usually at a specified price. Contract farming can then overcome imperfections in input and output markets or institutional deficiencies that constrain smallholder farmers in developing or transition economies.

Lease and management contracts refer to arrangements under which a farmer or farm management company farms on land that belongs to someone else. Holders of large estates (individuals, companies or state bodies) contracting an agribusiness company to manage their plantation are commonly used. In cases where smallholders and local communities hold land, the management company runs the farms on their behalf but does not acquire the land directly. These arrangements can bring in scale advantages such as bulk input supplies, as well as pooled value addition through quality control, processing and marketing. However, an enabling legal framework is needed to ensure equitable outcomes from these arrangements.

Joint ventures are arrangements, whereby two or more parties jointly run a business. Each party contributes to the joint business, whether in cash (capital) or in kind (e.g. land/natural resource rights, technology, know-how), and participates in any profits (or losses) made by it. Joint venture models between agribusinesses and smallholders are intrinsically attractive because it includes smallholders as full business partners in agribusiness activities, granting them shares of realized profits (rather than just one-off compensation, land rent or farm gate crop prices) and, in most cases, a legally recognized decision-making role in the business.

Upstream and downstream business models describe the set of business opportunities beyond direct agricultural production that exist for agribusinesses, small local enterprises and smallholders. They include agribusinesses working with smallholders in the upstream or downstream end of the agricultural supply chain, without direct involvement in agricultural production. Upstream links include supply of inputs (seeds, fertilizers, pesticides) as well as various business services such as micro-credit, insurance, advisory, certification, training (e.g. Farmer Field Schools). Unlike in contract farming, these services are often provided separately. Downstream links include specialized wholesale and retail. Upstream and downstream business links facilitate access to high-value niche markets and develop local enterprises that support smallholder farmers.

The "Driver Models"

Four "driver models" have been identified to describe the way smallholders are organized and linked to the market (Table 12.1): producer-driven, buyer-driven, intermediary-/facilitator-driven and integrated models (Kelly 2012; Miller 2012; Miller and Jones 2010; Vorley et al. 2009). Producer-driven models are motivated and owned by small-scale producers based on collective action for increased small farmers participation in markets. Buyer-driven models involve larger businesses organizing farmers into suppliers, which can also include the provision of inputs and technical advice based on the buyers' needs, also known as contract farming. Intermediary models that are commonly led by local NGOs involve the provision of technical assistance and support to identify and improve smallholder market linkages. Integrated model not only connects producers to input suppliers, intermediaries,

Table 12.1 Typical organizational smallholder business models in developing countries

Model	Driver	Motivation
Producer-driven	Smallholder groups, associations, cooperatives	Access to new markets, increased bargaining power, access to inputs, technical assistance, secure market position, farmer empowerment
Buyer-driven	Processors, retailers, exporters, traders, wholesalers	Access to land, supplies, increase volumes, supply niche markets
Intermediary-driven	NGOs, development agencies, governments	Local and national economic development, farmer empowerment
Integrated	Lead firms, supermarkets, multinationals	New and higher-value markets, low prices for good quality, market monopolies

Source Adapted from Kelly (2012), Miller (2012), Vorley et al. (2009)

processors, retailers and service providers (e.g. financial service providers) but integrates many of these through ownership and/or formal contractual relationships. It has many features of the other models.

Among the various agricultural business models described as inclusive of smallholder farmers, it is worth noting that no one is optimal for all smallholders in all situations (Vermeulen and Cotula 2010). Vorley et al. (2009) also argued that none of the driver models is inherently superior to smallholders since comparative evidence on benefits and impacts is weak. Organization of farmers into associations, market coordination and intermediation, business support and financial services, buyer behaviour, and enabling policies and infrastructure are important factors which influence sustained and equitable inclusion of smaller-scale farmers and SMEs (Vorley et al. 2009). Beyond the partnerships between smallholders and buyers, support services from governments, NGOs, etc. play a key role to promote more inclusive agribusiness models. They can help small farmers' cooperatives to gain negotiation power and access to funds, information concerning market trends, prices, royalties; they can also help to reduce the risk that poses obstacles to the partnership between large and small producers (Amador and Vera-Cruz 2013; Vermeulen and Cotula 2010).

To assure stakeholders about inclusive agribusiness practices and get reward for this, the adoption of social certifications such as Fairtrade is necessary. In adopting inclusive certification, agribusinesses demonstrate their commitment to the principle of corporate social responsibility (social value creation), allowing them to brand their business as socially responsible. This can allow them to increase their market share and get premium prices.

Inclusive Green Agricultural Business Model Innovation
Inclusive agricultural business models can provide socio-economic value for smallholder farmers. Some of them involve interesting green agricultural innovations (Seelos and Mair 2005, 2007). However environmental concerns are rarely a primary concern in most of the social business models described in the literature. Yet, along with economic and social value creation, environmental value creation is increasingly recognized as critical in all businesses including agribusinesses. Likewise, green business model may focus on environmental value creation while overlooking the inclusion of the poor as discussed above. The business model innovation should then be both inclusive and green.

Built on the definitions of innovation, inclusive business model innovation and green business model innovation, the following definition of an inclusive green business model innovation (IGBMI) is proposed: the implementation of new products, processes, marketing methods and/or organizational methods in business practices, which significantly improve the socio-economic conditions of low-income communities and reduce the ecological footprint of the business, while ensuring its economic viability. When IGBMI happens, the resulting new BM is an inclusive green business model (IGBM). With respect to agriculture specifically, inclusive green agricultural business model innovation (IGABMI) is the implementation of new products, processes, marketing methods and/or organizational methods in agricultural business practices, which significantly improve the socio-economic conditions of smallholder farmers and reduce the ecological footprint of the agribusiness, while ensuring its economic viability. Adopting green agricultural innovations in the inclusive agricultural business models described above, inclusive green agricultural business models (IGABMs) can be developed in a practical way. Green agricultural innovations start from sustainable agricultural practices (SAPs) at the farmer level (e.g. water and soil conservation practices, integrated pest management, organic farming) to green agricultural standards and certifications (Global G.A.P., Rainforest Alliance, Organic, etc.) that concern the entire value chain (Kassie et al. 2013; UNEP 2013). Green agricultural certifications (e.g. Organic certification) assure the commitment of agribusinesses to provide environmental values. They can be economically more rewarding than inclusive certifications, in terms of the branding of agribusiness, the market potential and price premium. However, the adoption of inclusive and green agricultural innovations is low in developing countries due to various factors (Kassie et al. 2013; Lee 2005).

12.5 Inclusive Green Agricultural Business Model Innovations in Africa: Potential Drivers, Barriers, Opportunities and Policy Role

The factors that drive or inhibit inclusive green business model innovations in Africa can be external and/or internal to the agribusinesses. Identifying these factors is important for policy purpose.

12.5.1 External Factors

External drivers include mainly environmental regulations and other government measures, market conditions and the increasing costs of resources and supply risk (Porter and van der Linde 1995; Hervani et al. 2005; Bisgaard et al. 2012; Beltramello et al. 2013; Economic and Social Commission for Western Asia (ESCWA) 2013). Pressure also arises from NGOs and other agencies that advocate for the protection of the environment, promotion of sustainable agricultural practices or linking smallholder farmers to markets.

International institutions and governments often issue regulations for the use and commercialization of some products with serious negative environmental and health impacts. These concerns are particularly critical in the agriculture and food sector. Therefore, in complying with environmental regulations to avoid penalties, firms adopt green innovations (Porter and van der Linde 1995; ESCWA 2013). But, beyond acting as constraints, innovation to meet regulations can bring economic benefits: better use of inputs, creation of better products or improvement in product yields (Porter and van der Linde 1995). However, environmental regulations should be innovation-friendly since bad regulation will damage competitiveness, while the right kind of regulation can enhance it (Porter and van der Linde 1995). In addition to regulations, governments and NGOs often initiate or support eco-innovation projects, enabling smallholder farmers to access profitably (green) markets. Therefore, a lack of environmental regulations as well as government and NGO support for change limits the adoption of green innovations.

Market pressures that make firms adopt green innovations include consumers' demand (market requirements) and the search for new competitive advantage and opportunities for increasing profits. On the one hand, consumer awareness about sustainability issues leads them to demand products that are safe, socially responsible and environmentally friendly; thus, while attempting to reduce their negative impacts, value chain firms also respond to market demand by producing and selling products that are branded as green (Bisgaard et al. 2012; Henriksen et al. 2012d; Beltramello et al. 2013; ESCWA 2013). A typical case is the increasing market for organic agricultural products. In doing so, companies differentiate their products and services from those of their competitors, thereby creating a competitive advantage and simultaneously improving their economic and environmental performance

(Porter and van der Linde 1995; Bisgaard et al. 2012). Again, as Porter and van der Linde (1995) pointed out, resisting innovation will lead to loss of competitiveness in global economy. However, market conditions can also be unfavourable to green innovations (Beltramello et al. 2013). Indeed, while a new environmental product or technology represents an improvement for many users, there is still a large group of customers who do not have enough knowledge about what sustainability is and who are too conservative to change their buying habits where price is the main purchasing incentive; this may result in a low uptake of beneficial innovations (Bisgaard et al. 2012; Beltramello et al. 2013). Competition from non-complying operators may also force firms to cut costs for greening in order to stay in business (ESCWA 2013).

Increasing costs of resources and supply risk as a result of non-renewable resource depletion, for example (Bisgaard et al. 2012; ESCWA 2013), as well as land degradation and climate risk especially in agriculture, have also forced value chain firms to consider alternative resources and strategies to adapt to changes. Thus, in agriculture, there is an increasing promotion of agro-ecological farming technologies, water and soil conservation practices, waste reuse as inputs, etc.

12.5.2 Internal Factors

Internal drivers and barriers to green innovations are mainly organizational factors that influence a firm's decision to adopt or not adopt innovative practices that lead to improvements in environmental outcomes and overall business performance. These include organizational resources and capacity such as human capital and financial resources along with organizational structure and monitoring system (Hervani et al. 2005; Bisgaard et al. 2012; Beltramello et al. 2013 ; ESCWA 2013).

Human capital enhances innovation through several channels: the generation of new knowledge; the adoption and adaptation of existing technologies and ideas; and the ability to adapt to changes and to learn new things (Beltramello et al. 2013). Education and training are key factors in forming human capital. Collaboration with external partners such as universities often provides innovators with additional sources of skills and knowledge that can lead to a successful outcome of innovations. Thus, authors argued that the knowledge-related barriers to green innovations include the traditional mindset among firms, the lack of competencies and knowledge of sustainability issues and about the economic benefits of green innovations and the lack of training for company managers and staffs (FORA 2010; Bisgaard et al. 2012 ; Beltramello et al. 2013).

Financial resources play an essential role in enabling firms to invest in innovation in general and green innovation more particularly. Indeed, green investments are costly and may bring in additional operating costs (Bisgaard et al. 2012; Beltramello et al. 2013; ESCWA 2013). Therefore, they are not easy especially for small and medium enterprises (e.g. farmers and middlemen/traders) because of their inability to meet the necessary investment required to comply with green standards (ESCWA 2013). Thus, limited financial resources constitute an important barrier to innovation in SMEs with respect to environmental strategy (Hervani et al. 2005). Yet, access to finance is more difficult for firms engaged in eco-innovation, due to the immaturity of the market, which increases the difficulties associated with accurately pricing the relative risk of investment, making it more difficult for such firms to obtain financing at reasonable costs than for firms involved in more established markets (Beltramello et al. 2013). Moreover, long payback periods and risk related to green investments, lack of long-term commitments and few financial incentives from green value chain buyers constrain many small operators in the adoption of green innovations (ESCWA 2013; Beltramello et al. 2013).

Other organizational factors that inhibit inclusive green innovation include the lack of integration between divisions in companies; insufficient R&D capabilities; the status of the environmental and social issues in the company; the lack of vision; and the lack of relationships with external stakeholders (Hervani et al. 2005; FORA 2010; Beltramello et al. 2013). Social and green entrepreneurship ideas may also be lacking.

12.5.3 Opportunities and Other Policy Challenges

There are increasing green commodity markets, mainly driven by demand from developed countries. Thus, traditional crops (cocoa, coffee, cotton) are progressively being sold on green export markets, while new export market opportunities are emerging in the horticultural sector (fruits and vegetables). Other green export market opportunities exist with cashew, sesame, spices, soy and palm oil. These opportunities can also be promoted in local markets especially for the horticultural sector by developing consumer awareness on environmental and health issues, while proposing green value to them through certified (inclusive and/or green) agricultural products. Consumers in urban areas are increasingly interested in such products. Various certifications including Global G.A.P., Rainforest Alliance, Fairtrade, Organic, etc., allow green entrepreneurs to differentiate their commodities from others, gain competitive advantage and reach green markets. However, existing standards and certifications are most often used for commodities oriented towards global markets.

Policy directed towards customizing existing standards for regional and local markets in Africa and/or developing awareness of environmental challenges could have a significant impact on the development of green local markets and agricultural value chains. Applying standards and certifications also requires good organization and coordination among value chain actors to reduce transaction costs. Public–private partnerships facilitate access to finance, green technologies, inputs and skills training on green practices. The role of policy-makers in facilitating the development of inclusive green agricultural businesses in Africa also includes support for research and information diffusion on inclusive and green markets; support for training on social and green agribusiness entrepreneurship; investment in rural and market infrastructures, etc. For this, African governments must renew their recognition of the need to increase investment in agriculture.

12.6 Green Agricultural Value Chain as an IGABMI

The green value chain approach is used to provide an illustration of how IGABMI can address many of the issues discussed above. There are many reasons for this choice. First, green value chain development is increasingly seen as the way to promote green growth, with a particular attention to agriculture (DCED 2012; ESCWA 2013; IDH 2013; Sinclair-Desgagné 2013). However, its link with the GBMI framework is not yet established. Green value chain is shown here as a particular case of IGBMI. Second, more and more products and services now pass through global value chains to reach the end consumers (DCED 2012), while there is an increasing interest of investors in Africa along the whole food supply chain (Deutsche Bank Research 2014). Third, a value chain approach allows an easy mapping of the actors involved in a business for a particular product, as well as the partnerships needed.

12.6.1 Value Chain as an Inclusive Business Model Innovation

Generally defined in reference to a product or service, value chain is often used interchangeably with various related concepts such as production chain, supply chain, marketing chains or distribution chains (Webber and Labaste 2010). Although these concepts are often overlapping and then difficult to be separated, their focus can be used to distinguish them. Value chains focus on value creation through innovation in products or processes, as well as marketing and also on the allocation of the incremental value, while the primary focus of supply chains is efficiency by reducing "friction" (e.g. delays), reducing outages or overstocks, lower transaction costs, etc (Webber and Labaste 2010). As value creation is at the heart of any business model (Zott et al. 2011; Osterwalder and Pigneur 2009), the term "value chain" is employed in this study.

Value chain refers to the full range of activities required to bring a product or service through the different phases of production, involving a combination of physical transformation, the input of various producer services, delivery to consumers and final disposal after use; it includes the vertically linked interdependent processes that generate value for the consumer (Webber and Labaste 2010). A value chain encompasses the actors (private and public, including service providers) and the sequence of value-adding activities involved in bringing a product from production to the end user/consumer. It is then clear that a value chain typically involves the upstream and downstream business links previously discussed as inclusive agribusiness models.

A value chain is traditionally viewed as a linear model, as shown below for agricultural products (Fig. 12.2). Indeed, an agricultural value chain shows the relations between input suppliers, farmers, processors, traders and a number of intermediaries and supporting actors (government agencies, NGOs, financial institutions, donors and certification bodies) who all work together to add value, improve the quality of

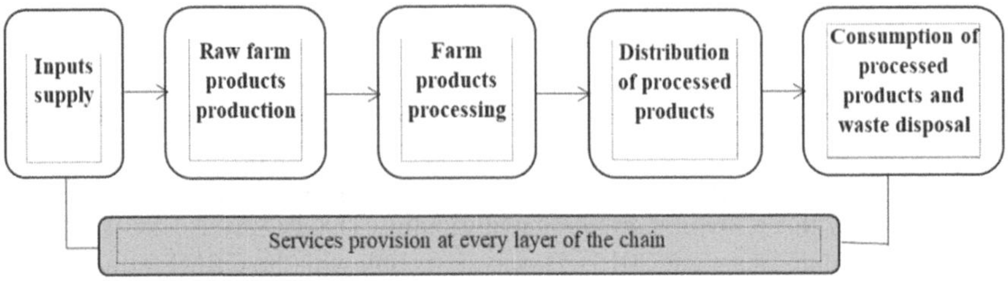

Fig. 12.2 Agricultural value chain

a product or increase efficiency (Miller 2012). Therefore, even if the value chain is a system composed of multiple enterprises, it can be considered as an economic unit of analysis (Miller 2012).

Doing business following the whole value chain perspective is an innovative approach of creating, delivering and capturing value. By its nature, this innovation is systemic for farmers and small and medium agribusinesses who previously used to organize their production and marketing activities alone, without formal partnership with other stakeholders. Working together instead of individually, actors in the product chain can overcome many constraints and take advantage of opportunities related to their individual activities. For example, the value chain business model allows tracing and linking geographically dispersed activities, identifying and analysing the roles of stakeholders and firms, as well as studying various other economic, social, technological, environmental and institutional issues; they can also be a source of economies of scale which are essential to achieve significant improvement in productivity, competitiveness and bargaining power (ESCWA 2013). Agricultural value chain development is increasingly recognized as a powerful tool for linking smallholder farmers to market, improving job opportunities in rural areas, alleviating poverty and enhancing food security among smallholders in Africa (Deutsche Bank Research 2014). Agricultural value chain management not only is a driver of green agricultural innovation, but can also be subjected to systemic green innovation and then result in a green agricultural value chain.

12.6.2 Value Chain as a Driver of Inclusive and Green Agricultural Innovations

Value chain functions as a source of its own product and process innovation. As such, many of the challenges related to the adoption of inclusive and green agricultural innovations in Africa can be addressed thanks to some important innovations driven by value chain development. These include various instruments of value chain financing (e.g. interlinked supplier–buyer–producer–bank financial arrangements), application of ICTs in mobile banking, mobile technical support, electronic networks, etc.

(Miller 2012). These innovations contribute to reduce cost and risk, while improving access to and use of financial services, information systems, contract farming systems, commodity exchange linkages and other aspects of value chain operations. Value chain development also drives policy innovations including extension services and investing in supportive infrastructure.

Value chain actors generally benefit from information sharing and training on various aspects of the product development and delivery, especially the application of Good Agricultural Practices (GAPs) in order to meet the requirements of consumers expressed through certification. Certification is generally costly and is not affordable for smallholder farmers and small processors who initially have very limited resources. Leading agribusinesses in the value chain, NGOs and government agencies, who work generally with certification bodies and banks, are the main providers of training and certification for them while facilitating their access to and use of financial services. Thus, an agricultural value chain business model allows a faster flow of environmental concerns from consumers to farmers and input suppliers, and a faster adoption of more green agricultural practices accordingly. By enhancing the whole value chain integration, innovations within the agricultural value chain improve the operating efficiency and minimize waste in the system (Wang and Liu 2014). In so doing, traditional agricultural value chains move towards greener ones; i.e. they undertake green agricultural value chain innovations.

12.6.3 Green Agricultural Value Chain as a Systemic Green Innovation

From a general perspective, the concept of green value chain has been more referred to as green supply chain management (GSCM) since the 1990s (Hervani et al. 2005; Srivastava 2007; Sarkis et al. 2011). The green component has been added to the conventional supply chain management in order to address the relationships between supply chain management and the natural environment. GSCM encompasses green purchasing, green manufacturing/material management, green distribution/marketing and reverse logistics (Hervani et al. 2005 ; Sarkis et al. 2011). The green value chain approach is increasingly seen as a promising way to promote green growth, with more focus on agriculture (DCED 2012; IDH 2013; Sinclair-Desgagné 2013). This new approach transforms the conventional linear view of value chains into a cyclical system view (Fig. 12.3). The environment is integrated into the value chain at its starting stage as a supplier of natural resources and at its final stage as recipient and recycler of end products (DCED 2012). Green value chain development aims to improve the sustainability of the entire chain through a strategic integration of resources to achieve economic, social and environmental objectives (DCED 2012; ESCWA 2013).

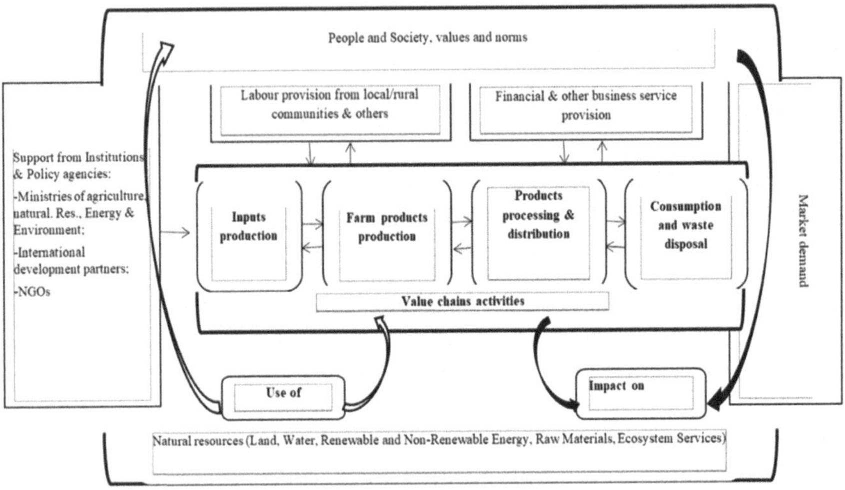

Fig. 12.3 Systemic perspective for developing green value chains. *Source* Adapted from DCED (2012)

Tools for implementing agricultural value chains in Africa have been extensively discussed in the literature, ranging from those for choosing priority sectors for interventions to tools for monitoring achievements in value chain performance (e.g. see Webber and Labaste 2010). Building on these, ESCWA (2013) put more emphasis on the following actions for developing green agricultural value chains in the Arab region including some Northern African countries: (i) prioritizing sectors for value chain interventions; (ii) assessing the status of value chains (business model appraisal); (iii) finding ways (critical success factors) for upgrading value chains; (iv) promoting vertical and horizontal linkages among value chain actors (input producers, smallholder farmers, processors, retailers, consumers); (v) applying standards and certifications; (vi) providing support services and improving the operating environment; and (vii) monitoring and evaluating achievements.

12.7 Framework for Assessing and Monitoring Achievement Towards IGABM

In promoting the transition from traditional business models to inclusive and green ones, it is important to evaluate the achievements using relevant indicators. From the triple bottom line (TBL) perspective of sustainability, an indicator can be defined as a measurement that shows the status of an environmental, economic or social system over time (Bae and Smardon 2011). Many indicators have been proposed to measure progress towards sustainable business (Hervani et al. 2005; Bae and Smardon 2011). International organizations have developed those that are internationally agreed for standardization and global reporting initiatives. This section first reviews the sustainable business indicators (SBIs) and their measurements, and uses the most relevant to this study to build a framework for assessing and monitoring the performance of agricultural businesses from the perspective of inclusive green innovation.

Sustainable Business Indicators

In accordance with their objectives and applications, indicators for sustainable business practices can be expressed in many different forms: qualitative or quantitative, general or specific, and absolute or relative (Bae and Smardon 2011). Quantitative indicators are measured in terms of mass, volume or number of environmental pollutants or physical materials, while qualitative indicators are expressed interpretively including social dimensions of a firm's activities, such as changes in cultural values or equity. General indicators can be used to measure issues that have already been discussed globally, including energy, water and material consumption, greenhouse gas emissions, carbon dioxide, methane and air emissions per unit of product, while specific indicators are defined differently and measured in accordance with characteristics of each industry or firm. Absolute indicators are used to measure a firm's quantitative environmental and social impact related to its activities, products and services, while relative indicators are expressed in terms of a ratio or proportion that compares an absolute indicator with another absolute indicator.

This last distinction between absolute and relative indicators is particularly important in analysing the inclusiveness and greenness of businesses which requires an analysis of absolute economic, social and environmental performance indicators, not only separately but also in an integrated manner. The challenge is how to identify these absolute indicators first and integrate them next. From a comprehensive review of ninety internationally agreed sustainable business indicators (i.e. general SBIs), Bae and Smardon (2011) identified seven groups including absolute and relative indicators. Absolute indicators include environmental indicators, economic indicators and social indicators. They can be used to build the following relative or composite indicators according to the type of absolute indicators involved: eco-efficient indicators (economic and environmental); socio-economic indicators (social and economic); socio-environmental indicators (social and environmental); and integrated indicators (environmental, economic and social) (Bae and Smardon 2011). However, the choice of the indicators depends on one's objective. Depending on the context, specific indicators (not internationally agreed) are even used.

Guidelines for Constructing a Performance Index for Inclusive Green Agricultural Business

A performance index for inclusive green agricultural business requires necessarily an integrated indicator built on absolute indicators of inclusiveness (social indicators), environmental indicators and economic indicators. The choice of absolute indicators of interest depends again on one's objective.

From the value chain perspective, value added (VA) is a good economic indicator. Life-cycle costing and common financial accounting approaches can be employed.

Environmental indicators can be identified in relation to the three broad environmental issues: depletion of resources (mineral resources, fossil fuels, water and land); human health hazards; and ecosystem quality degradation (Carvalho et al. 2014). The environmental life-cycle assessment (E-LCA) method is the most established approach for defining and calculating these indicators (Guinée et al. 2004). The social life-cycle assessment (S-LCA) method has emerged recently as the best approach for identifying and calculating social indicators (Benoît et al. 2010; Jørgensen et al. 2008) and is particularly relevant from the value chain perspective. Social indicators in S-LCA relate to broad social issues: human health as considered in E-LCA, and human dignity and well-being (Jørgensen et al. 2008). More specific indicators relate to social themes of interest to stakeholders and decision-makers: human rights, work conditions, cultural heritage, poverty, disease, political conflict, indigenous rights, discrimination, child labour, forced labour, etc.

In addition to well-established economic, environmental and social indicators, the following aspects (not exhaustive) should be given particular attention when assessing agricultural value chains in rural Africa with respect to inclusive and green innovations (Table 12.2).

For aggregation of the above absolute economic, environmental and social indicators, the following relative sustainability indicators can be explored: eco-efficiency indicator (Picazo-Tadeo et al. 2011), the green productivity index (Hur et al. 2004 ; Gandhi et al. 2006 ; Darmawan et al. 2014) or sustainable value added (Figge and Hahn 2004; Van Passel et al. 2009). In addition, an integrated inclusive green economic performance index can be constructed following the directions of Singh et al.

Table 12.2 Important specific indicators for monitoring agricultural business models in Africa

Proportion of smallholder farmers in the total number of farmers involved in the value chain	Perception of farmers involved about the governance of the agribusiness
Proportion of smallholder farmers in the value chain who are landowners	Perception of local community about the activities and products of the agribusiness
Proportion of farmers in the value chain who are in association	Poverty, food security and health status of farmers involved in the agribusiness
Existence of an inter-professional association in the value chain	Frequency of training provided to farmers on GAPs
Frequency of meeting of the inter-professional association	Frequency of communication using ICTs
Proportion of youth involved in the value chain	Proportion of land under GAPs
Proportion of women involved in the value chain	Proportion of farmers involved in inclusive certification (e.g. Fairtrade)
Contribution of the value chain activities to the whole farm income	Proportion of farmers involved in green certification (e.g. Organic)
Share of farmers' income in the value added of the product	Proportion of products certified inclusive
Farmers' involvement in price determination	Proportion of products certified green

Source Authors

(2012). For this, the absolute economic, environmental and social indicators should be normalized and weighted.

The aim of the normalization is to place indicators into a broader context and to adjust them to have common dimensions. The weighting factor represents the relative importance of the issue related to the corresponding indicator. Various approaches of weighting have emerged: subjective weighting, equal weighting, analytical hierarchy process (AHP) weighting, experts' knowledge, public opinions, optimization procedures, etc. (Singh et al. 2012 ; Carvalho et al. 2014 ; Picazo-Tadeo et al. 2011). Once the different indicators are weighted, they can be aggregated using various methods including simple summation, arithmetic mean, geometric mean, etc. (Singh et al. 2012).

12.8 Conclusion

Innovations needed for an effective green economy transition are different according to regions and economic sectors. This study proposed a framework of inclusive green agricultural business model innovation (IGABMI) for rural Africa. It adapts the green business model innovation framework that focuses on environmental issues in developed countries, to include social issues that are of primary interest to stakeholders in developing countries, particularly in rural Africa. The proposed framework is more consistent with the inclusive green economy concept that is agreed as the way for achieving sustainable development. It differs from the sustainable business model innovation framework by being more operational with respect to the social dimension (inclusiveness). Inclusive green agricultural business model (IGABM) provides a theoretical foundation for green agricultural value chain (GAVC) that is used in this study as an illustration. Moreover, a framework for assessing and monitoring achievement towards IGABM is proposed, with particular attention again to GAVC. Overall, this study provides guidelines for developing IGABMs in Africa and other developing regions.

However, the proposed framework is mainly conceptual. A review of practical GABMIs that exist in Africa will improve it by providing better illustrations and identifying success stories that other agribusiness models can learn from in order to go inclusive and green. Moreover, empirical assessments of agribusiness in Africa will be an opportunity for testing the applicability of the framework proposed regarding the performance index for IGABM.

References

Allen C, Clouth S (2012) A guidebook to the green economy: issue 1: green economy, green growth, and low-carbon development-history, definitions and a guide to recent publications. United Nations Division for Sustainable Development

Amador JF, Vera-Cruz AO (2013) Agricultural business models: a Mexican case. Porto: XV Ibero-American conference on management of technology, 27th–31 Oct 2013, Porto, Portugal. Retrieved from http://www.altec2013.org/programme_pdf/1136.pdf

Bae H, Smardon RS (2011) Indicators of sustainable business practices. In: Broniewicz E (ed), Environmental management in practice. InTech, pp 176–206. Retrieved from http://www.intechopen.com/books/environmental-management-in-practice/indicators-of-sustainable-business-practices

Beltramello A, Haie-Fayle L, Pilat D (2013) Why new business models matter for green growth. OECD green growth papers. OECD Publishing, Paris. doi:10.1787/5k97gk40v3ln-en

Benhin JKA (2006) Agriculture and deforestation in the tropics: a critical theoretical and empirical review. AMBIO: J Hum Environ 35(1):9–16. https://doi.org/10.1579/0044-7447-35.1.9

Benoît C, Norris GA, Valdivia S, Ciroth A, Moberg A, Bos U, Prakash S, Ugaya C, Beck T (2010). The guidelines for social life-cycle assessment of products: just in time! Int J Life Cycle Assess 15:156–163. doi:10.1007/s11367-009-0147-8

Benson-Rea M, Brodie RJ, Sima H (2013) The plurality of co-existing business models: Investigating the complexity of value drivers. Ind Mark Manage 42(5):717–729. https://doi.org/10.1016/j.indmarman.2013.05.011

Bilsen V, Blondiau T, Debergh P, Lukach R (2013) Exchange of good policy practices on innovation and green business models. European Commission, directorate-general enterprise and industry

Bisgaard T, Henriksen K, Bjerre M (2012) Green business model innovation: conceptualization, next practice and policy. Nordic Innovation, Oslo, Norway. Retrieved from http://www.nordicinnovation.org/publications

Bocken NMP, Short SW, Rana P, Evans S (2014) A literature and practice review to develop sustainable business model archetypes. J Clean Prod 65:42–56. https://doi.org/10.1016/j.jclepro.2013.11.039

Carvalho A, Mimoso AF, Mendes AN, Matos HA (2014) From a literature review to a framework for environmental processes impact assessment index. J Clean Prod 64:36–62. https://doi.org/10.1016/j.jclepro.2013.08.010

Cotula L, Leonard R (eds) (2010) Alternatives to land acquisitions: agricultural investment and collaborative business models. Food and Agriculture Organization of the United Nations (FAO)/International Institute for Environment and Development (IIED)/International Fund for Agricultural Development (IFAD)/Swiss Agency for Development and Cooperation (SDC)/Centro Terra Viva (CTV), London/Bern/Rome/Maputo

Dahan NM, Doh JP, Oetzel J, Yaziji M (2010) Corporate-NGO collaboration: co-creating new business models for developing markets. Long Range Plan 43(2–3):326–342. https://doi.org/10.1016/j.lrp.2009.11.003

Danish Enterprise and Construction Authority (FORA) (2010) Green business models in the Nordic Region: a key to promote sustainable growth. Copenhagen. Retrieved from http://www.motiva.fi/files/4878/Greenpaper_Green_Business_Models_in_the_Nordic_Region_A_Key_to_Promote_Sustainable_Growth.pdf

Danish Enterprise and Construction Authority (FORA) (2011) Project proposal green business model innovation. Danish Enterprise and Construction Authority, Copenhagen (FORA). Division for research and analysis. Retrieved from http://www.ebst.dk/file/170960/projectproposal.pdf

Darmawan MA, Islam Fajar Putra MP, Wiguna B (2014) Value chain analysis for green productivity improvement in the natural rubber supply chain: a case study. J Cleaner Prod, 1–11. doi:10.1016/j.jclepro.2014.01.098

Deutsche Bank Research (2014) Agricultural value chains in Sub-Saharan Africa. In: Lanzeni ML, (ed) From a development challenge to a business opportunity. Frankfurt am Main, Deutsche Bank AG, Deutsche Bank Research

Donor Committee for Enterprise Development (DCED) (2012) Green value chains to promote green growth. The Donor Committee for Enterprise Development. Retrieved from http://www.enterprise-development.org/page/ggwg

Economic and Social Commission for Western Asia (ESCWA) (2013) Green agricultural value chains for improved livelihood in the Arab Region. United Nations Economic and Social Commission for Western Asia (ESCWA), New York

FAO, IFAD, WFP (2012) The State of food insecurity in the world 2012: economic growth is necessary but not sufficient to accelerate reduction of hunger and malnutrition. Food and Agriculture Organization of the United Nations (FAO), Rome

Figge F, Hahn T (2004) Sustainable value added–measuring corporate contributions to sustainability beyond eco-efficiency. Ecol Econ 48(2):173–187. https://doi.org/10.1016/j.ecolecon.2003.08.005

Gandhi NMD, Selladurai V, Santhi P (2006) Green productivity indexing: a practical step towards integrating environmental protection into corporate performance. Int J Prod Perform Manag 55 (7):594–606. doi:10.1108/17410400610702179

Girotra K, Netessine S (2013) Business model innovation for sustainability (No. 2013/64/TOM). INSEAD: The Business School for the World. Retrieved from http://www.insead.edu/facultyresearch/research/doc.cfm?did=52400

Guinée JB, Gorrée M, Heijungs R, Huppes G, Kleijn R, de Koning A, Weidema BP et al (2004) Handbook on life cycle assessment: operational guide to the ISO Standards (ed: Guinée JB). Kluwer Academic Publishers, New York

Henriksen K, Bjerre M, Almasi AM, Damgaard-Grann E (2012a) Green business model innovation: conceptualization report. Nordic Innovation, Oslo. Retrieved from http://www.nordicinnovation.org/publications

Henriksen K, Bjerre M, Bisgaard T, Almasi AM, Damgaard E (2012b) Green business model innovation: empirical and literature studies. Nordic Innovation, Oslo. Retrieved from http://www.nordicinnovation.org/publications

Henriksen K, Bjerre M, Grann ED, Lindahl M, Suorti T, Friðriksson K, Muhlbradt T, Sand H (2012c) Green business model innovation: business case study compendium. Nordic Innovation, Oslo. Retrieved from http://www.nordicinnovation.org/publications

Henriksen K, Bjerre M, Øster J, Bisgaard T (2012d) Green business model innovation: policy report. Nordic Innovation, Oslo. Retrieved from http://www.nordicinnovation.org/publications

Hervani AA, Helms MM, Sarkis J (2005) Performance measurement for green supply chain management. Benchmarking: Int J 12(4):330–353. https://doi.org/10.1108/14635770510609015

Hur T, Kim I, Yamamoto R (2004) Measurement of green productivity and its improvement. J Clean Prod 12(7):673–683. https://doi.org/10.1016/j.jclepro.2003.08.004

Intergovernmental Panel on Climate Change (IPCC) (2014) Climate change 2014: impacts, adaptation, and vulnerability–summary for policymakers. Intergovernmental Panel on Climate Change

Jing H, Jiang BS (2013) The Framework of green business model for eco-innovation. J Supply Chain Oper Manag 11(1):33–46

Joller L (2012) Eco-innovation in business models–theoretical considerations. Paper to be presented at the DRUID academy conference 2012, 19–21 Jan, University of Cambridge, UK

Jørgensen A, Bocq A Le, Nazarkina L, Hauschild M (2008) Methodologies for social life cycle assessment. Int J Life Cycle Assess 13(2):96–103

Karnani A (2007) The mirage of marketing to the bottom of the pyramid: how the private sector can help alleviate poverty. Calif Manag Rev 49(4):90–111

Kassie M, Jaleta M, Shiferaw B, Mmbando F, Mekuria M (2013) Adoption of interrelated sustainable agricultural practices in smallholder systems: evidence from rural Tanzania. Technol Forecast Soc Chang 80(3):525–540. https://doi.org/10.1016/j.techfore.2012.08.007

Kelly S (2012). Smallholder business models for agribusiness-led development: good practice and policy guidance. Food and Agriculture Organization of the United Nations (FAO), Rome

Koohafkan P, Altieri MA, Gimenez EH (2012) Green Agriculture: foundations for biodiverse, resilient and productive agricultural systems. Int J Agric Sustain 10(1):61–75. https://doi.org/10.1080/14735903.2011.610206

Lee DRL (2005) Agricultural sustainability and technology adoption: issues and policies for developing countries. Am J Agr Econ 87(5):1325–1334

Lüdeke-freund F (2009) Business model concepts in corporate sustainability contexts: from rhetoric to a generic template for "Business models for sustainability." Centre for Sustainability Management (CSM), Leuphana University of Lueneburg, Lueneburg

Miller C (2012) Agricultural value chain finance strategy and design. International Fund for Agricultural Development (IFAD)

Miller C, Jones L (2010) Agricultural value chain finance: tools and lessons. Food and Agriculture Organization of the United Nations (FAO) and Practical Action Publishing, Rome

OECD and Eurostat (2005) Oslo manual: guidelines for collecting and interpreting innovation data, 3rd edn. OECD Publishing. https://doi.org/10.1787/9789264013100-en

Osterwalder A, Pigneur Y (2009) Business model generation: a hand book for visionaries, game changers, and challengers. Wiley, Hoboken, NJ

Picazo-Tadeo AJ, Gómez-Limón JA, Reig-Martínez E (2011) Assessing farming eco-efficiency: a data envelopment analysis approach. J Environ Manage 92(4):1154–1164. https://doi.org/10.1016/j.jenvman.2010.11.025

Porter ME, van der Linde C (1995) Towards a new conception of the environment-competitiveness relationship. J Econ Perspect 9(4):97–118

Prahalad CK (2010) The fortune at the bottom of the pyramid: eradicating poverty through profits, 5th edn. Pearson Education Inc., Upper Saddle River. https://doi.org/10.2139/ssrn.914518

Prahalad CK, Hart SL (2002) The fortune at the bottom of the pyramid. Strategy + Bus (26). Retrieved from www.strategy-business.com

Sarkar AN (2013) Promotion of eco-innovation to leverage sustainable development of eco-industry and green growth. Int J Ecol Dev 25(2):71–104

Sarkis J, Zhu Q, Lai KH (2011) An organizational theoretic review of green supply chain management literature. Int J Prod Econ 130(1):1–15. https://doi.org/10.1016/j.ijpe.2010.11.010

Schaltegger S, Lüdeke-freund F, Hansen EG (2011) Business cases for sustainability and the role of business model innovation: developing a conceptual framework. Centre for Sustainability Management (CSM), Leuphana University of Lueneburg, Lueneburg

Seelos C, Mair J (2005) Social entrepreneurship: creating new business models to serve the poor. Bus Horiz 48(3):241–246. https://doi.org/10.1016/j.bushor.2004.11.006

Seelos C, Mair J (2007) Profitable business models and market creation in the context of deep poverty: a strategic view. Acad Manag Perspect 21(4):49–63. https://doi.org/10.5465/AMP.2007.27895339

Sinclair-Desgagné B (2013) Greening global value chains: implementation challenges (No. 2013-04). OECD Publishing, Paris

Singh KR, Murty HR, Gupta SK, Dikshit AK (2012) An overview of sustainability assessment methodologies. Ecol Ind 15(1):281–299. https://doi.org/10.1016/j.ecolind.2011.01.007

Srivastava SK (2007) Green supply-chain management: a state-of-the-art literature review. Int J Manag Rev 9(1):53–80. https://doi.org/10.1111/j.1468-2370.2007.00202.x

Teece DJ (2010) Business models, business strategy and innovation. Long Range Plan 43(2–3):172–194. https://doi.org/10.1016/j.lrp.2009.07.003

The Sustainable Trade Initiative (IDH) (2013) Building green global value chains: committed public-private coalitions in agro-commodity markets (No. 2013-03). IDH. OECD Publishing, Paris. https://doi.org/10.1787/5k483jndzwtj-en

Thompson JD, MacMillan IC (2010) Business models: creating new markets and societal wealth. Long Range Plan 43(2–3):291–307. https://doi.org/10.1016/j.lrp.2009.11.002

United Nations Development Programme (UNDP) (2008) Creating value for all: strategies for doing business with the poor. The Growing Inclusive Markets Initiative, United Nations Development Programme, New York. Retrieved from http://growinginclusivemarkets.org/media/gimlaunch/Report_2008/GIM Report Final August 2008.pdf

United Nations Development Programme (UNDP) (2010) The MDGs: everyone's business-how inclusive business models contribute to development and who supports them. The Growing

Inclusive Markets Initiative, United Nations Development Programme, New York. Retrieved from http://www.growinginclusivemarkets.org/media/mdgreport/mdgreport_full.pdf

United Nations Environment Programme (UNEP) (2011) Towards a green economy: pathways to sustainable development and poverty eradication. United Nations Environment Programme, Geneva. Retrieved from www.unep.org/greeneconomy

United Nations Environment Programme (UNEP) (2013) Green economy and trade: trends, challenges and opportunities. United Nations Environment Programme, Geneva. Retrieved from http://www.unep.org/greeneconomy/GreenEconomyandTrade

Van Passel S, Van Huylenbroeck G, Lauwers L, Mathijs E (2009) Sustainable value assessment of farms using frontier efficiency benchmarks. J Environ Manage 90(10):3057–3069. https://doi.org/10.1016/j.jenvman.2009.04.009

Vermeulen S, Cotula L (2010) Making the most of agricultural investment: a survey of business models that provide opportunities for smallholders. Food and Agriculture Organization of the United Nations (FAO)/International Institute for Environment and Development (IIED)/International Fund for Agricultural Development (IFAD)/Swiss Agency for Development and Cooperation (SDC), London/Rome/Bern

Vorley B, Lundy M, Macgregor J (2009) Business models that are inclusive of small farmers. In: da Silva CA, Baker D, Shepherd AW, Jenane C, Miranda-da-Cruz S (eds) Agro-industries for development. Food and Agriculture Organization of the United Nations (FAO), Rome, pp 186–226

Wang X, Liu N (2014) The application of internet of things in agricultural means of production supply chain management. J Chem Pharm Res 6(7):2304–2310

Webber CM, Labaste P (2010) Building competitiveness in Africa's agriculture: a guide to value chain concepts and applications. The World Bank, Washington. https://doi.org/10.1596/978-0-8213-7952-3

Yunus M, Moingeon B, Lehmann-Ortega L (2010) Building social business models: lessons from the Grameen Experience. Long Range Plan 43(2–3):308–325. https://doi.org/10.1016/j.lrp.2009.12.005

Zott C, Amit R, Massa L (2011) The business model: recent developments and future research. J Manag 37(4):1019–1042. https://doi.org/10.1177/0149206311406265

Chapter 13
Insights on Africa's Future in its Transition to the Green Economy

Dorothe Yong Ngondjeb, Calvin Atewamba, and Moussa Macalou

13.1 Introduction

After having generated a lot of excitement both during the preparatory process and at the time of the Rio+20 conference, the concept of green economy continues to raise debate about its content and its link with sustainable development. Within the context of multiple crises including food, economic, financial, climatic and environmental crises, the green economy is perceived by some as an answer. Despite abundant natural resources and good economic growth over the past decade, Africa is still struggling to effectively and sustainably exploit the economic potential it has. From fisheries resources to non-renewable energy resources, there is indeed an extraction that is both anarchic and non-exploitative, sometimes at the artisanal scale. In addition, Africa is the continent with the lowest industrialization, and all too often, the benefits of exploiting natural resources are of little benefit to local populations (UNECA 2015).

In order to overcome this, it is useful for Africa to seize the current opportunities for green growth by developing a well-defined strategy transforming the continent's economic structures to better integrate sustainable and effective management into policy. Indeed, if Africa's economic growth has not led to the reduction of poverty and hunger, or the problems of employment, it is because there is no structural and

D. Y. Ngondjeb (✉) · C. Atewamba
United Nations University Institute for Natural Resources in Africa, Acrra, Ghana
e-mail: njedora@gmail.com

C. Atewamba
e-mail: atewamba@gmail.com

D. Y. Ngondjeb
Faculty of Economics and Management, University of Yaoundé II, 1365, Yaoundé, Cameroon

M. Macalou
University of Ghana, Acrra, Ghana
e-mail: macaloumoussa29@yahoo.fr

C. Atewamba and D. Yong Ngondjeb (eds.), *Inclusive Green Growth*, Advances in African Economic, Social and Political Development,
https://doi.org/10.1007/978-3-030-44180-7_13

sustainable transformation of African economies that remain vulnerable to external shocks (AfDB et al. 2013). With a low rate of industrialization, the continent also remains dependent on the outside, especially with regard to food products. To reverse this trend, Africa should develop its agribusiness and especially manufacturing sector to increase its production of high added-value goods.

A number of emerging countries have not considered the impact of development on environmental protection today and therefore face environmental problems including overexploitation of resources, water scarcity, pollution, waste, land degradation that are impediments to sustainable growth. Recent studies show that both aspects can be reconciled through the green economy, which offers opportunities for African countries to attract investment in environmental resources and renewable energy, which are ultimately beneficial for development and reduce poverty and create jobs according to numerous analytical studies conducted. The question that arises in this chapter is how to reconcile seemingly antagonistic interests and structurally transform Africa through industry while preserving the environment.

In light of various studies in this book, this chapter aims to examine the feasibility of transforming the current industrial development challenges into green growth opportunities in rural Africa. To achieve this goal, we discuss the main results of various studies in this book to highlight some empirical evidence explaining the factors determining the emergence or otherwise of green entrepreneurship in Africa. It is important to note that Chap. 1 of this book looks at the factors of the emergence of the green economy in Africa. Chapter 2 explores the potential opportunities for innovative small- and medium-sized enterprises (SMEs). Chapter 3 analyzes the factors influencing green entrepreneurship activities in Africa. Chapter 4 examines the effect of access to credit and the use of technological inputs on crop yields. Chapter 5 examines the factors that influence the demand for environmentally friendly products. Chapter 6 identifies the determinants of SME engagement in sustainable development and corporate responsibility. Chapter 7 examines the environmental and economic burden of dredging on artisanal fisheries. Chapter 8 assesses the impact of grazing on cereal and tuber crops. Chapter 9 examines the impact of climate change-induced migration and its implication for the development of green SMEs. Chapter 10 assesses the factors that determine green growth in the forest sector in Africa. Chapter 11 examines the barriers that can limit the capacity of manufacturing SMEs in peripheral regions. Chapter 12 proposes a conceptual framework for transforming traditional agribusiness in Africa into a green and inclusive agribusiness through an inclusive innovation business model and a green innovation business model. These studies show that the transition to the green economy in Africa offers both opportunities and challenges.

The rest of the chapter is organized as follows: The second section presents a synthesis of the results of the research. In the third section, the results are discussed in detail to highlight the book's contributions in filling some gaps in the green economy literature for rural Africa. The fourth section examines the limitations of research and the prospects for future research. A conclusion is made in the last section

13.2 Opportunities and Challenges of the Green Economy in Africa

The green economy is a new paradigm necessary for the implementation of sustainable development, which remains a vital human development project. Its implementation should necessarily consider the specific strengths and comparative advantages of the regions. The African context shows that natural resource-based sectors such as agriculture, forestry, mining, fisheries, renewable energy, water control and other transport and waste sectors are the most important sectors for job creation and green growth niches. Indeed, agriculture, forestry, mining and fisheries sectors generate 80% of employment in Africa (UNECA 2013). Assessing the impact of the emergence of the green economy in these sectors for African countries, we have seen in the various studies presented in this book that green entrepreneurship can help promote innovation, diversify business, empower businesses at the local level, enhance social cohesion, support environmental protection and so on. It has also emerged that, in some cases, the emergence of the green economy in Africa can face several constraints, including those related to financing, opening of markets and so on. This section discusses the challenges and opportunities for developing green entrepreneurship in various economic sectors, including agriculture, energy and forestry in rural Africa, while summarizing the main findings of the studies.

13.2.1 Green Small and Medium Enterprises

Innovation in Green SMEs

Many studies have been done on innovation in SMEs. The results of these studies identify several factors that determine a company's innovation capacity. These include: an organization's internal capacity to acquire information, absorb knowledge, engage with teams and learn from the organization (Oxborrow and Brindley 2013). For Cuerva et al. (2014), the propensity for innovation depends on the availability or lack of resources and innovation capacity of companies. Pinget et al. (2015) reveals the need for effective research, education and training as major drivers of eco-innovation for small businesses. To these factors, Lewis and Cassells (2010) add government legislation, the need to expand market share, employee motivation and eco-efficiency.

In an African context, Chap. 2 examines the factors that hinder innovation and innovation opportunities in green enterprises in rural Uganda. A descriptive analysis of the secondary data of 423 companies was made. In addition, a comparative analysis of the case studies of three green firms was also conducted to allow for more in-depth and contextualized investigation. The results of this analysis of the SME sector in Uganda suggest that some of the factors highlighted above may not apply in the context where SMEs operate informally. Therefore, any effort to influence the positive behavior of SMEs vis-à-vis environmental performance should focus on the relevant factors for sub-Saharan Africa. For example, Uganda, like many countries in sub-Saharan Africa, does not have a specific eco-innovation policy, and measures to promote eco-innovation are often framed by national environmental policy. Informally, operating SMEs may not easily respond to government regulations, especially when there is no specific policy to guide implementation. As a result, the use of environmental laws to initiate eco-innovation is doomed to fail. For this reason, Uganda's legal environment may not be effective in influencing the behavior of SMEs with respect to environmental practices.

SMEs in Uganda, particularly in the agricultural sector, lack financial resources and, as a result, rarely employ skilled labor and have difficulty accessing public extension agents. In addition, SMEs in Uganda do not perform research and development (R&D), and their human capacities as well as technological capabilities are lacking. SMEs often depend on technologies already developed by universities, public research centers and foreign companies, but most of the time, these technologies are not easily accessible, and the adoption rate is often negatively affected.

This chapter also explores additional information to understand the nature of opportunities to accelerate the adoption of eco-innovation for SMEs. The most discouraging factor for innovation is the "cost factor," specifically the lack of funding from outside sources. The financial system in Uganda imposes serious barriers on SMEs that need financing for innovation. In addition to cost factors, market factors also pose serious challenges for Ugandan SMEs. These include: the market dominated by established companies; the uncertain demand for innovative goods or services; an innovation that is easy to imitate; knowledge factors, such as difficulty finding partners for innovation and lack of qualified staff.

The results of Chap. 11, which examines the factors that prevent SMEs in the infrastructure sector from introducing eco-innovation in the peripheral areas of Lagos, Nigeria, are similar to those in Uganda. In terms of barriers to eco-innovation, the results of the correlation analysis suggest that barriers such as the high cost of innovation, insufficient technological capabilities, inaccessibility to the market, inadequate infrastructure and regulations are unfavorable for innovation by SMEs in outlying areas of Lagos metropolis. However, a more in-depth analysis using advanced analytical technique (probit ordinal regression) suggests that most of these barriers are important but not sufficient to drive or limit the introduction of eco-innovation among SMEs. The results revealed that eco-innovation among SMEs showed a significant positive link with the technological barrier and the regulatory barrier. At the same time, the company's characteristics, such as in-depth research of external knowledge and the business sector, also have a significant impact on the introduction of eco-innovation among companies.

Entrepreneurship in Green SMEs

Entrepreneurship was introduced as a catalyst for economic development. However, green entrepreneurship in Africa is still very unsatisfactory and faces challenges. Identifying the factors influencing green entrepreneurship activities would help inform greening African business policies.

The study on entrepreneurship in Chap. 3 examines and analyzes the factors affecting green entrepreneurship activities in South Africa using qualitative and quantitative methods. 103 green entrepreneurs were asked about the factors that influence green entrepreneurship activities. In-depth interviews, key informants, observations and a comprehensive review of the literature were conducted for triangulation purposes. The results of the study indicate that the availability of opportunities to move to the green economy was one of the most influential factors on green entrepreneurship activities. This factor is followed by incentives, then public policies, access to finance, green education and finally institutions. The reason why 61% of respondents ticked off opportunities as the reason they were moving toward implementing sustainable business practices could be attributed to the number of programs and initiatives that have been unveiled in the country over the past five years. This result is also explained by the fact that initiatives that encourage entrepreneurs to meet the socioeconomic and environmental challenges in sub-Saharan Africa are more and more common.

Social Responsibility in Green SMEs

In developing countries, ownership of Corporate Social Responsibility (CSR) by companies remains a major concern for governments, researchers and business leaders, despite the existence of a legislative framework (in most countries) and the gradual adoption of sustainable development strategies by the various UN Member States. The question here is what are the determinants of the commitment of African SME leaders to adopt a favorable posture for sustainable development. Work on the appropriation of CSR practices in other countries shows that SME managers who adopt a CSR-friendly posture benefit their companies in terms of reducing costs, increasing employee loyalty, better relations with the territorial authorities, acquiring new knowledge and, incidentally, a better reputation (Bonneveux and Saulquin 2009). While despite these positive repercussions, the mobilization of SME leaders in sustainable development remains problematic, Berger and Douce (2008) believes that it is because of the lack of time but especially of knowledge and information of these SMEs on CSR.

What about the African context? Chapter 6 identifies the determinants of the adoption by business leaders of Ivorian SMEs of sustainable development practices. A methodological approach combining qualitative and quantitative methods was adopted, and the data were collected from fifty-seven (57) Ivorian SMEs from various sectors of activity. Of these, 24 are commercial, nine are industrial, 14 are services and 10 are multi-business enterprises. In addition, 37 of these enterprises are located in Abidjan, seven in the suburbs of Abidjan and the rest (13) in the interior of the country. Despite the existence of some CSR practices in SMEs, the analysis of the data highlights a number of elements identified by managers as hindering their commitment to SD: the "absence of a regulatory framework and support from the State," "absence of a management structure," "lack of information on the concept of SD/CSR." These data show that for most SME managers in the sample, these three elements are the main causes of non-engagement of Ivorian SMEs business leaders in the CSR approach.

Green SMEs and Migration
Migration can be defined as the movement of people from one place to another for economic, social or political reasons (Sinha 2005; Jolly and Reeves 2005). In Africa, migration is often the end result of a process and can be induced by an event such as extreme climate shocks. A question that is pertinent in the African context is how green SMEs can be introduced as an alternative option to climate change adaptation and thus stem the wave of climate change-induced migrants while improving climate change rural livelihoods and development.

Chapter 9 aims to examine how green SMEs should be implemented in rural Africa in the presence of increasing migration induced by climate change. The results of the study show that green SMEs are a potentially viable alternative for labor, migration and income generation, especially in the most vulnerable farming systems. Policy interventions using green SMEs as an alternative to migration should consider promoting economically important crops for households and focusing on entrepreneurial and economically viable farming systems; these systems should not be demanding in terms of manpower. In addition, these agricultural system setup in rural areas should be identified and defined by characteristics or variables such as the level of education, especially the head of the household and the spouse; the level of resources they own and control and assets that are both sustainable and unsustainable that can easily be converted into cash. These variables summarized as education, resources, assets, households and market define the degree of independence from resource owners, income generation, standard of living and the adoption of innovation.

The study also shows that migrants are connected to their fellows back home through the transfer of funds. This is an indication that climate change-induced migration could be used as a potential market development strategy for green SMEs.

13.2.2 Green Business in the Agricultural Sector

For Rosen et al. (2012), the specific challenges for the adoption of green technologies in agriculture are numerous and relate to: the identification of appropriate green technologies for income generation, that is, ecological agriculture, renewable energy and so on; review of the impact and implications of national policies related to the extension of appropriate green technologies; the diagnosis of the political impact of such green technologies on rural income in the context of sustainable agriculture development and the constraints and policy options available for a sustainable adoption of green technologies.

Access to Credit

Chapter 4 on the instability of crop yields among small-scale rural farmers discusses the challenges of adopting green technologies in Ghana. Data from the study come from the Ghana Living Conditions Survey collected in 2012/2013. The results of the regression analysis revealed that, in general, the difficulty of farmers' access to credit is a constraint for the adoption of green technologies. Indeed, access to credit is an important element of agricultural production systems. Credit allows producers to meet the financial and resource needs of the production cycle, which is characterized by crucial activities such as land preparation, planting, cultivation, harvesting, storage and transportation, usually over several months. The availability of credit would thus allow for greater investment in soil conservation techniques, inputs, tools and infrastructure, which in turn leads to increased farmer performance and protect the environment.

The study used three separate ordinary least squares (OLS) models. The pooled model showed that smallholders with access to credit achieve 35.5% higher yields per acre than their counterparties who do not have access to credit at a 1% level of significance, other variables remaining constant. This significant effect is driven by increased yields among rural farmers due to access to credit. Rural farmers thus have a chance to increase their production by about 38.5% if they have access to credit. In terms of opportunities, the adoption of new technologies allows smallholder farmers to increase the yields of their plots and thus income, but also to improve soil conservation, market green products, and to practice agroforestry and cover crop.

Level of Education and Experience

The overall objective of Chap. 7 is to determine the factors influencing environmental efficiency in fisheries. The study used primary and secondary data. Primary data were collected in July 2014 from two local governments in Lagos State, Nigeria, namely Ikorodu and Epe, known for artisanal fishing and sand dredging. Both areas of the local government have rivers flowing into Lagos Lagoon. Using cost/benefit analysis, the study shows that education, fishing experience and sand dredging are factors that affect the environmental effectiveness of fishermen in sand dredging and non-dredging zones.

The Agricultural System put in Place

One of the challenges identified by Rosen et al. (2012) is the identification of an appropriate agricultural technology and/or farming system for income generation for the transition to a green economy.

Chapter 8 highlights the impact of the extensive system on crop production in Abuja, Nigeria. An extensive animal production system refers to a system that allows animals to roam and search for food without restriction (Ezeibe 2010). On the other hand, a semi-intensive system allows good control of feeding, management and protection of animals. As part of this study, questionnaires were used to survey small-scale crop farmers in agrarian communities about the benefits of the extensive system. The analysis of the survey reveals that the extensive system causes the destruction of crops, which causes conflicts between farmers and breeders. This disastrous scenario is detrimental to sustainable agricultural development. The study also highlights several plot-level issues that require the implementation of viable policies by governments to reduce the negative impacts of agricultural systems and thus encourage sustainable development.

13.2.3 Green Business in the Energy Sector

Usman (2012) reported that rural people use solar energy for cooking, water heating, refrigeration, domestic lighting, distribution of domestic water pumps, ironing, television and radio. Green infrastructure is particularly useful for the agricultural economy, providing solar technology for drying, processing and storage of agricultural products, especially those that are perishable; pumping and distribution of water for irrigation; grinding flour; threshing; expulsion of oil; drying of crops and use of various agricultural tools. Green infrastructure is also useful for commercial purposes. The question that arises here is what factors affect the demand for and use of energy products and infrastructure in Africa?

Chapter 5 examines the constraints and opportunities associated with the use of energy infrastructure, such as energy-saving light bulbs, solar panels and inverters and energy storage devices in rural areas of western Nigeria. The study is based on primary data obtained from a field survey of 160 rural households in the states of Ekiti and Ondo in Nigeria.

In terms of constraints, the study identifies several factors that determine the use of products and infrastructure in Nigeria. First, financial problems are a huge constraint to the development of green infrastructure. In developing countries with very limited financial resources such as in Africa, green energy facilities, such as solar panels, inverters and energy-saving light bulbs, are more expensive than traditional electric generators and incandescent bulbs. Because of the higher initial cost of installing solar panels, even if in the long run it becomes cheaper, most people cannot afford to install it. Then, there is little knowledge of environmentally friendly infrastructure in the context of climate change. The lack of awareness of these products means that a good proportion of the African population, particularly Nigeria, is very ignorant of the green economy concept. Furthermore, the lack of robust policies and research on these technologies are constraints related to their adoption. Finally, the study shows that the factors that determine the use of green technologies in rural areas include the level of awareness of climate change, age and income level.

In terms of opportunities, the study shows that green infrastructure improves the quality of life in rural Africa, creates jobs, alleviates poverty and encourages innovation.

13.2.4 Green Business in the Forestry Sector

The question of how different green growth strategies and green investments interact with the current pattern of forest use and management at the level of rural communities in Africa is rarely raised (UNECE 2009). As a result, factors that could facilitate or hinder the transition to green growth in the forest sector at the level of rural communities remain largely unknown.

Chapter 10 was designed to address this challenge by assessing the current mode of forest use and management at the rural community level in South Africa to find spaces for green growth initiative development and factors facilitating or hindering these initiatives. Rural tributary communities in Vhembe District were deliberately selected as a case study.

The results of the study show that the formal forest sector does not offer good prospects for transition to green growth in the short term in the communities studied. The current contribution of the formal forest sector to household incomes and employment opportunities in the communities studied is significantly lower than that of the informal sector. In addition, the contribution of the formal forest sector is highly dependent on the availability of successful tree plantations and associated forest industries. It should be noted that the expansion of a community's core forest resources is therefore a necessary prerequisite for the development of a vibrant formal forest sector in a community and, therefore, a prerequisite for the development of formal forest management initiatives for green growth in the forest sector. However, this expansion of a community's core forest resources can only be beneficial for long-term green growth, simply because afforestation is a long-term project.

The current pattern of forest use and management in these communities is also one of the most obvious constraints to the transition to green growth in the informal forest sector at the rural community level. Forests in most rural communities in South Africa are treated as open access resources with little or no regulation on their use and management (DAFF 2010). As a result, unsustainable forest use and management practices leading to forest degradation are increasing in most rural communities in South Africa (Berliner 2005). This condition limits the ability of forests to contribute sustainably to the security of household livelihoods (Berliner 2005; Dovie 2003) and can hinder the transition to green growth in the forest sector at the level of rural communities (Berliner 2005). This study also showed that the promotion of the forest enterprise based on firewood and edible insects (for example, Mopani worms) offers good prospects for profitable green forest growth initiatives in the studied communities.

13.3 Policies that Facilitate a Successful Transition to a Green Economy in Africa

Favorable and specific conditions for transition to a green economy are needed and focus on national regulations, policies, subsidies and incentives, international markets, legal infrastructure and trade and aid agreements. The creation of these conditions requires certain actions to be undertaken. These actions include changes in fiscal policy; political, economic, fiscal, legislative and financial reforms; the reduction of environmentally harmful subsidies; the use of new economic instruments; targeting public investments in key ecological sectors; taking the environment into account in public procurement; improving environmental regulations and laws and strengthening their application and stakeholder training (UNEP 2011, 2015). This section identifies some policies to be put in place in the African context to facilitate the transition to the green economy.

13.3.1 Promote Eco-innovation in SMEs

In order to enable SMEs to change their environmental behavior, we must strengthen the driving forces or weaken the resistance to eco-innovation, or a combination of both.

Understanding barriers is the key to develop appropriate eco-innovation policies. The main barriers to innovation in SMEs identified in Chap.2 are the high costs of innovation, inadequate knowledge and market factors. In addition to these factors, Chap. 11 refers to the lack of infrastructure, and unfavorable regulation affects the introduction of eco-innovation by SMEs in the peripheral areas of Lagos metropolis. It is therefore essential to develop policies such as the subsidy in investments to assist SMEs in innovation. It is also imperative to develop infrastructure and encourage SMEs to opt for the least costly types of innovation. The government should also consider using its existing research institutions to help SMEs innovate. In addition, legislators must adopt policies that support and strengthen SMEs that engage in innovative activities. Given that the sector in which SMEs belong is important for the introduction of eco-innovation, policymakers should avoid putting in place comprehensive eco-innovation policies for all sub-sectors. In addition, according to Cooke et al. (1997), firms located in outlying areas may need specific policy interventions that may be different from those required by businesses located in central business districts.

Chapter 2 also reveals that customer needs are driving innovation in SMEs. To reinforce this driver, it is necessary to encourage the training of clients and the promotion of eco-literacy as a potential option to stimulate eco-innovation among SMEs. Chapter 11 highlights the importance of learning-by-doing as a means of generating knowledge for eco-innovation. In other words, SMEs should focus on the learning-by-doing models and seek to interact with various external sources of knowledge to complement their internal innovation strategies.

In addition, SMEs lack the resources to invest in eco-innovation. It would be crucial to put in place policies to facilitate their collaborations with partners to invest in innovation. One solution to this problem of low investment could be to encourage public–private partnerships (Koppenjan and Enserink 2009).

13.3.2 Regulate the Informal Forest Sector and Facilitate Participatory Forest Management

The forest sector is seen as one of the key priority sectors identified in the green growth landscape because of its potentialities especially in terms of mitigation strategies and adaptation to climate change. Opportunities for transition to green growth in the forest sector come in different forms: market growth for the existing products, technical innovation for new product and market development, increased efficiency in product use, best practices in forest management, improved trade arrangements for small commercial enterprises (Bishop et al. 2015; FAO 2009; UNECE 2009).

The case study of South African communities in Chapter 10 shows that the formal forest sector offers few prospects for transition to green growth in the short term. Therefore, for the immediate adoption of green growth initiatives and the provision of green growth benefits to rural communities, the informal forest sector is the one that offers a good option. Indeed, the informal forest sector is crucial for rural development and livelihood sustainability and for the transition to green growth in rural communities. The forest products that the informal sector offers have potential in terms of promoting green entrepreneurship in these communities. This potential comes in the form of prospects for immediate and widespread impact through job creation, income generation and the achievement of food security. However, open access to forest resources due to poor or lack of regulation in the informal sector is one of the most obvious obstacles to the transition to green growth (DAFF 2010). Implementing a policy to regulate the informal forest sector could facilitate the transition to rural green growth. The development of technical skills needed through capacity-building initiatives and access to credit to develop forest enterprises could also be a catalyst for regulation.

Implementing a forest management strategy that effectively links the informal forest sector to green growth initiatives requires an understanding of the informal forest sector that goes beyond the assumptions that informal activities are disorganized and chaotic in form or practice (Smit and Musango 2015). Participatory forest management involving local populations and the development of forest enterprises is essential for the sustainability and cost-effectiveness of forest development initiatives. Thus, the government and local authorities must actively encourage all segments of the community to participate without hindrance in the management of forests and the development of forest enterprises. A hybrid forest management approach open to such duality is needed; it embraces both "the traditional and the modern, the small and the large scale, the informal and the formal," correct power imbalances and protects vulnerable people (Smit and Musango 2015).

13.3.3 Strengthen Entrepreneurship Education and Enhance Indigenous Knowledge

Education for green entrepreneurship plays an important role in the development of green culture and green aspirations and promotes the development of green entrepreneurs in countries. More specifically, the development of employee skills in technology helps improve the environmental performance of SMEs by reducing the negative impact of production and consumption and also promotes growth (Adeoti 2000). In order to mitigate knowledge constraints, such as lack of skills and information on technologies, SMEs should establish partnerships with universities, consultants and research institutes, where they could access new knowledge.

Chapter 2 reveals that green SMEs in the agricultural sector rely on indigenous knowledge and technologies to undertake innovation. This is mainly because local knowledge and technologies are cheaper to adapt. Indigenous knowledge can be a major component of process and product innovation because it reduces production costs and improves environmental performance; therefore, innovation among green SMEs in low-income countries can be supported by exploiting this existing knowledge and know-how. Strategies should also be developed to promote the exploitation of this local and indigenous knowledge.

In addition, potential suppliers, especially foreign firms, would have the opportunity to tap into the existing knowledge systems, including indigenous knowledge to develop more affordable and efficient technologies, such as degradable polymers to replace plastic.

13.3.4 Develop Pastoral Trails

An equally salient fact stemming from the case study of Nigeria in Chap. 8 is the conflict between small farmers and their fellow breeders. Smallholders cultivate small portions of land without fencing, while their livestock counterparts raise animals in an extensive system, allowing animals to move and feed with little or no restriction. This animal production system causes the destruction of crops, which causes conflicts between farmers and breeders.

This disastrous scenario is detrimental to sustainable agricultural development. Farmers know that at any vegetative stage of their crops, animals can devastate them, hence the demotivation of not investing in green technologies that are sometimes expensive to be destroyed without profitability. The implementation of a policy to separate farmers and breeders is necessary, either by the determination of animal passageways such as advocating by the West African Economic and Monetary Union (UEMOA) or the adoption of an intensive system of animal production.

13.3.5 Facilitate Access to Credit and Diversify Sources of Financing

Access to financial resources including credit remains a brake on green economic activities. African agriculture is dominated by smallholders who play an important role in the African economy. However, access to credit remains a major constraint for the adoption of environmentally friendly agricultural technologies for these smallholders. The case study on Ghana in Chap. 4 confirms that access to credit remains essential for the adoption of technologies that are essential for the development of green agriculture, especially in rural areas. The success of green agriculture depends on policies to facilitate access and the cost of credit for rural and urban smallholders.

The quest for a healthy environment, energy resources and sustainable and clean energy use has become a recurring problem in global development. However, financial problems are a huge constraint to the development of green energies. In developing countries with very limited financial resources like Africa, green energy facilities, such as solar panels, inverters and energy-saving light bulbs, are more expensive than traditional electric generators and incandescent bulbs. Because of the higher initial cost of installing solar panels, even if in the long run it becomes cheaper, most people cannot afford to install it. It is therefore necessary for African development partners and governments to develop modalities to provide financing or subsidize the costs associated with the transition to green energy. The use of these clean energy sources essentially requires increased investment and a change of attitude with regard to ecological infrastructure (Mkpado 2013a, b).

As a financial measure for promoting green energy, Chap. 5 shows that government and non-government organizations should inject money by first providing these green energy products and services for free or at subsidized prices on the market so that they reach a higher scale.

13.3.6 Promote and Develop Green Technologies to Facilitate the Transition to the Green Economy

Lack of awareness about available green energy products is common in African countries. As noted in Chap. 5 above, a good proportion of the African population is very ignorant of the green economy concept. To alleviate this state of affairs, it is strongly recommended to create an interactive environment for experts in green economy to enable them to communicate with the masses. It should also encourage the establishment of credible platforms through which real changes in the green economy should be generated.

According to the study by Wai et al. (2008), it is necessary to develop a communication and awareness action plan to raise awareness of the benefits of using renewable energy and energy efficiency compared to other forms. Marketing and promotion are very critical to the growth of any business in the industry. Green energy should be promoted and products widely marketed. The digital and physical marketing interaction is crucial at this stage. There is a growing popularity, expressed in social media, on green energy products, such as the use of energy-saving light bulbs, inverters and others. However, their uses and benefits are rarely advertised on social media, as well as on media such as newspapers, magazines, radio and television.

13.3.7 Put in Place a Regulatory Framework to Guide Investment Toward Environmentally Friendly Activities

Practical policies must be put in place to encourage the use of green energy and allow investors to adhere to it. A healthy and profitable economy is only realistic if there are laws and a basic order in society. It is imperative to develop legislative and regulatory frameworks to stimulate the use of renewable energy (Wai et al. 2008). Policies can stipulate the percentage of financial participation from the government budget, measurement and goal evaluation strategies, tax incentive structures and mass campaigns and education systems to enhance green energy development (Mkpado 2013a).

An adequate policy framework is needed to stimulate public spending on green technologies. NEPAD has advised African countries to invest at least 10 percent of their annual budget in agriculture (CSAC 2008; Mkpado 2013b). A similar model can be followed by African countries by investing a certain percentage of their budget in green technologies in order to transition to a green economy.

13.4 Limitations of Research and Prospects for Future Research

The studies in this book examine different green business issues in different sectors and countries. The main sectors examined in this book are small and medium-sized enterprises, agriculture, energy and forestry. The study covers several countries such as Uganda, South Africa, Cote d'Ivoire and Nigeria. Chapters 2 and 11 examine the constraints and opportunities of green SMEs for innovation. Chapter 3 examines the factors that influence entrepreneurship in SMEs. Chapter 6 examines the obstacles to SMEs's commitment to sustainable development. Chapter 9 examines the potential of green SMEs in the fight against migration. In the case of agriculture, Chap. 4 reviews the role of credit in the adoption of agricultural technologies. Therefore, the interpretation of the results should not necessarily be generalized to the whole of rural Africa. Indeed, African countries are diverse and what works in one country may not work in another. The study of the same sector in many African countries would make it possible to compare the results of research between countries, which could reveal more interesting results. In addition, examining the constraints and opportunities of green enterprises in many African countries would contextualize the policy implications of research findings for wider applicability in rural Africa.

The availability of data is a major problem when researching in Africa. Most of the studies in this book are based on primary data collected through small-scale fieldwork. In addition, various methods are used in this book. For example, some studies are more quantitative in nature, while others have used a more qualitative approach. While both approaches are commendable, this indicates that the results are not strictly comparable, and caution must be used in interpreting the results. There is a need for large-scale surveys of the opportunities and constraints of green enterprises in rural Africa with good quality data and sound methodology. This may be a necessary condition for testing the robustness of some of the findings in this book.

Although the studies in this book have identified many constraints and opportunities for the development of green enterprises, they have paid less attention to the opportunities offered by new green commodity markets, mainly driven by demand from developed countries. These include new export opportunities in the horticultural sector (fruit and vegetables), cashew, sesame, spices, soybeans and palm oil. Another area that needs attention is the development of the green value chain, which is increasingly seen as a way to promote green growth. Chapter 12 develops a conceptual framework for an inclusive green agricultural business model innovation for rural Africa and demonstrates that a green agricultural value chain is a special case. There is, however, a need to document challenges and opportunities of the existing green value chains in agriculture and other sectors in Africa in order to inform policy development and strengthening of the value chain. Value chain operates as a source of its own product and innovation process. As such, many challenges related to adopting inclusive and green agricultural innovations in Africa can be addressed through important innovations driven by value chain development. These include various value chain finance instruments (e.g., financial arrangements between suppliers, buyers, producers and banks), the application of ICT in mobile banking, mobile technical support, electronic networks (Miller 2012). These innovations help reduce costs and risks, while improving access to and use of financial services, information systems, contract farming systems, product exchange links and other aspects of value chain operations. Value chain development also brings about policy innovations, including extension services and investment in support infrastructure for green agriculture.

For future studies, it would therefore be interesting to extend green technologies to other sectors of activity and specially to embrace other African countries. This will enable the design and implementation of specific policies for each business sector as well as for each African country to ensure the transition to a green African economy. In addition, larger-scale studies that address the issues raised in this book will help contextualize the findings of this book and inform policymaking in rural Africa.

13.5 Conclusion

One of the major concerns throughout this study was to assess the constraints and opportunities of the transition to green economy in rural Africa, through the Green Business Note. This issue is particularly important for Africa as green business is a development opportunity for Africa. Building on growth sectors such as agroforestry, renewable energies, agriculture, industry, ecotourism and carbon sequestration, green business is likely to create wealth and jobs while preserving the environment and the balance of ecosystems. This is why Africa must accelerate and succeed in the transition to this new paradigm of business development. The continent can count on the vastness of its potential with regard to the richness of biodiversity, the diversity of soils and streams, ideal sunshine, etc. But resource potential is not enough to revolutionize business development in Africa. It still needs a supportive human and material environment.

As noted throughout this chapter, the transition to this new business model faces many challenges related to innovation, entrepreneurship, access to finance, education, regulation and so on. In terms of opportunities, studies show that green businesses improve the quality of life in rural Africa, create jobs, alleviate poverty and encourage innovation.

The results obtained in the chapters have strong political implications for the transition from rural Africa to a business-driven green economy. If the main objective of African governments is to create jobs, stimulate growth and reduce poverty for the rural population, then encouraging the development of green enterprises appears as an entry point to a greener world. In light of these results, several recommendations can be made to African governments:

- Put in place mechanisms for monitoring and evaluating country progress toward the green economy;
- Work for the promotion of endogenous technologies and technology transfer;
- Develop common standards to incentivize green production by, for example, subsidizing clean technologies;
- Encourage public–private partnerships for the promotion of the green economy;
- Continue to encourage African countries to put in place institutional, legal and participatory mechanisms for the development and promotion of the green economy by developing land use and investment plans in key sectors such as forestry, energy, infrastructure, agriculture, industry and so on;
- Encourage the initiative for the rapid implementation of sectoral development programs for this economy, such as, among others, the support program for the

development of renewable energies and the program for the development of non-timber forest products and so on;

- Encourage the creation by the African Development Bank and the regional central banks of a fund dedicated to the financing of activities related to the green economy in Africa. Urge these banks to put in place a fund mobilization strategy at the subregional and international levels to feed the Green Fund, and to use diversified financing mechanisms and capture the most innovative.

On an international level, it would be good for the international community to support the efforts of African countries in the field of development and the promotion of the green economy through financing and transfer of the necessary technologies.

References

Adeoti JO (2000) Small enterprise promotion and sustainable development: an attempt at integration. J Develop Entrep 5.1 (April 2000):57–71

AfDB, OECD, UNDP, UNECA (2013) African economic outlook 2013: structural transformation and natural resources. UNECA, Addis Ababa, Ethiopia

Berger-Douce S (2008) Rentabilité et pratiques de RSE en milieu de PME: Premiers résultats d'une étude française. Revue Manage et Avenir 15:9–29. http://dx.doi.org/10.3917/mav.015.0009

Berliner D (2005) Systematic conservation planning for the forest biome of South Africa: approach, methods and results of the selection of priority forests for conservation action. Department of Water Affairs and Forestry (DWAF), Pretoria, South Africa

Bishop R, Eis J, Harrison P, Shukla P, Awafo EA, Herweijer C, Hourcade JC (2015) Assessing and communicating benefits of green growth. In: G. G. practice, green growth in practice: lessons from country experiences. The Green Growth Best Practice (GGBP), London, pp 79–102

Bonneveux E, Saulquin JY (2009) L'appropriation de la RSE par les dirigeants de PME. Le réseau comme vecteur de l'apprentissage managérial, Management and Avenir 23

Cooke P, Uranga MG, Etxebarria G (1997) Regional innovation systems: institutional and organizational dimensions. Res Policy 26(4):475–491

CSAC (2008) (civil society input unto the African commission on effective development cooperation with Africa) Agriculture, growth and employment in Africa—need for institutions, infrastructure, information, innovation, investments and integration, discussion paper number 2, April 2008. http://www.friisbach.dk/fileadmin/cfb/images/Africa/Agriculture-final.pdf

Cuerva MC, Triguero-Cano Á, Córcoles D (2014) Drivers of green and non-green innovation: empirical evidence in Low-Tech SMEs. J Clean Prod 68:104–113

DAFF (2010) Policy principles and guidelines for control of development affecting natural forests. Department of Agriculture, Forestry and Fisheries, Pretoria, South Africa

Dovie DB (2003) Rural economy and livelihoods from the non-timber forest products trade. Compromising sustainability in southern Africa? Int J Sustain Dev World Ecol 10(24):7–262

Ezeibe ABC (2010) Profitability analysis of pig production under intensive management systems in Nsukka local government area of Enugu state, Nigeria'. Int J Econ Dev Res Invest 1(2):48–54

FAO (2009) Green Jobs. An Int J For For Ind. Unasylva, 60(3), ISSN 0041-6436. ftp://ftp.fao.org/docrep/fao/012/i1025e/i1025e01.pdf. 28 Nov 2013

Jolly S, Reeves H (2005) Gender and migration. Bridge Publishers, ISBN 1858648661, 9781858648668

Koppenjan J, Enserink B (2009) Public-private partnerships in urban infrastructures: reconciling private sector participation and sustainability. Public Admin Rev 69(2): 284–296

Lewis K, Cassells S (2010) Barriers and drivers for environmental practice uptake in SMEs: a New Zealand perspective. Int J Bus Stud 18(1):7
Miller C (2012) Agricultural value chain finance strategy and design. International Fund for Agricultural Development (IFAD)
Mkpado M (2013a) Green technologies for African agriculture and energy and possible strategies for incubation and development. J Agric Sustain 2(1):43–65. http://infinitypress.info/index.php/jas/article/view/41
Mkpado M (2013b) The status of African Agriculture and capacity development challenges for sustainable resilience from global economic shocks. J Sustain Dev Stud 3(1):45–79
Oxborrow L, Brindley C (2013) Adoption of "eco-advantage" by SMEs: emerging opportunities and constraints. Eur J Innov Manag 16(3):355–375
Pinget A, Bocquet R, Irege C (2015) Barriers to environmental innovation in SMEs: empirical evidence from French Firms. Manage 18(2):132–155
Rosen MA, Kishawy HA (2012) Sustainable manufacturing and design: concepts, practices and needs. Sustain 4:154–174
Sinha BRK (2005) Human migration: concepts and approaches. Foldrl Ert, Geogr Bull 3–4:403–414
Smit S, Musango JK (2015) Towards connecting green economy with informal economy in South Africa: a review and way forward. Ecol Econ 116:154–159
UNECA (2013). "A Regional Perspective on the Post-2015 United Nations Development Agenda," United Nations, E/ESCW A/OES/2013/2
UNECA Report (2015) Industrializing through Trade
Usman M (2012) Rural solar electrification in Nigeria: renewable energy potentials and distribution for rural development. http://www.sases.conference-services.netresources/2522859pdf SOLAR2012_0232_full%20paper.pdf electric rural
Wai RJ, Wang WH, C-Y Lin (2008) High-performance stand-alone photovoltaic generation system. In: Proceedings of IEEE transactions on industrial electronics, vol 55, no 1

The manufacturer's authorised representative in the EU is Springer Nature Customer Service Centre GmbH, Europaplatz 3, 69115 Heidelberg, Germany. If you have any concerns regarding our products, please contact ProductSafety@springernature.com

Printed and bound by CPI Group (UK) Ltd, Croydon, CR0 4YY
05/10/2026
02228124-0010